Digital and Discrete Geometry

Li M. Chen

Digital and Discrete Geometry

Theory and Algorithms

 Springer

Li M. Chen
University of the District of Columbia
Washington
District of Columbia
USA

ISBN 978-3-319-34862-9   ISBN 978-3-319-12099-7 (eBook)
DOI 10.1007/978-3-319-12099-7
Springer Cham Heidelberg New York Dordrecht London

Printed on acid-free paper

Springer is part of Springer Science+Business Media (www.springer.com)

To the researchers and their supporters in Digital Geometry and Topology.

Preface

Discrete geometry is the study of the geometric properties of discrete objects including lines, triangles, rectangles, circles, cubes, and spheres. These shapes are usually subsets of Euclidean space. On the other hand, digital geometry has two meanings: (1) The objects are formed by digital or integer points, more narrow digital geometry; and (2) The objects are computerized formations of geometric data. Sometimes we can view digital geometry as a subcategory of discrete geometry.

While discrete geometry has a long history, it has recently garnered much attention due to its large role in fulfilling computer vision and image processing needs. Such a need is the motivation behind the creation of digital geometry. The subject provides tremendous new research areas within discrete geometry. In the past, geometric tiling and counting were the primary research topics in discrete geometry.

Digital geometry mainly comes from two areas: image processing and computer graphics. A digital image in 2D is in the form of digital grid points; it is a natural treatment of using geometry in image processing including segmentation, recognition, and reconstruction. On the other hand, computer graphics use geometric design, object dynamics, and modification.

Computerized geometry must deal with efficient algorithms for many applications including classifications of digital objects, which also uses topological properties and geometry processing. It can be applied to a vast number of areas including biomathematics, medical imaging, the film industry, etc.

Digital geometry is also highly related to algorithmic geometry (computational geometry), which is more focused on algorithm design for discrete objects in Euclidean space. However, digital geometry has its own set of problems and challenges including those involving distance measure and the formatting of digital objects, which are different than that of discrete objects. Digital geometry also has some advantages since sampling the data can usually be directly applied in its digital form. There is no need to do a conversion from discrete forms.

This book provides detailed methods and algorithms in discrete geometry, especially digital geometry. We also provide the necessary knowledge in its connections to other types of geometry such as differential geometry and algebraic topology. In addition, there is much discussion on the recent development of applications in variety of methods of image processing, computer vision, and computer graphics.

This book is intended to offer comprehensive coverage of the modern methods for geometric problems in the computing sciences. We also discuss concurrent topics in BigData and data science as well.

This book is written to be suitable to different groups of readers. Chapters 1–6 are for junior and senior college students in computer science and mathematics; Chaps. 7–12 are for graduate students. Chapters 13–15 are written for researchers or students with advanced knowledge in geometry and topology.

This book can also be categorized into three parts: (a) Chaps. 1–9 are introductions to digital and discrete geometry, (b) Chaps. 10–12 mainly deal with geometric processing for readers interested in applications, and (c) Chaps. 13–15 present topics in high level mathematics that are related to discrete geometry. The sections marked with "*" may require some advanced knowledge. The book is also self-contained.

Acknowledgments: Many thanks to my daughter Boxi Cera Chen who helped me correct my grammar for the whole book.

Many thanks to my wife Lan Zhang and my son Kyle Chen for their patience and support while I was working on this book. I never thought that I had to put increasingly more effort into completing this book. Many thanks to my colleagues Professors Feng Luo, Petra Wiederhold, and Sherali Zeadally for their continued support and encouragement. Many thanks to my colleagues in digital topology for their help and support, Professors Reinhard Klette, Reneta Barneva, Jacques-Olivier Coeurjolly, Tae Yung Kong, and Konrad Polthier, just name a few. Thanks also to UDC for giving me one semester of sabbatical to work on this project.

My goal was set to write a complete introductory and comprehensive book to digital and discrete geometry. As I was reviewing my writing today, I found that it is still too far from reaching this initial vision. I hope that this book has laid a good foundation for learning digital and discrete geometry, as well as linking to various topics as a stepping stone to future research in this relatively new discipline of computer science and mathematics.

Aug., 2014

Li M. Chen
Washington, DC

Contents

Acronyms

N	The natural number set
I	The integer number set
R	The real number set
$G = (V, E)$	A graph G with the vertex set V and the edge set E
$|A|$	The number of elements in set A
$d(x, y)$	The distance between x and y
GVF	Gradually varied functions
E_m	m_dimensional Euclidean space
Σ_m	m_dimensional digital space
N_p	The $3 \times 3 \times 3$ neighborhood of point p in Σ_3
B^k	k-dimensional unite ball
S^k	k-dimensional unite sphere
∂M	The boundary of M
$O(f)$	The time cost of an algorithm which is smaller than a constant $c \cdot f$
f'	The derivative of function f
$C^{(n)}$	The class of functions having up to nth continuous derivatives
f_x	The partial derivative of function f with respect to the variable x
$\partial f / \partial x$	The same as f_x
∇f	The gradient vector of f : $\frac{\partial f}{\partial x} i + \frac{\partial f}{\partial y} j$
K_G	The Gaussian curvature

Part I
Basic Geometry

Chapter 1
Introduction

Abstract Geometry comes from measurements of visible shapes, sizes, patterns, and positions. It is one of the most basic human subjects in civilization. Through thousands of years, humans have developed different types of geometry including: elementary geometry, analytic geometry, differential geometry, topology, algebraic geometry, and many other related research areas. Along with the fast development of digital computers, in recent years, people have been studying digital geometry.

Digital geometry focuses on digital objects, which are usually represented by a finite number of integer points or vectors. However, in a much larger sense, digital objects could be digital data saved in computers or data sets in electronic form. Digital geometry comes from two primary sources: digital images and digital data displays of computer graphics. Computer memory is the primary domain of digital geometry. Since computer memory is arranged by arrays, its location (called address) in the arrangement is similar to n-dimensional grid points in Euclidean space. Therefore, digital geometry can also be viewed as geometry of grid space. The main difference between digital space and Euclidean space is how we measure the distance between two points. Digital geometry is a branch of discrete geometry that mainly consists of the study of geometric relationships among discrete objects. In this chapter, we will give a brief introduction to the different types of geometry, especially digital and discrete geometry.

Keywords Geometry · Digital geometry · Discrete geometry · Computer graphics · Image processing · Introduction

1.1 What is Geometry

In general, geometry meaning measurement of the earth, is the study of spatial figures and their relationships. The figures can be lines, surfaces, and solids in space. In a flat surface (plane), describing the relationship between two lines introduces the concept of angles. If the space is a sphere, a "line" that is the shortest path on the surface of sphere, becomes an arc or a circle. Therefore, different geometric systems exist.

Geometry is one of three major branches of mathematics. The other two are algebra and analysis. In geometry, Euclidean geometry is the most common [17]. Some rules (also called axioms or postulates) are first assumed. In other words, in

© Springer International Publishing Switzerland 2014

L. M. Chen, *Digital and Discrete Geometry,* DOI 10.1007/978-3-319-12099-7_1

Euclidean geometry, the axioms are made to describe relationships among points, lines, and circles in a flat surface or plane. Based on these axioms, we can define basic objects such as points, lines, triangles, etc. The induced relationship of those objects are called theorems, which are self-consistence conclusions. For instance, the sum of the interior angles in a triangle is 180°.

Analytic geometry was first introduced by Rene Descartes. A figure can be specified relative to the coordinate systems by a set of numbers or equations. They usually involve a system of linear equations for geometric transformation. Projective geometry was established by Jean-Victor Poncelet. It is a modification of Euclidean space by including points at infinity. Differential geometry was started by Karl Gauss who treated the local curvature as the central concept of a surface. Naturally, the length, area, and volume of randomly shaped objects are also topics of differential geometry. However, in most of cases, these topics are covered in calculus, i.e. analysis since these are classical geometries. Canadian mathematician H.S.M. Coxeter's classic book, Introduction to Geometry, includes a concise and comprehensive introduction to those topics [14].

The existence of Non-Euclidean geometry was first observed a long time ago as the geometry on a sphere. However Non-Euclidean geometry is based on axioms that differ from those of Euclidean geometry. This branch of geometry was founded by Nikolai Lobachevski, Janos Bolyai, and G. F. B. Riemann. They verified the existence of self-consistent systems based on all of Euclidean's axioms, except the one concerning parallel lines. It is very interesting that the classification of different geometries is dominated by how many parallel lines we can make at a point that is not on a target line. In 2D Euclidean geometry, we can just make one. In elliptical geometry, we cannot make any since any two big circles on the surface of a sphere must intersect. In hyperbolic geometry, we can make more than two parallel lines.

Due to the fact that non-Euclidean geometry exists, the basic measurement of Euclidean distance has also changed in that the space can be bent. In general, a space can be made by any shape, circle, ellipse, donut, etc. These are examples of manifolds, which is the extension of arbitrary shapes.

The goal of geometry is to study the manifold. A manifold could be as big as an entire universe or as small as a unit circle. Similar to Euclidean space, a manifold is usually assigned a dimension. Basically, a k-dimensional manifold means that for any point in the manifold, there is a neighborhood that is the "same" as a k-dimensional cube in Euclidean space.

Topology is a special type geometry that concerns visual objects that remain unchanged upon deformation. Topology first began with combinatorial topology by Henri Poincare [2], where the key concept is to use triangles to represent a surface and then extend the surface using k-simplexes to represent k-manifolds.

Graph theory can also be viewed a type of geometry that deals with the relationship among vertices and the edges between vertices [23].

There are many other types of geometry such as differential geometry, differential topology, algebraic geometry, and algebraic topology [2, 28]. The most impressive result in resent years is the positive solution for the 3D Poincare conjecture concluded by Grigori Perelman [33]. In this book, we mainly discuss the theory and problems in computer related geometry: digital and discrete geometry.

The basic references for classical geometric theories are: H.S.M. Coxeter, Introduction to Geometry, Wiley; 2nd edition, 1989. P.S. Alexandrov, Combinatorial Topology, New York: Dover, 1998. E. Kreyszig, Differential Geometry, Dover, 1991.

For graph theory, we recommend the book: F. Harary, Graph theory, Addison-Wesley, Reading, Mass., 1969.

1.2 Contemporary Geometries in Modern Computer Times

Digital geometry can be traced back over 120 years to the geometry of numbers. It was initiated for number theory by Hermann Minkowski [31]. Minkowski's theorem states that an integer grid point other than the origin point must be involved in a special convex region if it has a certain amount of sufficient volume.

Modern digital geometry is due to the development of digital image processing and computer graphics. It overlaps with numerical geometry and algorithmic geometry (also called computational geometry).

In image processing, digital geometry is usually used in image segmentation, edge detection, thinning, and transformation. The basic display methods in computer graphics use digital methods such as Bresenham's line algorithm, polygon clipping, and 3D rendering.

The measurement of digital objects is also a research topic in those geometries. Classic topics, such as the shortest path, are part of graph theory, but the shortest path on a surface is a topic in differential geometry. It is sometimes difficult to separate these categories for special problems in the real world. This book will cover all the essential background materials while focusing on newer concepts and methodologies.

1.2.1 Discrete Geometry

Discrete geometry mainly deals with the geometry among discretely represented objects. Before the computer age, discrete geometry was called combinatorial geometry [9]. For instance, *the geometry of numbers* has combinatorial properties used in number theory. Discrete geometry also deals with combinatorial problems, especially counting and tiling problems.

A famous example of discrete geometry is how to calculate the area for irregular polygon shapes. The simple formula is called Peak's theorem on grid or lattice planes. We will discuss this result in Chap. 4. Another interesting result is called the Catalan number for counting the number of triangulations if given n points on the edge of a convex. The answer is $\frac{1}{(n-1)}C(2n-4, n-2)$ (called the $(n-2)$-th Catalan number) [6], where $C(n, m) = n!/((n-m)!m!)$. This formula tells us we can not practically go through all triangulation to pick a best one. We can only look for the most reasonable triangulation when it is needed. This fact also gives a good reason for why we want to develop digital geometry.

1.2.2 Digital Geometry

Digital geometry focuses on the properties and applications in digital, or grid, space [9, 25, 29]. It was created due to the development of image processing, computer vision, and computer graphics [10, 21, 36]. A good example is how to draw a digital line.

Digital geometry also deals with the modeling and recognition of curves and surfaces that are stored in computers, since objects can only be represented digitally in computers. Because the memory address system is assigned as an array, this performs best if we use an integer grid system as the foundation. In this case, space usually means memory space, which is why we call it digital space.

The main difference between digital space and Euclidean space is how we measure the distance between two points. In Euclidean space, for points $u = (x_1, y_1)$, $v = (x_2, y_2)$, we are mainly interested in $d(u, v) = \sqrt{(x_1 - x_2)^2 + (y_1 - y_2)^2}$ as the metric. In digital geometry, we are more interested in $d_1(u, v) = |x_1 - x_2| + |y_1 - y_2|$ and $d_\infty(u, v) = \max\{|x_1 - x_2|, |y_1 - y_2|\}$. Digital geometry is a branch of discrete geometry that mainly consists of the study of geometric relationships among discrete objects.

Another challenge is that digital space does not always hold the Jordan curve theorem: A closed simple curve always separate a plane into to components [2, 35].

1.2.3 Computational Geometry and Numerical Geometry

Another type of modern geometry is called computational geometry, the purpose of which is to consider the computational cost to design an algorithm to complete a geometrical task, such as finding the maximum circle in a polygon. It is sometimes called algorithmic geometry [15, 22].

The name computational geometry was first used by Hyman Minsky and Seymour Papert in 1969 [32]. Afterwards, Forrest (1972) and Su and Liu (1981) used it for curve and surface fitting in numerical methods [19, 45]. Today, the name of computational geometry should covers a vast area. The curve and surface fittings using numerical methods, especially in computer graphics, which can be viewed as part of computational geometry.

Numerical geometry refers to using numerical methods in geometry problems. This usually means using constructive methods for curves and surfaces, many of which are used in computer graphics. Bezier and B-spline methods are popular examples [38].

Fig. 1.1 Example of an image and its components made by P. Arbelaez et al [1]: **a** Original image, **b** human segmentation, and **c** machine segmentation

1.3 Geometry and Topology in Image Processing and Computer Graphics

In order to transfer a continuous picture or image to be stored in computers, we usually partition the image into a number of small, square regions. This process is called digitization. Each of the small regions is called a pixel element, or pixel as it is known colloquially today. A pixel is the combination of a small region and its value.

In three-dimensional (3D) space, we use the volume element instead of the picture element, so we have voxels. Sometimes people call the voxel a 3D pixel. In other words, these pixels are arranged as 2D or 3D arrays.

The goal of image processing is to extract information from digital images, which is why digital geometry is one of the fundamental research areas in digital image processing. See Fig. 1.1.

A basic problem of digital geometry related to image processing is identifying a region that indicates an object in a digital image and extracting this region as a set of pixels. This problem relates to image segmentation. Another popular example is when we recognized an object and we want to find the orientation of the object. The principal component analysis in statistics can be used to find the major and minor axis of the object.

Another very interesting and useful information about an image is the number of holes in a region of the image. This refers to the topological properties of digital spaces. This problem becomes more complex when we deal with 3D digital images. There are many applications of this in medical imaging and informatics [11]. This research area is called digital topology as suggested by Rosenfeld [41, 42].

Fig. 1.2 Example of
computer graphics made by
Krystina Dizayas

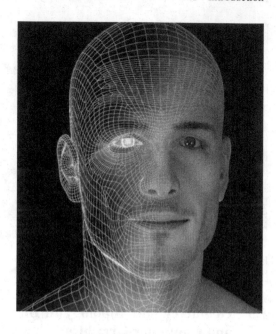

On the other hand, in computer graphics, we want to display 2D or 3D pictures
using digital display devices such as TVs or computer monitors. See Fig. 1.2. A
display device normally has the pixel dimensions 1280×1024 for monitor resolution.
This means that the monitor contains 1280×1024 pixels. We then have to arrange a
picture inside of the array. Some parts of the picture will not be seen and some parts
are shaded so there are many geometry technologies involved. For instance, how is a
3D object displayed in an 1280×1024 digital array? This is called image rendering
[18].

To solve a problem that cannot be represented as a formula, we usually use several
instruction-type steps, called an algorithm [13]. The first famous algorithm related
to computer graphics is to find the best algorithm to draw a line in 2D digital space.
This algorithm was found by Bresenham and is called the Bresenham line drawing
algorithm.

Objects in computer graphics are usually represented by triangles and not individual image points. This is because the zoom-in and zoom-out display functions would
take up too much space when we store an object in its pixel format. In computer
graphics, when two objects in 3D space are viewed from an angle, the problem of
finding the intersection line of two or more triangles (or polygons) becomes a basic
problem as well. This issue is called clipping.

Researchers today use differential geometry for deformation of objects in graphics. The process called morphing is also a profound topic. In this book, we will give
brief introductions to discrete differential geometry in Chap. 13.

Researchers in computer graphics referred to digital geometry as the fourth wave
of computer graphics. Note that the digital geometry they refer to is also the general
geometric method for computer graphics, called discrete geometry in this book, and
is not limited to geometry in grid space.

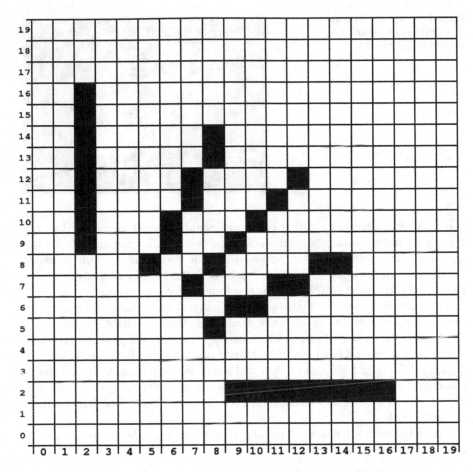

Fig. 1.3 Digital lines: only few can be considered to be straight lines drawn by Tim Bell

1.4 Problems and Concepts of Digital and Discrete Geometry

As we discussed above, an object in image processing and computer vision is a set of pixels that is used to approximate the original object [25, 26, 29].

How to describe and distinguish different digital objects is a difficult task. For instance, what is a digital line? What is the distance between two digital points or two pixels? The answers will be based on the specific problems. We still do not have perfect or unique answers to those questions. This is because there are only few perfect digital lines in a plane: vertical, horizontal, and diagonal. All other digital lines do not look like straight lines at the pixel level, meaning that we examine each of the pixels in the "line." See Fig. 1.3.

Therefore, we can only define digital curves. Rosenfeld first studied digital curves and its topological properties in the early 70's [39–44]. For digital images, the

Fig. 1.4 A medical CT image
of brain. (From standard
medical imaging library)

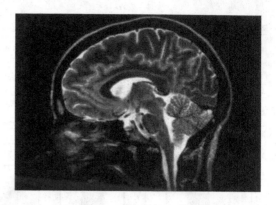

boundary of an object is a digital curve, or several curves if the object does not contain any holes. One can count the area of the object and extract this object using algorithms. It is interesting that the best algorithm for extracting such an object is called breadth-first-search or depth-first-search, the famous methods in graph-theoretic algorithms [13]. This fact was first found in [36].

This example tells us that digital geometry has great deal of connections to graph theory, along with the fast development of medical imaging technology such as CT and MRI. In the early 1980s, researchers started to consider the problems related to digital surfaces since a boundary of 3D objects is a surface in 3D images (Fig. 1.4).

Artzy, Frieder and Herman [3] designed a tracking algorithm in medical image processing. One year later, Morgenthaler and Rosenfeld [34] gave a first definition of digital surfaces in 3D array mathematically. Unlike digital curves, the digital surface is much complex.

In terms of the developments of digital geometry, two major directions have gained considerable interest with researchers:

(1) Some practical methods and algorithms have developed in digital geometry and topology such as object labeling, thinning, shrinking, and boundary tracking and extraction. The two good examples are the 2D and 3D thinning algorithms [47].
(2) Special properties and theorems are found for digital objects in digital space. The development of the theory includes the finite topological spaces studied by Khalimsky, Kopperman, and Meyer [24], and Kovalevsky [27]. The definitions of digital manifolds given by Chen and Zhang [9, 12].

On the other hand, discrete geometry that mainly use graph and triangulations has raised importance in recent years due to the cloud data processing (random scattered data samples). We can ask: Is there a topological structure of the massive cloud data set?

Researchers are also interested in finding the lower dimensional information of a data set in a high dimension space. This research relates to manifold learning, dimension reduction, and persistent analysis. So, manifold learning is to find the shape of an unknown data set. For instance, we want to know if the data points

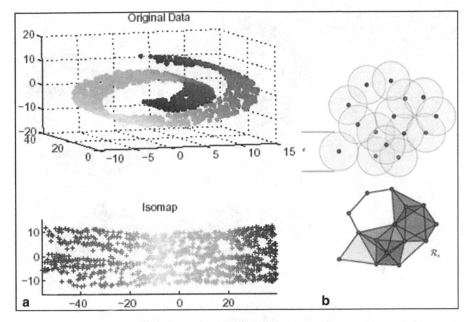

Fig. 1.5 Modern geometric data processing: **a** Manifold learning, and **b** Persistent analysis

represent a sphere or a spiral shape [30, 46]. The example of manifold learning is shown in Fig. 1.5a [30].

To learn a data set that will most likely represent a random shape is a very difficult task. So there are many tasks to be complete in manifold learning.

Persistent analysis is used to find out how many holes and possibly locations of these holes consistently appear when we enlarge the coverage of each data point [7, 20], see Fig. 1.5b [20]. These topics relate to geometric processing, and we will discuss them in Chap. 9.

Even though, considerable significant results have been discovered in digital and discrete geometry. There are still many concerns and new developments for the future, we just list a few here.

(1) More significant discoveries including the problem of digital form of the Poincare conjucture.
(2) More dispensable technologies include image processing and computer vision, such as 3D thinning.
(3) Mathematical fascination means that more beautiful results can be found when comparing to the existing Pick's theorem and hole counting.

For the first problem, if one can solve a very famous mathematical problem such as "four color problem" in digital geometry and topology, then digital geometry and topology will make a great impact. For the second problem, if we can find an efficient digital geometry method to track and register moving objects in computer vision, it will help to build the reputation of digital geometry and topology.

Fig. 1.6 An example in
discrete differential geometry
where the curvature drawn on
a triangulated surface in [37]

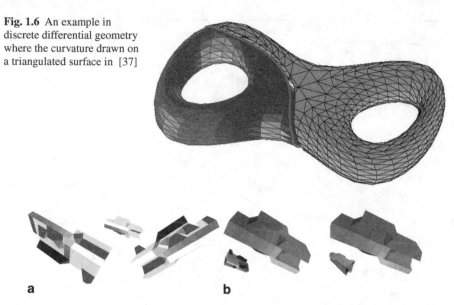

a   b

Fig. 1.7 Data fitting on manifolds: **a** A curve function on a 3D digital manifold, **b** another view of
the curve, **c** a digital function fitting, and **d** the harmonic function fitting

For the third problem, if we could simplify the terminology of digital geometry and topology, make its methods easier to understand, to use, and to implement,
along with making it closer to traditional research fields such as discrete and computational geometry, then digital geometry and topology will surely be more attractive
to scientists and engineers in terms of scientific study.

Recent developments in discrete differential geometry has merged profound
knowledge and results in differential geometry with computer graphics. Researchers
found great potential in image science research. The research relates to minimal
surface computing, circle packing, and Ricci flow [5, 37]. We will discuss them in
Chap. 13 (Fig. 1.6).

In data reconstruction, merging digital methods such as digital functions with
classical harmonic analysis for surface fitting on discrete manifolds has also been investigated, and the results are very promising. See Fig. 1.7 [10]. A brief introduction
is presented in Chap. 11.

1.4.1 Some Developments in Digital Geometry

Since this book mainly deals with digital geometry, the following is a list of important
developments in digital geometry and topology.

1. The need for digital geometry, especially digital curves in digital image processing, inspired work by Rosenfeld, Mylopoulos, and Pavlidis in 1970–1971, along with Herman's initiative on digital surfaces in 1980.
2. The definition of digital surfaces by Morgenthaler and Rosenfeld was the first mathematical definition of digital surfaces.
3. Research by Kong and Roscoe on plate surface points and compound cells provided a detailed articulation for most of the cases of digital surfaces.
4. Kovalevsky's finite topology became a type of digital topology based on the analog of classical topology.
5. Chen and Zhang's work on the six types of digital surface points in 3D and Chen and Rong's theorem on genus in 3D were solid and definitive results in digital topology.
6. Chen and Zhang's definition of digital manifolds using parallel-moves was a more practical definition of digital manifolds and storage-saving methods compared to classical topology, including simplicial and cell complexes.
7. Thinning algorithms, considered one of the best practical uses of digital topology, were the work of Zhang and Sun's methods, which consisted of methods in 2D and 3D.
8. Kong and Rosenfeld's survey paper on digital topology and Klette and Rosenfeld's book on digital geometry popularized digital geometry and topology.
9. Rosenfeld's digital continuous functions and Chen's extension theorem on the gradually varied extension of digital functions connected digital geometry to classical calculus [8, 41].

In terms of discrete geometry in computer graphics, remarkable developments include circle packing, minimal surface calculation, and Ricci flow calculation.

References

1. P. Arbelaez, M. Maire, C. Fowlkes and J. Malik. Contour detection and hierarchical image segmentation. IEEE TPAMI, 33(5), 898–916, 2011.
2. P. S. Alexandrov, Combinatorial Topology, New York: Dover, 1998.
3. E. Artzy, G. Frieder and G.T. Herman, The theory, design, implementation and evaluation of a three-dimensional surface detection algorithm, Comput. Vision Graphics Image Process. 15, 1981, 1–24.
4. E. Bishop and D. Bridges (1985) *Constructive Analysis,* Springer Verlag, 1985.
5. A. Bobenko, Y.U. Suris, Discrete Differential Geometry: Integrable Structure, AMS, 2008.
6. R.A. Brualdi, Introductory Combinatorics, 4th ed. New York: Elsevier, 1997.
7. G. Carlsson and A. Zomorodian, Theory of multidimensional persistence, Discrete and Computational Geometry, Volume 42, Number 1, July, 2009.
8. L. Chen, The necessary and sufficient condition and the efficient algorithms for gradually varied fill, Chinese Science Bulletin, 35:10 (1990).
9. L. Chen, *Discrete Surfaces and Manifolds: A theory of digital-discrete geometry and topology,* 2004. SP Computing.
10. L. Chen, Digital Functions and Data Reconstruction, Springer, NY, 2013.

11. L. Chen, and Y. Rong, Digital topological method for computing genus and the Betti numbers, Topology and its Applications, Volume 157, Issue 12, 2010, Pages 1931–1936.
12. L. Chen and J. Zhang, Digital manifolds: A Intuitive Definition and Some Properties, Proceedings of the Second ACM/SIGGRAPH Symposium on Solid Modeling and Applications, Montreal, 1993, 459–460.
13. T. H. Cormen, C.E. Leiserson, and R. L. Rivest, Introduction to Algorithms, MIT Press, 1993.
14. H.S.M. Coxeter, Introduction to geometry, John Wiley, 1961.
15. de Berg, M., VAN KREVELD M., OVERMARS, M., and SCHWARZKOPF, O. Computational Geometry: Algorithms and Applications. Springer-Verlag, New York, 1997.
16. C. de Boor A Practical Guide to Splines, Springer-Verlag, 1978.
17. Euclid's Elements, Green Lion Press, 2002.
18. J. D. Foley, A. Van Dam, S. K. Feiner and J. F. Hughes, Computer Graphics: Principles and Practice. Addison-Wesley. 1995.
19. A. R. Forrest, Interactive interpolation and approximation by Bezier polynomials, Computer J. 15:71–79 (1972).
20. R. Ghrist, Barcodes: the persistent topology of data, Bull. Amer. Math. Soc., 45(1), 61–75, 2008.
21. R. C. Gonzalez, and R. Wood, *Digital Image Processing*, Addison-Wesley, Reading, MA, 1993.
22. Goodman, O'Rourke, Handbook of Discrete Geometry, CRC, 1997.
23. F. Harary, Graph theory, Addison-Wesley, Reading, Mass., 1969.
24. E. Khalimsky, R. D. Kopperman and P. R. Meyer, Computer graphics and connected topologies on finite ordered sets, Topology and its Applications, Vol. 36, 1990, 1–17.
25. R. Klette and A. Rosenfeld, Digital Geometry, Geometric Methods for Digital Picture Analysis, series in computer graphics and geometric modeling. Morgan Kaufmann, 2004.
26. T. Y. Kong and A. Rosenfeld, Digital topology: introduction and survey, Computer Vision, Graphics and Image Processing, Vol. 48, 1989, 357–393.
27. V. A. Kovalevsky, Finite topology as applied to image analysis, Computer Vision, Graphics and Image Processing, Vol. 46, 1989, pp. 141–161.
28. E. Kreyszig, Differential Geometry, University of Toronto Press, 1959.
29. L.J. Latecki, Discrete Representation of Spatial Objects in Computer Vision, Kluwer Academic Publishers, 1998.
30. T. Lin and H. Zha, Riemannian Manifold Learning, IEEE Trans. Pattern Analysis and Machine Intelligence, vol. 30, no. 5, pp. 796–809, May 2008.
31. H. Minkowski, Geometrie der Zahlen, Leipzig and Berlin: R. G. Teubner, 1910.
32. M. Minsky and S. Papert, Perceptrons: An Introduction to Computational Geometry, The MIT Press, Cambridge MA, 1969.
33. J. Morgan, G. Tian, Ricci flow and the Poincare conjecture, Clay Mathematics Monographs, Cambridge, MA, 2007.
34. D.G. Morgenthaler and A. Rosenfeld, Surfaces in three-dimensional images, *Imform. and Control* 51, 1981, 227–247.
35. M. Newman, Elements of the Topology of Plane Sets of Points, Cambridge, London, 1954.
36. T. Pavilidis, Algorithms for Graphics and Image Processing, Computer Science Press, Rockville, MD, 1982.
37. K. Polthier, Polyhedral Surfaces of Constant Mean Curvature, Habilitationsschrift Technische Universitt Berlin (2002).
38. W. H. Press, et al. *Numerical Recipes in C: The Art of Scientific Computing*, 2nd Ed., Cambridge Univ Press, 1993.
39. A. Rosenfeld, Connectivity in digital pictures, Journal of the ACM, Vol. 17, 1970, pp. 146–160.
40. A. Rosenfeld, Arcs and curves in digital pictures, Journal of the ACM, Vol. 20, 1973, pp. 81–87.
41. A. Rosenfeld, "Digital topology," Amer. Math. Monthly, Vol 86, pp 621–630, 1979.
42. A. Rosenfeld, Three-dimensional digital topology, Inform. and Control 50, 1981, 119–127.

43. A. Rosenfeld, Continuous' functions on digital pictures, Pattern Recognition Letters 4 (1986), 177–184.
44. A. Rosenfeld and A.C. Kak, *Digital Picture Processing*, 2nd ed., Academic Press, New York, 1982
45. B. Su and D. Liu, Computational geometry, Shanghai Science and Technology Press, 1981.
46. J. B. Tenenbaum, V. de Silva, J. C. Langford, A Global Geometric Framework for Nonlinear Dimensionality Reduction, Science 290, (2000), 2319–2323.
47. T. Y. Zhang, C. Y. Suen, A fast parallel algorithm for thinning digital patterns, Communications of the ACM, v. 27 n. 3, p. 236–239, 1984.

Chapter 2
Discrete Spaces: Graphs, Lattices, and Digital Spaces

Abstract The best way to describe discrete objects is to use graphs. A graph consists of vertices and edges. The vertex usually represents a part of an object, an whole object, or the location of an object; the edge represents a relationship between two vertices. A graph can be defined as $G = (V, E)$ where V is a set of vertices and E is a set of edges, each of which links two vertices. For a certain geometric object, e.g. a rectangle, one can draw four points on the corners and link them using four edges. The drawback of using edges is that the edge is not a geometric line and it usually does not carry a distance. A geometric space also requires a measurement of distance (the length between two points), called a metric. Therefore, people prefer to use specialized graphs such as triangulated graphs and grid graphs, to represent an object in discrete space. In this chapter, we introduce the discrete spaces made by graphs, lattices, and grid points. We briefly review some of the basic concepts related to discrete objects and discrete spaces.

Keywords Graph · Lattice · Space · Digital space · Discrete space · Algorithms

2.1 Objects in Discrete Spaces

Objects are things that are visible or tangible. An object usually has a form that is relatively stable. Geometry is the study of objects and their properties in space. Due to the fact that computers can only take a discrete or finite number of objects, discrete geometry has become more important in recent years.

There are always two ways to view an object: (1) A contiguous interpolation of discrete points to from a continuous object, (2) A discretizing or sampling of a continuous object to get its discrete representation. In this chapter, we assume that our space is discrete. In the next chapter, we discuss spaces and objects that are continuous.

The recent motivation behind studies in discrete geometry is due to the need of computer graphics and computer vision. There are a great deal of applications, especially in medical imaging, that require a high level of math including topology in a digital format. Images are stored in digital spaces, discrete spaces, and even with finite topologies.

The simplest way to describe a discrete object is to use graphs. A graph consists of vertices and edges. The vertex usually represents a part of an object, a whole object,

© Springer International Publishing Switzerland 2014

L. M. Chen, *Digital and Discrete Geometry,* DOI 10.1007/978-3-319-12099-7_2

or the location of an object; the edge represents a relationship between two vertices. For instance, a rectangle can be represented by drawing four points on the corners and linking them using four edges.

In this chapter, we start with an introduction of an undirected graph $G = (V, E)$, where V denotes the set of vertices and E denotes the edges between the vertices. Then we introduce the lattice that has geometrically regular assigned locations for each vertex. The edge in lattices is typically assumed. Note, the lattice we deal with here differs from an algebraic lattice, which has a partial order assigned on each vertex.

Then, we focus on the grid space that is the simplest lattice, similar to arrays in computers. This grid space is called digital spaces.

Another popular discrete space is called the triangulated space in two-dimensional spaces (2D). This space contains only triangles as 2D elements. In mathematics, a triangle is called a two-dimensional simplex (2-simplex), while a point is called a 0-simplex and a line-segment is called 1-simplex.

In undirected graphs, we can view that vertices are 0-simplices, and edges are 1-simplices. However, to describe 2-simplices, one needs special structures to define it. On the other hand, edges in a graph cannot just be viewed as line-segments because edges in graphs can have more general meanings.

2.2   Graphs and Simple Graphs

A graph G consists of two sets V and E, where V is the set of vertices and E is the set of pairs of vertices called edges. An edge is said to be incident to the vertices if it joins [2, 11, 16]. In this book, assume that $G = (V, E)$ is an undirected graph, which means that if $(a, b) \in E$ then $(b, a) \in E$, or $(a, b) = (b, a)$. a, b are also called ends, endpoints, or end vertices of edge (a, b). It is possible for a vertex in a graph to not belong to any edge. V and E are usually finite, and the order of a graph is $|V|$, which is the number of vertices. The size of a graph is linear to $\max\{|V|, |E|\}$, meaning that it requires this much memory to store the graph.

The degree of a vertex is the number of edges that incident with (or link to) it. A loop is an edge that links to the same vertex. In such a case, the degree would be counted twice. Figure 2.1 shows two examples of graphs. Figure 2.1a shows a directed graph, where the edge has an arrow. Figure 2.1b shows an undirected graph.

2.2.1   Basic Concepts of Graphs

Graph $G = (V, E)$ is called a simple graph if every pair of vertices has at most one edge that is incident to these two vertices and there is no loop $(a, a) \in E$ for any $a \in V$. See Fig. 2.2.

If (p, q) is in E, then p is said to be adjacent to q. Let $p_0, p_1, ..., p_{n-1}, p_n$ be $n + 1$ vertices in V. If (p_{i-1}, p_i) is in E for all $i = 1, ..., n$, then $\{p_0, p_1, ..., p_{n-1}, p_n\}$ is

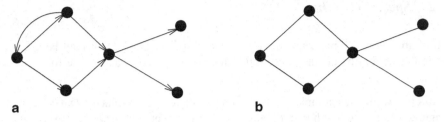

Fig. 2.1 Example of graphs: **a** A directed graph, and **b** An undirected graph

Fig. 2.2 Paths: A path
$\{A, C, D, B, C, E\}$; A simple
path $\{A, B, D, F\}$; and A loop
at F

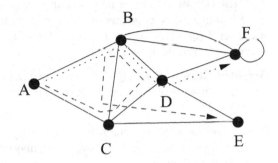

called a path. If $p_0, p_1, ..., p_{n-1}, p_n$ are distinct vertices, the path is called a simple path.

A simple path $\{p_0, p_1, ..., p_{n-1}, p_n\}$ is closed if (p_0, p_n) is an edge in E. A closed path is also called a cycle. Two vertices p and q are connected if there is a path $\{p_0, p_1, ..., p_{n-1}, p_n\}$ where $p_0 = p$ and $p_n = q$. G is called connected if every pair of vertices in G is connected. In this book, it is always assumed that G is connected (unless otherwise specified).

Let S be a set. If S' is a subset of S, then their relationship is denoted by $S' \subseteq S$. If S is not a subset of S', then S' is called a proper-subset of S, denoted by $S' \subset S$.

Suppose $G' = (V', E')$ is a graph where $V' \subseteq V$ and $E' \subseteq E$ for graph $G = (V, E)$. Then, $G' = (V', E')$ is called a subgraph of G.

If E' consists of all edges in G whose joining vertices are in V', then the subgraph $G' = (V', E')$ is called a partial-graph of G and their relationship is denoted by $G' \preceq G$. If V' is a proper-subset of V, then the relationship is denoted by $G' \prec G$. It is noted that for a certain subset V' of V, the partial-graph G' with vertices V' is uniquely defined. [1] A path is a subgraph. For example in Fig. 2.2, $\{A, B, D, C, E\}$ is a path. Let $V' = \{A, B, D, C, E\}$ and $E' = \{(A, B), (B, D), (D, C), (C, E)\}$. $G' = (V', E')$ is a subgraph, but it is not a partial graph because it does not include the edge (D, E). So, $G'' = (V', E' \cup \{(D, E)\})$ is a partial graph.

---

[1] In [6], we made an error in revising the meanings of the two concepts: Subgraphs and Partial graphs.

2.2.2   Special Graphs

In this subsection, we present some examples of special graphs that may be used later in this book. The detailed description for these concepts can also be found in [2, 11].

Complete Graph   In a complete graph, each pair of vertices is joined by an edge. A triangle is a complete graph with three vertices. A complete graph with five vertices contains 10 edges. See Fig. 2.3a. A complete graph with n vertices is denoted as K_n.

Bipartite Graph   In a bipartite graph, the vertices can be divided into two sets, X and Y, so that every edge has one vertex in each of the two sets, i.e. no edge is inside set X (or Y) alone. See Fig. 2.3b. A Bipartite graph with n vertices in X and m vertices in Y is denoted as $K_{n,m}$.

Weighted Graph   In a weighted graph, each edge can be assigned a weight. The weights are usually real numbers that could indicate distance if the vertices are cities. See Fig. 2.3c.

Planar Graph   A planar graph is a graph that can be drawn in a plane with no crossing edges. A graph with crossing edges may or may not be a planar (plane-able) graph. In the following subsection, we present a theorem related to planar graphs. See Fig. 2.3d.

Tree   A tree is a connected graph with no cycles. A forest is a graph with no cycles. See Fig. 2.3e.

2.3   Basic Topics and Results in Graph Theory

Graph theory was established by Euler, who solved the well-known —Seven Bridges of Konigsberg problem in his time. See Fig. 2.4a. In this map, Euler was asked to draw a path where each bridge would be traveled just once. Euler found that it was impossible, since he represented this problem as a 4 vertices and 7 edges graph, Fig. 2.4b. When one passes a vertex, he must go through two edges, one in and one out. That is to say, if such a path exists, each vertex must contain even number of edges. This property became the first theorem in graph theory.

Two of the most famous problems in graph theory are the Four Color Problem and the Traveling Salesman Problem. The former deals with a well-known "rule" in map printing. Only four colors are needed in a map, a planar graph, in which all adjacent points are colored by different colors. It is believed that this problem was solved positively by a computer program. However, even now, no one is able to verify the correctness of the computer program [2].

As for the Traveling Salesman Problem, its goal is to find the shortest path (for n cities) when a salesperson only travels to each city once. The path finding procedure exists but any of these solutions require extremely long computation time to complete

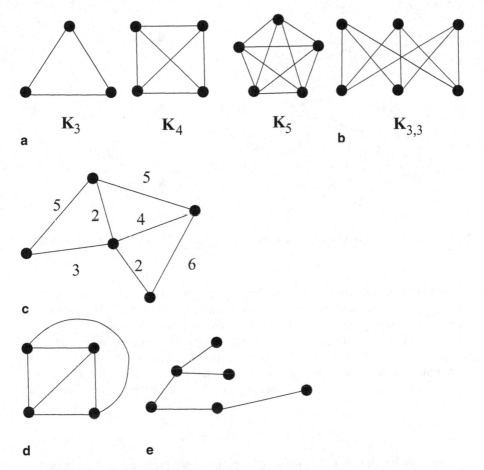

Fig. 2.3 Some special graphs: **a** A complete graph K_3, K_4, K_5, **b** A bipartite graph $K_{3,3}$, **c** A weighted graph, **d** A planar graph K_4, and **e** A tree

the job. Today, no one knows if there is an efficient algorithm to solve this problem. It is related to the $P =?NP$ problem [8].

2.3.1 Graph Representation, Searching Graph, and Graph Coloring

To solve a complex problem in graphs, for example the graph contains hundreds or even thousands of vertices, we must rely on computers. But how do we represent a certain graph in the computer? There are two ways to do this: the adjacency matrix and the adjacency list.

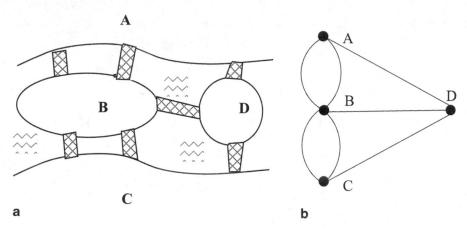

Fig. 2.4 Seven Bridges of Konigsberg: **a** An original map, and **b** The equivalent representation in graphs

In the adjacency matrix, we assume Graph G has n vertices. An $n \times n$ $\{0, 1\}$ matrix $M = \{m_{ij}\}$ is used to hold the information of adjacency: $m_{ij} = 1$ if and only if vertex i and j are adjacent. This is the simplest way, but if there are not many edges in G, then there needs to be a large amount of storage space to represent matrix M. For instance, if $n = 100$, we need 10,000 memory units to represent M.

Another way that could save storage space is called the adjacency list. For vertex v in G, we just attach all adjacent points of v to v. So each link becomes a linked list that is lead by v.

$$v_i \rightarrow u_{i1} \rightarrow u_{i2} \cdots \rightarrow u_{ik}$$

v_i represents the ith vertex. The length of each list is not the same, so we can save a lot of space. However, the process of calculating may become a little more difficult.

To find whether a graph G is connected, we need to search the graph. The best way is called the depth-first search or breadth-first search technique. In mathematics and computer science, if one needs a procedure to solve a problem, the procedure is called an algorithm. An algorithm usually contains several steps of instructions that solves a problem. The basic idea of the depth first search algorithm is to find the set of connected vertices until no more vertices can be found. Then, we go back to each vertex we visited to see if there are any other ways to go. We continue, where possible, until there are no possible ways left. This procedure requires a special data structure to hold the vertices we visited. We discuss this further in Chaps. 4–6.

Graph coloring assigns different colors to adjacent vertices. It usually tries to assign the minimum number of colors. The most famous problem is called the four color problem for planar graphs, as we have mentioned before. This problem was believed to be solved in 1975 with the help of computer programs. Since some errors were found and fixed in the program, some mathematicians are still looking for pure,

Fig. 2.5 Find a minimum
spanning tree T

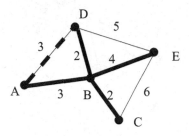

mathematical proofs. However, a famous theorem states: Five colors are enough for
a planar graph [11].

2.3.2   The Minimum Spanning Tree

Given a connected graph $G = (V, E)$, a spanning tree is a subgraph of G, which is
a tree and contains all the vertices of G. This special tree is called a spanning tree
since it spans every vertex.

A graph may have several different spanning trees. If G is a weighted graph, a
minimum spanning tree (MST) is the one that has the minimum total weight. The
simplest method for finding a MST is called Kruskal's Algorithm. Its principles are
as follows (Fig. 2.5):

Algorithm 2.1: Kruskal's Algorithm. Find a minimum spanning tree T for graph G.

Step 1   Sort the edges based on the weights of the edges from smallest to largest.
Step 2   Set initial tree T to be empty.
Step 3   Select an edge from the sorted edge list and add it to T if such an added
             edge does not generate a cycle.
Step 4 T would be the minimal spanning tree if $|V| - 1$ edges are added to T.

The proof of Algorithm 2.1 can be found in [8].

2.3.3   The Shortest Path*

Finding the shortest path for each pair of vertices in a graph is one of the most
common problems in the real world. A method of finding the shortest paths from a
single source vertex to all of the other vertices in a weighted directed graph is called
the Bellman–Ford algorithm.

Let $G = (V, E)$ and $|V| = n$. We use W(e)=W(u,v) to represent the weight on
an edge $e = (u, v)$. The principle of the algorithm is to reach a vertex v using k edges
from the source vertex S and maintain the shortest path using at most k edges. In

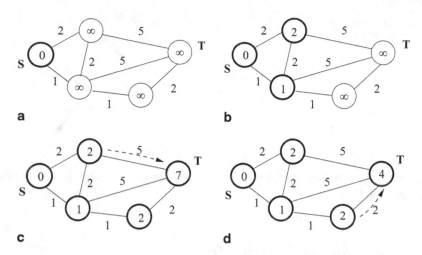

Fig. 2.6 Bellman-Ford Algorithm for finding the shortest paths: **a** Original graph, **b** Move to the direct neighbors and no change in relaxation, **c** Move to the next neighbors, and **d** Value changed in relaxation

other words, from S, if we use one edge, we can only get to the neighboring cities. If we use two edges, we can get to the neighboring cities of the neighboring cities. Then, we update the shortest distance on all vertices we can reach on the path with at most two edges. Continuing this idea, we can reach all vertices by using at most $n - 1$ edges and we can finish our task.

In the detailed algorithm, we always mark or record the distance from the source to every other vertex at the vertex. It is obvious that we always mark the smallest value (the shortest one, by using k edges). When we have used $(n - 1)$ edges, we would have the solution.

Algorithm 2.2 : (The Bellman–Ford Algorithm) Find the shortest paths from a vertex S to all vertices in graph G (Fig. 2.6).

Step 1   Mark all vertices other than S as ∞, where $d(v)$ is the distance from S to v and $d(S) = 0$.

Step 2   For each vertex, follow an iterative procedure called relaxation: for each v where u is adjacent to v, check if $d(v) > d(u) + W(u, v)$. If so, then $d(v) = d(u) + W(u, v)$. Repeat Step 2 for $(n - 1)$ times.

Another algorithm is called Dijkstra's algorithm. Dijkstra's algorithm runs faster than the Bellman–Ford algorithm, but Dijkstra's algorithm is unable to deal with graphs that have some negative weight edges. The Bellman–Ford algorithm is also easier to understand. It checks all possible links $n - 1$ times. During each iteration, we get the shortest path passing through k edges. Until we pass $n - 1$ edges, we naturally get the shortest path from S. The idea behind this algorithm is called dynamic programming, which means dynamically using the results already calculated.

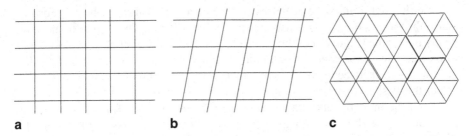

Fig. 2.7 Lattice graphs: **a** Squares, **b** Parallelograms, and **c** Triangles

2.3.4 Graph Homomorphism and Graph Isomorphism*

Homomorphism and isomorphism are related to functions between to two graphs. A graph homomorphic mapping is a function from the vertex set of G to the vertex set of G' so that if (a, b) is an edge of G, then $(f(a), f(b))$ is an edge of G'. If such a mapping exists, we say that G, G' are homomorphic.

If f is a 1-to-1 mapping (bijection), then f is called isomorphic. Homomorphism is called an edge-preserving mapping whereas isomorphism is an edge-preserving bijection.

Graph homomorphism has some applications to graph coloring problems. However, graph isomorphism mainly describes two graphs having "the same structure."

When the preserving edge can be allowed to shrink into a vertex, the function/mapping is called immersion and embedding. Graph immersion is an important concept to discrete surface reconstruction in this book. See Chap. 11.

A famous unsolved problem in computer science states: Given two graphs with the same number of vertices, is there a polynomial time algorithm to decide if these two graphs are isomorphic? This problem is called the graph isomorphism problem [1, 5].

2.4 Lattice Graphs, Triangulated Space, and Grid Space

To view a general graph as a discrete space is sometime too general. There are much more specific graphs that are better for the purpose of defining discrete spaces. These are called lattice graphs. A lattice graph is a simple graph with a distance measurement (called metric) of a geometric object. Lattice graphs are regular graphs, meaning that each edge has the same weight or represents the same distance in Euclidean space as in other spaces.

For instance, in regular triangles, each edge has the same length as in hexagons and parallelograms. (See Fig. 2.7) A more general 2D shape is called a polygon, which is formed by multiple edges as a closed boundary. In Chap. 5, we show more examples of these shapes.

A metric used to measure distances is dependent on geometric objects. For instance, the metrics on the plane and the sphere are different. The weight of an edge usually means the length of the shortest path between the two lattice points. In other words, the edge is usually the minimum distance curve in the space, called the geodesic curves between two adjacent lattice points. In computer graphics, we call lattices the meshes. Therefore, meshes on surfaces are good examples of lattices.

Two of the most popular lattice graphs are the triangulated space and the grid space. See Fig. 2.7b and c. Since the triangle is the simplest shape containing 2D information. It is called a 2D simplex. Adding another point in a new dimension, we will have a 3D simplex that contains four end points. Any geometric shape can be partitioned using simplexes in general. So mathematicians treat the simplex as the most significant discretization unit. However, this is difficult to represent in computers since the computer memory or disk storage are arranged as arrays, which is similar to grid spaces.

A grid space is a special lattice in which each point is at the integer coordinate location in Euclidean space and the edges are usually parallel to the coordinate lines. In other words, a grid graph is a graph whose vertices correspond to the points with integer coordinates. For instance, in a 2D plane, x-coordinates are in the range $1,...,n$, and y-coordinates are in the range $1,...,n$. A grid space is similar to a TV screen or a mathematical 2D array.

The grid space is also called the grid-cell space, which is the main topic of concern in this book—digital space. However, digital space has more inner-meanings than a grid graph, and we discuss it in further details in the following sections of this chapter.

In summary, we can usually view a lattice as a graph embedded in Euclidean space. Lattices only contain points and edges, where edges are usually straight lines. Examples, including meshes for computer graphics, are very popular. The lattice graph differs from the algebraic lattice, which is defined on a partially ordered set where any two elements have a supremum and infimum in the algebraic lattice [15].

2.5 Basic Concepts of Digital Spaces

Digital space has two definitions. First, in the narrow sense, a digital space is a discrete space in which each point can be defined as an integer vector, i.e. each component of the vector is an integer. Second, in the general sense, the space is a digitized space or discretely sampled space that is saved in digital form. In this book, we usually reference the first definition of grid space when discussing digital space.

2.5.1 2D and 3D Digital Spaces

Let us consider a two-dimensional digital space Σ_2. It contains all integer points of a Euclidean plane, $\mathbf{E}_2$. A point P (x, y) in Σ_2 has two horizontal $(x, y \pm 1)$ and two

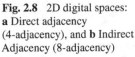

Fig. 2.8 2D digital spaces:
a Direct adjacency
(4-adjacency), and **b** Indirect
Adjacency (8-adjacency)

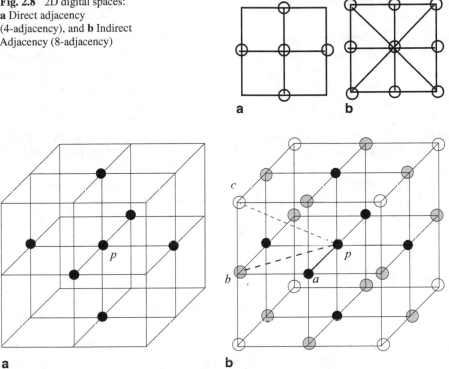

a   **b**

a   **b**

Fig. 2.9 3D digital spaces: **a** Direct adjacency (6-adjacency), and **b** Indirect Adjacency (26-adjacency)

vertical neighbors $(x \pm 1, y)$. These four neighbors are called directly adjacent points of p called 4-adjacency. p also has four diagonal neighbors: $(x \pm 1, y \pm 1)$. These eight (horizontal, vertical and diagonal) neighbors are called general (or indirect) adjacent points of p, which is called 8-adjacency. (Fig. 2.8 shows the digital points and their neighborhood.) In 2D, a connected component based on 4-adjacency is called 4-connected component while a component based on 8-adjacency is called 8-connected. Digital spaces can be represented as a graph. However, digital space has its own meaning in Euclidean space, it contains all meanings of the digitization of Euclidean space. Indirect adjacency also includes the property of non-Jordan space (a closed curve might not separate a plane into two parts), and is therefore more complicated than it appears.

We can extend the definition of Σ_2 to Σ_3. Σ_3 is formed by all integer points of 3D Euclidean space, $\mathbf{E}_3$.

In Σ_3, a point has 6 directly adjacent points and 26 indirectly adjacent points. Therefore, saying that two points in Σ_3 are connected has two definitions for a path: (1) Using directly adjacent points is called 6-connected, and (2) Using indirectly adjacent points is called 26-connected (Fig. 2.9).

2.5.2   mD Digital Spaces*

In general, Σ_m represents a special graph $\Sigma_m = (V, E)$. V contains all integer grid points in the m dimensional Euclidean space. The edge set E of Σ_m is defined as $E = \{(a, b) | a, b \in V \& d(a, b) = 1\}$, where $d(a, b)$ is the distance between a and b. In fact, E contains all pairs of adjacent points. Because a is an m-dimensional vector, $(a, b) \in E$ means that only one component, the i-th component, is different in a and b, $|x_i - y_i| = 1$, and the rest of the components are the same where $a = (x_1, ..., x_m)$ and $b = (y_1, ..., y_m)$. This is known as direct adjacency. One can define indirect adjacency as $\max_i |x_i - y_i| = 1$. Σ_m is usually called an m-dimensional digital space.

Formally, two points $a = (x_1, x_2, ..., x_m)$ and $b = (y_1, y_2, ..., y_m)$ in Σ_m are directly adjacent points, or we say that p and q are direct neighbors if

$$d_D(a, b) = \sum_{i=1}^{m} |x_i - y_i| = 1.$$ (2.1)

a and b are indirectly adjacent points if

$$d_I(a, b) = \max_{1 \leq i \leq m} |x_i - y_i| = 1.$$ (2.2)

Note: "Indirectly adjacent points" include all directly adjacent points here. We use general adjacency instead of indirect adjacency in most cases. Since direct adjacency is more stringent than indirect adjacency, in this book, when we discuss adjacency without further specifications, we are referring to direct adjacency [7].

In real world problems, digital space cannot just be viewed as a graph or a digitization. The meanings of and how to choose connectivity of digital space is much more complex.

2.5.3   Points, Line-cells, and Surface-cells in Digital Space

In digital space, the point (0-cell) is the basic element. This is because our devices can only collect points, but other information such as connectivity (edges) are human's interpretations.

As we can see, a line-segment is a line-cell or 1-cell, and a triangle is a 2-cell. In topology, cells are the basic unit in forming a complex shape or object. A k-dimensional cell is usually called a k-cell. In order to consider more complex shapes or objects in digital space, we need to define cells in each dimensions. In this section, we only include point-cell, surface-cell, and 3-cell. We discuss general k-cells in Chap. 5.

A surface-cell is a set of 4 points which form a unit square parallel to the coordinate plane. A 3-dimensional cell (or 3-cell) is a unit cube which includes eight points.

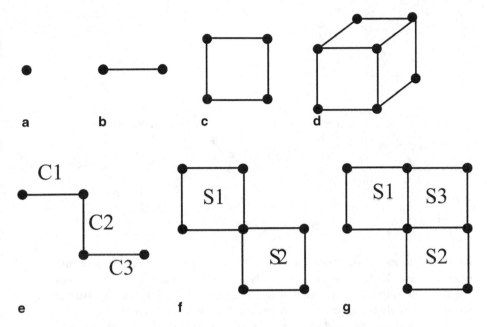

Fig. 2.10 Examples of basic unit cells and their connections: **a** 0-cells, **b** 1-cells, **c** 2-cells, **d** 3-cells, **e** Point-connected 1-cells, **f** Point-connected 2-cells, and **g** Line-connected 2-cells

By the same reasoning, we may define a k-cell. Fig.-2.5(a–d) shows a point-cell, line-cell, surface-cell, and 3-cell, respectively.

Now, let us consider the concepts of adjacency and connectedness of (unit) cells. Two points p and q (point-cells, or 0-cells) are connected if there exists a simple path $p_0, p_1, ..., p_n$, where $p_0 = p$ and $p_n = q$, and p_i and p_{i+1} are adjacent for $i = 1, ..., n - 1$.

Two cells are point-adjacent or 0-adjacent if they share a point. For example, line-cells $C1$ and $C2$ are point-adjacent in Fig. 2.10e, and surface-cells $s1$ and $s2$ are point-adjacent in Fig. 2.10f. Two surface-cells are line-adjacent or 1-adjacent if they share a line-cell. For example, surface-cells $s1$ and $s3$ in Fig. 2.10g are line-adjacent.

Two line-cells are point-connected or 0-connected if they are two end elements of a line-cell path in which each pair of adjacent line-cells is point-adjacent. For example, line-cells $C1$ and $C3$ in Fig. 2.10e are point-connected. Two surface-cells are line-connected or 1-connected if they are two end elements of a surface-cell path in which each pair of adjacent surface-cells are line-adjacent. For example, $s1$ and $s2$ in Fig. 2.10g are line-connected.

2.5.4 Points in Digital Space and Data in Real World

What are digital objects such as curves and surfaces in digital space? Not only we see and recognize when they are drawn out. but we also want the computer or algorithms

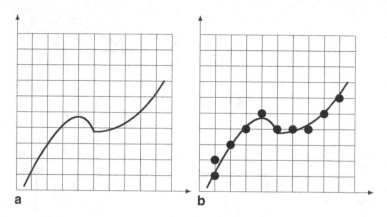

Fig. 2.11 **a** Curves in continuous space, and **b** curves in Σ_2 represented by *dots*

to be able to recognize them automatically. This is a key motivation to this new research area, digital geometry: Not only to digitize a continuous object, but also use procedures to generate a set that has the properties of digital curves/surfaces or more importantly recognize them when an arbitrary set is given.

Curves and surfaces in discrete spaces are different than they are in continuous spaces. Most of the related research work deals with curves and surfaces in Σ_m, the m-dimensional grid space. This space contains all integer grid points, we usually call such a space a digital space.

In Fig. 2.11, without looking at the curve (a), how does one know that digital points (b) is a digitization of the curve (a) and not some curve from elsewhere? In other words, there are hundreds of continuous curves have the same digitization. No one can answer this question. We can only say that a digital object is a set of points. What makes the set look a certain way is just human interpretation. The interpretation is the topological geometric structure of this point set. This argument shows us the demand of studies on digital objects in mathematics, not only the digitization of the continuous object.

2.6 Characteristics of General Discrete Spaces

In 2D, the simplest 2D discrete space is triangulated space, where the graph is made by a collection of triangles. We call the process of making the decomposition (triangulation).

Therefore, a 2-cell can be a triangle. It can also be a rectangle and also be any type of polygon. A polygon can be viewed as a simple closed path in a graph, but each edge is a straight line segment in 2D Euclidean space. So the most general type of 2D discrete space is the graph that consists of a number of polygons that may have different sizes (Fig. 2.12).

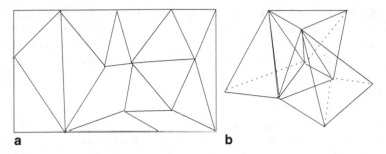

Fig. 2.12 Decompositions of 2D space and 3D Space: **a** 2D decomposition, and **b** 3D decomposition

In this section, we investigate the basic characteristics of a general discrete space. First we show the triangulation in Fig. 2.12.

In this example, we can see that each 1-cell is contained by two 2-cells. There are no 3-cells. If a 1-cell is contained by only one 2-cell, then this 1-cell must be on the boundary. This is true for any polygonal representation of a surface. A discrete surface is closed if each 1-cell in the object is contained within exactly two 2-cells.

When we fill the closed surface with water or other substances, we will have a 3D object, or 3-cell if it is small enough such as a unit ball or a cube.

Let us use the following definition for digital or discrete space (in 2D) as the conclusion of this section. One of the major topics of this book is the general discrete space in relation to discrete manifolds. We only present preliminary consideration here.

Definition 2.1 Let M be a set of vertices. M is connected in terms of using point-paths. A 1-cell is an edge or line-segment. A 2-cell is a simple polygon (that does not contain any other polygons). M is said to be a 2D space if each 1-cell is contained within one or two 2-cells.

Definition 2.2 Let M be a set of vertices. M is connected in terms of using point-paths. A 3-cell is a polyhedron. M is said to be a 3D space if each 2-cell is contained within one or two 3-cells.

The above definition can be extended to define kD discrete spaces in a recursive way. We discuss this in Chap. 7. We need to note that the above two definitions are not very strict. We give more precise definitions in Chaps. 5 and 7. A discrete manifold can also be described by simplicial complexes [3, 10] and finite topologies introduced by Kovalevsky [13]. We discuss this in Chap. 9.

2.7 Historical Remarks on Digital Space

A new branch of mathematics is usually established or invented to fit the needs of the time, especially the need to solve real world problems. Digital images and computer graphics are both based on digital pixels or voxels. The pixels and voxels are arranged in two and three dimensional spaces that usually correspond to two and three dimensional arrays. The geometric relationship among pixels and voxels forms a new geometry: digital geometry. More generally, digital geometry mainly deals with finite sets in grid space within Euclidean space. It has a closer connection to discrete (or combinatorial) geometry, which studies discrete sets in Euclidean or measured space.

Why is a geometric object in digital space important to us? In computer vision, an image (or picture) is stored in the memory. To analyze the image, one usually needs to find each object in the image. The easiest way to describe an object is to find its boundary, which is a set of dots. Since every object in digital space is a set of dots, we refer the boundary dots as digital boundary curves. A problem occurs: what is a digital curve mathematically or formally?

For instance, a binary image is a $\{0, 1\}$-function. It can be stored in an array. So, studying geometric and topological properties for a connected "0" or "1" point set in a 2D or 3D array has become an interesting research topic [12, 20].

In the early 1970s, Rosenfeld started the study of digital curves [17, 18]. In 1979, Rosenfeld suggested the study of digital topology [19]. He tried to use topological properties of digital spaces to build a solid foundation for image processing and computer vision [9, 15, 21]. However, in the early days of digital geometry and topology, at least in image processing and computer vision, researchers studied some useful properties of geometry and topology in digital space. The research was mostly isolated in the image processing community; it was not closely related to advanced geometry and topology.

Due to the fact that most researchers in image processing were more familiar with discrete mathematics rather than continuous mathematics, the methodology of digital space was more related to set theoretical methods and not the simulation of continuous methods. In other words, an object is viewed as a set of digital points. Therefore, there were no attempts to define digital curves and digital surfaces directly, tracking boundaries of a 3D solid digitally and finding the thinning of a digital object became the central topics.

Because of this, the independent work on digital surfaces and digital manifolds were remarkable compared to the ordinary digitization of a surface or manifold in continuous space. These definitions lead to some of the fast tracking algorithms and recognition algorithms, which were impossible when based on only their continuous counterparts.

To establish a new theory, there must be considerable independent work that compliments existing research areas. In Chaps. 4, 5, and 6, we examine the digital curve, surface, 3D solid, and k-manifolds in digital space.

Of course, to define a k-manifold, one must first define k-cells. A kD digital space is formed by individual k-cells. Therefore, this chapter gives an introduction to the basic concepts.

However, for general discrete space, and not digital space, the k-cell is much more difficult to deal with. In continuous space, we use k-simplex. We know a k-simplicial complex or CW-complex were defined by topologists many years ago. However, the k-simplicial complex cannot just be viewed as a discrete manifold. In Chap. 7, we mainly focus on discrete manifolds.

The key for both computer scientists and mathematicians is to find the best triangulation. Basically, a discrete manifold means the discretization of a continuous manifold. For example, the simplex decomposition of a manifold is a discrete manifold. However, a manifold may have thousands of different discretizations. For computer scientists, it is important to define what a discrete manifold is; otherwise, there is no way to track or search a discrete manifold using computers.

The definition of a discrete manifold must not contradict unreasonably that of a continuous manifold. However, their differences are obvious. For example, a neighborhood of a point in a continuous manifold contains an infinite number of points, but a discrete manifold itself may only contain a finite number of points. There are many real discrete manifolds. For instance, the set Σ_m containing all integer coordinate points of an m-dimensional Euclidean space is a discrete manifold [14]. We sometimes call this manifold a digital manifold. A simplex decomposition Δ_n of an n-dimensional manifold is also a discrete manifold.

There is a recursive way of defining k-cells. Roughly speaking, we can see that a discrete k-cell is a set of $(k-1)$-cells. Those $(k-1)$-cells are joined in a closed way, meaning that each of the $(k-2)$-cells contained by the $(k-1)$-cells is only exactly two $(k-1)$-cells. This k-cell also maintains a minimal volume. Plus, there is no hole inside the k-cell. This is the idea of defining the k-cell.

A discrete k-manifold can be defined by k-cells: Each $(k-1)$-cell is contained by one or two k-cells. These k-cells are connected, and there are no $(k+1)$ cells in the union of the k-cells.

Generally, a discrete manifold is a graph with some additional geometric and topological structures. The general definition of discrete manifolds was given by Chen [4–6].

In summary, the strict mathematical definition of discrete manifolds must not depend on continuous manifolds, because it is impossible to determine an infinite number of elements by an algorithm that only uses a finite set of actions.

References

1. G. Agnarsson and L. Chen, On the extension of vertex maps to graph homomorphisms, Discrete Mathematics, Vol 306, No 17, 2021–2030, Sept. 2006.
2. G. Agnarsson and R. Greenlaw, Graph theory: modeling, applications, and algorithms. Pearson Prentice Hall, Upper Saddle River, NJ, 2007.
3. P. S. Alexandrov, Combinatorial Topology, New York: Dover, 1998.
4. L. Chen, Generalized discrete object tracking algorithms and implementations, Melter, Wu, and Latecki ed, Vision Geometry VI, SPIE 3168, pp 184–195, 1997.
5. L. Chen, Note on the discrete Jordan curve theorem, Vision Geometry VIII, Proc. SPIE Vol. 3811, 82–94. 1999.
6. L. Chen, *Discrete Surfaces and Manifolds: A theory of digital-discrete geometry and topology*, 2004. SP Computing.

7. L. Chen and J. Zhang, Digital manifolds: A Intuitive Definition and Some Properties, Proceedings of the Second ACM/SIGGRAPH Symposium on Solid Modeling and Applications,
   Montreal, 1993, 459–460.   .
8. T. H. Cormen, C.E. Leiserson, and R. L. Rivest, Introduction to Algorithms, MIT Press, 1993.
9. R. C. Gonzalez, and R. Wood, *Digital Image Processing*, Addison-Wesley, Reading, MA,
   1993.
10. Goodman, O'Rourke, Handbook of Discrete Geometry, CRC, 1997.
11. F. Harary, Graph theory, Addison-Wesley, Reading, Mass., 1969.
12. R. Klette and A. Rosenfeld, Digital Geometry, Geometric Methods for Digital Picture Analysis,
    series in computer graphics and geometric modeling. Morgan Kaufmann, 2004.
13. V. A. Kovalevsky, Finite topology as applied to image analysis, Computer Vision, Graphics
    and Image Processing, Vol. 46, 1989, pp. 141–161.
14. D.G. Morgenthaler and A. Rosenfeld, Surfaces in three-dimensional images, *Inform. and
    Control* 51, 1981, 227–247.
15. T. Pavilidis, Algorithms for Graphics and Image Processing, Computer Science Press, Rockville
    MD, 1982.
16. K. H. Rosen, Discrete Mathematics and Its Applications McGraw-Hill Higher Education, Jan
    2007.
17. A. Rosenfeld, Connectivity in digital pictures, Journal of the ACM, Vol. 17, 1970, pp. 146–160.
18. A. Rosenfeld, Arcs and curves in digital pictures, Journal of the ACM, Vol. 20, 1973, pp. 81–87.
19. A. Rosenfeld, "Digital topology," Amer. Math. Monthly, Vol 86, pp 621–630, 1979.
20. A. Rosenfeld, Three-dimensional digital topology, Inform. and Control 50, 1981, 119–127.
21. A. Rosenfeld and A.C. Kak, *Digital Picture Processing*, 2nd ed., Academic Press, New York,
    1982

Chapter 3
Euclidean Space and Continuous Space

Abstract This chapter introduces Euclidean spaces, topological spaces, and their relationships to discrete spaces. We first introduce the concept of metrics, the distance measure of Euclidean spaces. Then, we introduce general continuous spaces—topological space. At the end, we discuss the relationship between continuous spaces and discrete spaces.

In continuation of the previous chapter, but in the opposite direction, we present the basic formulas for Euclidean space, functions, and linear transformations in this space. For the relationship of discrete and continuous space, triangulation (simplicial decomposition) plays an important role. We give a brief introduction to the method. In addition, we also discuss some other decomposition methods. Decomposition is the method of making continuous spaces into discrete spaces. Changing from discrete spaces to continuous spaces is called fitting or reconstruction, which we discuss in Chap. 11.

Keywords Euclidean space · Metric · Distance · Function · Calculus · Topology

3.1 Euclidean Space and Properties

In mathematics, Euclidean space came from Euclidean geometry, which is usually planar and three-dimensional geometry. It is named after the Greek mathematician Euclid of Alexandria who authored the book *Elements*. Euclidean geometry uses certain postulates or axioms. Then, based on those postulates, the other properties, called theorems, can be deduced. Modern mathematics uses Cartesian coordinates to define Euclidean space. This was because mathematicians found that the most important thing in Euclidean space was the measure of distance. This allows mathematicians to use algebra to solve problems and extend Euclidean space to any dimensions. Euclidean space is the most important continuous space.

© Springer International Publishing Switzerland 2014

L. M. Chen, *Digital and Discrete Geometry,* DOI 10.1007/978-3-319-12099-7_3

3.1.1   Euclidean Spaces

Let R be the real number set. An n-tuples of real numbers, $(x_1, x_2, ..., x_n)$, is called a vector on R. An n dimensional (nD) Euclidean space is defined on R^n, an nD real number vector space, where

$$R^n = R \times R \times \cdots \times R = \{(x_1, x_2, ..., x_n) | x_i \in R\}, \tag{3.1}$$

is called the Cartesian product on R. The measure of distance between two points $x = (x_1, x_2, ..., x_n)$ and $y = (y_1, y_2, ..., y_n)$ is defined as

$$d(x, y) = \sqrt{\sum_{i=1}^{n} (x_i - y_i)^2}. \tag{3.2}$$

x and y can also be viewed by two vectors from origin to points $(x_1, x_2, ..., x_n)$ and $(y_1, y_2, ..., y_n)$. d is called the metric (of the space), or Euclidean metric. Thus, the n-D Euclidean space can be written as $\mathbf{E}_n = (R^n, d)$, or simply R^n with default distance measure.

Based on the metric, we can define the length of a vector as the distance from the origin (i.e. $(0, 0, ..., 0)$) as

$$\|x\| = \sqrt{\sum_{i=1}^{n} (x_i)^2},$$

which is called the Euclidean norm, i.e.

$$\|x\|_2 = \left(\sum_{i=1}^{n} x_i^2 \right)^{1/2}.$$

Based on these simple definition, the angle of two vectors x and y, θ, can be determined by

$$cos(\theta) = \left(\frac{\sum_{i=1}^{n} x_i y_i}{\|x\| \|y\|} \right). \tag{3.3}$$

For further simplicity, we define the inner product (also called dot product) of x and y as

$$x \cdot y = x_1 y_1 + \cdots + x_n y_n. \tag{3.4}$$

Therefore,

$$\cos(\theta) = \left(\frac{x \cdot y}{\|x\| \|y\|} \right). \tag{3.5}$$

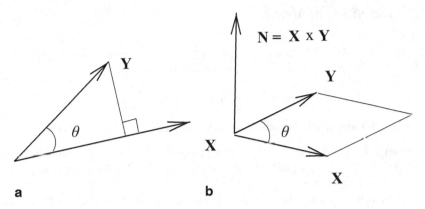

Fig. 3.1 The inner product and the vector product of vectors: **a** The inner product, and **b** The vector product

The meaning of the dot product is a projection from x to y, in particular, $\frac{x}{\|x\|}$ is a unit vector on the x axis, so $y \cdot \frac{x}{\|x\|} = \|y\| \cos(\theta)$ is the length of the project from y to x. See Fig. 3.1a.

Since $\cos(\theta) \leq 1, x \cdot y \leq \|x\| \|y\|$. This is called the Cauchy–Schwartz inequality, a very important inequality in Euclidean space. Using this inequality, we can prove the triangle inequality:

$$\|x\| + \|y\| \geq \|x + y\|. \tag{3.6}$$

The vector product (cross product) of two vectors x and y is a vector that is perpendicular to the plane determined by x and y. Let $\mathbf{n}$ be the unit vector of this vector product, then

$$x \times y = \|x\| \|y\| \sin(\theta) \cdot \mathbf{n}. \tag{3.7}$$

The norm of $x \times y$, $\|x \times y\|$ is the positive area of the parallelogram having x and y as sides. See Fig. 3.1b. In 3D, we can use a matrix and its determinant to represent the vector product.

The cross product can also be expressed by the following determinant:

$$\mathbf{x} \times \mathbf{y} = \begin{vmatrix} \mathbf{i} & \mathbf{j} & \mathbf{k} \\ x_1 & x_2 & x_3 \\ y_1 & y_2 & y_3 \end{vmatrix} \tag{3.8}$$

where $\mathbf{i}, \mathbf{j}$, and $\mathbf{k}$ are three unit vectors in three dimensions. That is,

$$\mathbf{x} \times \mathbf{y} = ((x_2 y_3 - x_3 y_2), (x_3 y_1 - x_1 y_3), (x_1 y_2 - x_2 y_1)). \tag{3.9}$$

3.1.2   Definition of Metrics

The standard Euclidean distance in R^n can also be given by $d(p,q) := \|p - q\|$.
Another form of the triangle inequality is the following.

$$d(p,q) + d(q,r) \geq d(p,r)$$

A metric for any space, not only Euclidean Space, can be defined as follows:

Definition 3.1 A metric on a set X is a mapping $d : X \times X \to R$ such that:
(1) $d(p,q) \leq 0$, with equality if and only if $p = q$. (2) $d(p,q) = d(q,p)$. (3)
$d(p,q) + d(q,r) \geq d(p,r)$.

A metric usually means the measure of distance between two points. It can also
mean the shortest path between two vertices in a graph as defined in Chap. 2. In other
words, the distance between two vertices can be defined as the shortest path.

3.1.3   Spheres and Distance on Spheres

The sphere is a special geometric shape that can be in any dimension. A circle is a
1D sphere. We define an arc as part of a circle. It has two end points p and q on
the circle. Let o be the centre (origin) of the circle. The angle of the arc is the angle
of two line-segments $\bar{o}p$ and $\bar{o}q$ The length of the arc on the circle is the distance
between two points when we travel on the circle. We have,

$$arcLength = r \cdot \theta$$

where θ is the angle of the arc.

A sphere in 3D Euclidean space can be described as a set of points where every
point on the sphere has the same distance to its origin, namely r, the radius of the
sphere. A great circle of a sphere is defined as the circle that cuts the sphere into two
equal parts.

The shortest path between two points on the sphere is always along a great circle.

In other words, distance on the sphere is defined as the shortest length of all arcs
from one point to another on the sphere. See Fig. 3.2. This equals the length of the
arc in the great circle of the sphere.

The length of the arc is $d_{sphere}(X, Y) = R\theta$, where R is the radius of the great cir-
cle. According to the cosine formula given in the last subsection, $\theta = \arccos \frac{X\dot{Y}}{\|X\|\|Y\|}$.
Since $\|X\| = \|Y\| = R$, therefore,

$$d_{sphere}(X, Y) = R \cdot \arccos \left(\frac{x_1 x_2 + y_1 y_2 + z_1 z_2}{R^2} \right) \tag{3.10}$$

Fig. 3.2 The shortest path
between two points on the
sphere

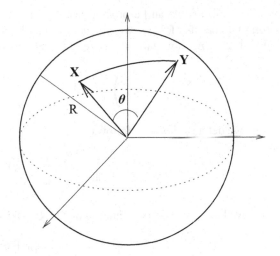

This distance measure, a metric on sphere $d_{sphere}(X, Y)$, is one of the general forms of the Riemann metric. The distance on the sphere, here a 2D sphere, is an example of non-Euclidean space. It also satisfies Definition 3.1.

3.1.4 Two Inequalities of Euclidean Space*

Two inequalities are important in Euclidean space. First, the Cauchy−Schwarz inequality states

$$\left(\sum_{i=1}^{n} x_i y_i \right)^2 \leq \left(\sum_{i=1}^{n} x_i^2 \right) \left(\sum_{i=1}^{n} y_i^2 \right). \tag{3.11}$$

We can see that $\sum_{i=1}^{n} x_i y_i$ is the inner product of two vectors x and y in Euclidean space. It is denoted as $x \cdot y$ or $< x, y >$. To understand this formula, we can see the definition of the angle of two vectors x and y as we have defined above:

$$|x \cdot y| = \|x\| \|y\| |\cos \theta|.$$

Because $|\cos \theta|$ is always less than or equal to 1, therefore,

$$|x \cdot y| \leq \|x\| \|y\|.$$

Second, the Minkowski's inequality is:

$$\left(\sum_{i=1}^{n} |x_i + y_i|^2 \right)^{1/2} \leq \left(\sum_{i=1}^{n} |x_i|^2 \right)^{1/2} + \left(\sum_{i=1}^{n} |y_i|^2 \right)^{1/2} \tag{3.12}$$

This inequality can be proven based on the Cauchy–Schwarz inequality. The complete proofs of these two inequalities can be found in [11]. Generally, for any $p = 1, 2, ..., n, ...$, we have the general triangle inequality:

$$\left(\sum_{i=1}^{n} |x_i + y_i|^p \right)^{1/p} \leq \left(\sum_{i=1}^{n} |x_i|^p \right)^{1/p} + \left(\sum_{i=1}^{n} |x_i|^p \right)^{1/p} \tag{3.13}$$

If we define p-norm as follows:

$$\|x\|_p = \left(\sum_{i=1}^{n} |x_i|^p \right)^{1/p} \tag{3.14}$$

then we have a short representation of Minkowski's inequality:

$$\|x + y\|_p \leq \|x\|_p + \|y\|_p. \tag{3.15}$$

It is interesting to indicate that the direct adjacency in digital space is equivalent to 1-norm and the general (indirect) adjacency is $p = \infty$-norm.

3.2   Functions on Euclidean Space

A function is a mapping from one set to another. For example, the polynomial $f(x) = x^2 + x + 1$ is a function that maps E_n to R if x is an n dimensional vector. A function is called a *one-to-one* function if no two elements in X map to the same element in Y. A function is called *onto* if all elements in Y have been mapped, meaning that for $y \in Y$, there exists an $x \in X$, such that $f(x) = y$. A function f that is both one-to-one and onto is called invertible, its inversion is denoted by f^{-1}. So we can see that arccos is $\cos^{-1}$.

If a function does not have any jumps or gaps, it is called a continuous function. In mathematics, we use *limits* to define a continuous function.

$f(x)$ is said to be continuous at $x_0 \in R$ if

$$f(x_0) = \lim_{\delta \to 0} f(x_0 + \delta) = \lim_{\delta \to 0} f(x_0 - \delta). \tag{3.16}$$

The differentiation of a function is introduced by the derivatives as defined below:

$$f'(x) = \lim_{y \to x} \frac{f(y) - f(x)}{y - x} \tag{3.17}$$

$f(x)$ is said to be differentiable in the first order if $f'(x)$ is continuous. We can further define nth order differential functions. A function is said to be infinitely differentiable if it has any order of differentiation. Such a function is called a smooth function. We use C^k to represent all functions whose kth derivative is continuous.

We usually use $\frac{dy}{dx}$ or $\frac{\Delta y}{\Delta x}$ to represent $f'(x)$. dy and dx are symbolic representations of differentiation. The second order of the derivative can be represented as $\frac{d^2 y}{dx^2}$, which is the short representation of $d(\frac{dy}{dx})/dx$.

For the function in n-dimensional space, $f(x_1, x_2, ..., x_n)$, we will have a derivative in each direction, that can be represented by $\frac{\partial f}{\partial x_i}$. These are all very basic concepts in calculus [16, 18].

In common sense terms, a curve is a function with only one variable on the X-axis and a surface is a function on the XY-plane. However this type of function cannot be used to represent a sphere since, for a domain point in XY-plane, a spheres has two values that share this domain point. A curve (or surface), that can be written as $f(x)$, is called a non-parametric curves (or surface). The more precise definition of curves in parametric form is presented in Chap. 4.

When we view a function in Euclidean space, such as a surface, then the surface is also a continuous space. The definition of metric on a surface (function) has particular importance since it will lead a new type of geometry called Riemannian geometry. We discuss this further in Chap. 13.

3.2.1 Geometric Transformation, Linear Transformation, and Matrix Algebra

The simplest geometric transformation is used to transform an object from one coordination system to another. Translation and rotation are the two basic moves. This is called isometric (congruent) transformation since the distance between two points in the original object will not be changed after the transformation. This type is popular in computer graphics since people want to observe an object from different angles.

The formula of translation is simple. It can be represented by

$$\begin{cases} u = x - u_0 \\ v = y - v_0 \end{cases} \tag{3.18}$$

To rotate by an angle θ clockwise at from the origin, the equation is

$$\begin{cases} u = x \cos \theta + y \sin \theta \\ v = -x \sin \theta + y \cos \theta \end{cases} \tag{3.19}$$

In matrix form, the same equation is

$$\begin{pmatrix} u \\ v \end{pmatrix} = \begin{pmatrix} \cos \theta & \sin \theta \\ -\sin \theta & \cos \theta \end{pmatrix} \begin{pmatrix} x \\ y \end{pmatrix}$$

We can see examples of the translation and rotation in Fig. 3.3:

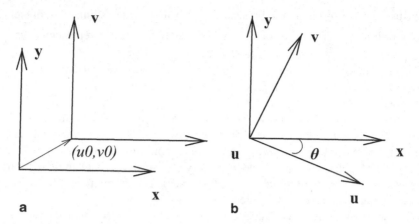

Fig. 3.3 The shortest path between two points on the sphere

Translations and rotations are linear transformations. A linear transformation between two Euclidean spaces X and Y is a function $f : X \rightarrow Y$ satisfying:

(1) $f(u + v) = f(u) + f(v)$ for $u, v \in X$, and
(2) $f(\lambda u) = \lambda f(v)$ for any scalar number λ (i.e. $\lambda \in R$).

This definition can be extended to vector spaces. In general, a linear transformation $f : X \rightarrow Y$ can be done using matrix multiplication. Let $\mathbf{u} \in X$ be an m-dimensional vector and $\mathbf{v} \in Y$ be an n-dimensional vector. We can define an $n \times m$ matrix A as

$$\begin{pmatrix} a_{1,1} & a_{1,2} & \cdots & a_{1,m} \\ a_{2,1} & a_{2,2} & \cdots & a_{2,m} \\ \vdots & \vdots & \ddots & \vdots \\ a_{n,1} & a_{n,2} & \cdots & a_{n,m} \end{pmatrix} \tag{3.20}$$

Then,

$$\mathbf{v} = A\mathbf{u}, \tag{3.21}$$

We transform vector $\mathbf{u}$ with m coordinates into vector $\mathbf{v}$ with n coordinates.

One of the most important concepts of linear or matrix algebra is that of the eigenvalues and eigenvectors, In this book, we will need to use these concepts multiple times. A brief introduction is as follows. For more details, refer to [8].

For an $n \times n$ matrix A, if we can find a real number λ and a vector $\mathbf{x} = (x_1, x_2, \cdots, x_n)$ such that,

$$A\mathbf{x} = \lambda\mathbf{x}, \tag{3.22}$$

then the multiplier λ is called an eigenvalue of A and $\mathbf{x}$ is an eigenvector. This formula is equivalent to

$$(A - \lambda I)\mathbf{x} = 0, \tag{3.23}$$

where I is the identity matrix, an $n \times n$ matrix with all diagonal elements assigned as 1 and all other elements as zero. Therefore, the determinant of $(A - \lambda I)$ is zero. Theoretically, $|(A - \lambda I)|$ is a polynomial of λ with n number of roots possible. For each root λ_i, we can solve the Eq (3.23) to get the eigenvectors.

We introduce other transformations such as the Fourier transform and the Radon transform in Chap. 11 when we discuss data analysis and reconstruction.

3.3 Topological Spaces and Manifolds

Topological space is a generalization of Euclidean space. Topology refers to the most general structure in geometry. A 2D or 3D object can be described using the terminology of a topological structure or topology.

The mathematical definition of topological space is abstract. The definition of a topological space is very general. It is based on the definition of topology on sets [1, 2].

Definition 3.2 Let X be a set and τ be a collection of subsets of X. (X, τ) is said to be a topological space if (X, τ) satisfies the following axioms: (1) The empty set and X are in τ. (2) The union of an arbitrary number of elements of τ is in τ (τ is closed under arbitrary union). (3) The intersection of any finite number of elements of τ is in τ (τ is closed under finite number of intersections).

In mathematics, τ is called a topology on X. The elements of X are called points, the elements of τ are called open sets. The complement of an open set A in X, denoted $X - A$, is called a closed set.

Almost all geometric objects or spaces we deal with are topological spaces. For a finite set X, the open set is also the closed set. A topological space can also refer to a function space in which each element of X is a function.

Functions on topological spaces usually mean that the function is on the base set X. We can also define a function between two topological spaces (X, τ) and (Y, τ'). For instance, $f : X \to Y$.

Intuitively, we say that two objects are topologically equivalent if there is a process that can continuously change one object into another. It can be defined as a continuous one-to-one onto function. We also say that these two objects have homeomorphism.

Definition 3.3 (X, τ) and (Y, τ') are said to be homeomorphic or topologically equivalent if there exists a continuous and invertible function f.

A (topological) n-manifold is a topological space $M = (X, \tau)$. Each of element (point) of X has an open nD neighborhood U_x that is continuously equivalent or homeomorphic to an nD Euclidean space. In other words, there is a continuous function $f_x : U_x \to E_n$ where its inversion f_x^{-1} is also continuous.

A smooth n-manifold is a manifold where for any two open sets U_x and U_y in M, $f_y \cdot f_x^{-1}$ on $f_x(U_x \cap U_y)$ is smooth or C^k-continuous. Smoothness is defined on E_n and through M. The intuitive meaning of this definition is that the local homeomorphic functions f_y and f_x on intersection $U_x \cap U_y$ guarantees that it can be smoothly put on $U_x \cap U_y$ without creating a bent angle in each local space.

Almost all geometric objects or spaces we deal with are topological spaces. Euclidean space is a topological space if we use natural open sets as elements in τ.

Example 3.1. In R, a line segment $[a, b]$ is a close set. $(0, 1) = [0, 1] - \{0, 1\}$ is an open set. If we have $E_1 = (R, \tau)$, then $\tau = \{s | s$ is an open set$\}$ is a topological space.

In general, a neighborhood in E_n can be defined as follows: for an element $a \in E_n$, if we define a neighborhood of a in E_n as all elements $x \in E_n$ such that $\|x - a\| < d$, where d is any number in R, then such a neighborhood is an open set. We also can define the union and the intersection of two balls are in τ.

Example 3.2 Euclidean metric induces a topology in (R, d). We can examine all three criterion of topology as follows: $< R, \tau >$ where τ is the collection of all open sets. According to the definition 3.2.:

(1) The union of any collection of open sets in τ is open.
(2) A finite intersection in τ is open. (The intersection of infinitive number of open sets can be converge to a single point. It is not an open set.)
(3) The empty set $\emptyset$ and R are open sets.

3.4   Decomposition: From Continuous Space to Discrete Space

This section introduces the most common types of discrete approximation of the continuous space: triangulation or simplicial decomposition. Then we extend the special space to general finite point space.

The commonly called triangulation was first introduced by H. Poincare. Simplicial complexes are also the basic foundations for topology and even for modern geometry.

The relationship between continuous space and discrete space is one of decomposition and fitting. To transform from continuous space to discrete space, we use decomposition, and changing from discrete space to continuous space requires fitting or reconstruction.

Decomposition is a partition. Partitioning a 2D plane into small squares is the easiest way to discretize a continuous space to a set of corner points, edges, and small square units. These points, edges, and unit-squares are discrete objects.

In theory, triangulation is the best method for decomposing a manifold into triangles, called 2-simplex. It was first used in combinatorial topology.

The common assumption is this: Any 2D smooth space has triangulation, and any smooth manifold has a representation using a collection of simplexes. This collection is called a simplicial complexes. The condition of smoothness is required here because a local smooth area can be triangulated. A non-smooth area is hard to evaluate.

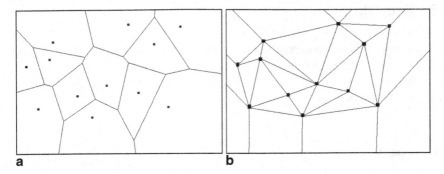

Fig. 3.4 Voronoi and Delaunay diagrams: **a** A Voronoi diagram; **b** Delaunay triangulation

A point is a 0-simplex, a line unit is a 1-simplex, and a triangle is a 2-simplex. Therefore, we can define a tetrahedron as a 4 point, 3-simplex (See Fig. 2.12b). Simplexes may not always be convenient in practice. However, as long as a cellular shape, or cell, is used, such as squares or cubes, we can still define topology using these objects. In modern topology, cellular complexes are very popular (especially CW−complexes) [12].

As a result, the decomposition of a space can be made by a cell complex. Since digital k-dimensional cube is special k-cell, it is obvious that k-cubes can be used and it is called digitization. It is the same as commonly used in sampling technology in electronic sensors, especially used in medical imaging and other industries. The sampled data is saved to a digital device as 2D or 3D arrays [6]. The characteristics of these samplings are regulated and each sample has equal distance to next sample.

When the samples collected in a random manner, the domain decomposition is different. In 2D, a popular method is called the Voronoi diagram. The Voronoi diagram partitions a plane into polygons, each polygon containing a sample point (also called a site); a point x inside a specific polygon containing site p if x is closer to p than to any other site [5, 9, 14].

The Voronoi diagram method has particular importance in science and engineering since the partition is made based on the closest distance to the particular site comparing to other sites. The dual diagram of the Voronoi decomposition is a triangulations of the domain, it is called Delaunay triangulation.

Delaunay triangulation is the most popular form among different types of triangulation. See Fig. 3.4 from [4].

We will discuss the detailed algorithms for Delaunay triangulation and the Voronoi diagram in Chap. 10.

3.5   Remark

Triangulation, polygon, and polyhedron decompositions have long history in mathematics [9, 13]. Some algorithms were also found by computer scientists [3, 7, 17]. Numerical analysis is the area that fits and reconstructs discrete data to continuous functions. We discuss this in Chap. 11.

References

1. P. S. Alexandrov, Combinatorial Topology, New York: Dover, 1998.
2. M. A. Armstrong, Basic Topology, rev. ed. New York: Springer-Verlag, 1997.
3. B. Chazelle, Triangulating a Simple Polygon in Linear Time. Disc. Comput. Geom. 6, 485–524, 1991.
4. L. Chen, *Digital Functions and Data Reconstruction*, 2013. Springer, New York.
5. T. H. Cormen, C. E. Leiserson, and R. L. Rivest (1993), Introduction to Algorithms, MIT Press, Cambridge, MA, 1993.
6. Euclid's Elements, Green Lion Press, 2002.
7. Garey, M. R.; Johnson, D. S.; Preparata, F. P.; and Tarjan, R. E. "Triangulating a Simple Polygon." Inform. Process. Lett. 7, 175–179, 1978.
8. I. M. Gelfand, Lectures on Linear Algebra, Dover, New York, 1989.
9. J. E. Goodman, J. O'Rourke, Handbook of discrete and computational geometry, CRC Press, Inc., Boca Raton, FL, 1997.
10. F. Harary, *Graph Theory,* Addison-Wesley, Reading, 1972.
11. G. H. Hardy, Littlewood, J. E.; Pólya, G.. Inequalities. Cambridge Mathematical Library (second ed.). Cambridge: Cambridge University Press. 1952.
12. A. Hatcher, *Algebraic Topology*, Cambridge University Press, 2002.
13. N. J. Lennes, "Theorems on the Simple Finite Polygon and Polyhedron." Amer. J. Math. 33, 37–62, 1911.
14. F. P. Preparata, M. I. Shamos, Computational geometry: an introduction, Springer-Verlag New York, Inc., New York, NY, 1985
15. K. H. Rosen, Discrete Mathematics and Its Applications McGraw-Hill Higher Education, Jan 2007.
16. J. Stewart, Calculus, Brooks/Cole Publishing Company, Pacific Grove, CA, 4th ed, 1999.
17. Tarjan, R. and van Wyk, C. "An Algorithm for Triangulating a Simple Polygon." SIAM J. Computing 17, 143–178, 1988.
18. G. B. Thomas, Jr., Calculus and Analytic Geometry, 4th ed. Addison-Wesley, Reading, Mass., 1969.

Part II
Digital Curves, Surfaces, and Manifolds

Chapter 4
Digital Planar Geometry: Curves and Connected Regions

Abstract In this chapter, we introduce basic 2D digital geometry. The main topic in 2D geometry is curves. A 2D digital curve is a simple path in Σ_2. A simple closed digital curve is usually the boundary of a connected component. We first discuss how we precisely define a curve in a graph and Euclidean space, then we discuss how we represent digital curves. Digital curves have two important applications in computer graphics and computer vision: (a) Construction of a digital line when two end points are given, and (b) Determination of a closed digital curve to identify a connected region in computer vision. At the end of this chapter, we present two classic theorems related to 2D digital geometry: Pick's theorem and Minkowski's theorem. In addition, we discuss the basic concept of image segmentation, one of the major applications of 2D digital planes.

Keywords Curve · Plane · Digital curve · Simple path · Curve Representation · Connectivity · Component · Image

4.1 General Continuous Curves and Discrete Curves

What is a continuous curve? A moving point generates a curve. In other words, the trace of a moving point forms a curve. Another example is if we have an open rubber band with length 1, we can extend and bend it into any shape in 2D, 3D, or an even higher dimension. Since bending must be continuous, a curve is also continuous.

Mathematically, we define a curve as a continuous function from $[0, 1]$ to R^3 or E_n as an n-dimensional curve. For example, a 3D curve f is

$$f : [0, 1] \to R^3 \tag{4.1}$$

Thus, a 3D curve is in a triple form: $f(t) = (x(t), y(t), z(t)), t \in [0, 1]$. This is called parametric form. (t is used as the variable for curves here to indicate the time for a moving point from $t = 0$ to $t = 1$.). See Fig. 4.1a.

In Euclidean space, the distance metric is defined as:

$$d(f(t), f(s)) = \sqrt{(x(t) - x(s))^2 + (y(t) - y(s))^2 + (z(t) - z(s))^2}$$

We can see that when t moves to s, $t \to s$, $d(f(t), f(s)) \to 0$. So, $f(t)$ is continuous if and only if $x(t)$, $y(t)$, and $z(t)$ are continuous [17]. This definition was found

© Springer International Publishing Switzerland 2014

L. M. Chen, *Digital and Discrete Geometry*, DOI 10.1007/978-3-319-12099-7_4

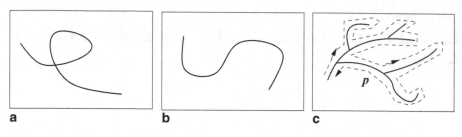

a   **b**   **c**

Fig. 4.1 Example of continuous curves: **a** a curve, **b** a simple curve, **c** an ordinary curve that is a tree

by C. Jordan [12]. In Chap. 13, we will discuss more from differential geometry perspective.

A curve is called a simple curve if no $f(t) = f(t')$ when $t \neq t'$ except at the two end points(e.g. $t = 0$ and $t' = 1$). See Fig. 4.1b. More generally,

Definition 4.1 A union of a finite collection of simple curves is called an ordinary curve if the union is connected.

In addition, since an ordinary curve only contains a finite number of simple curves, we can use Jordan's definition to go through all points on the curve. A tree is an ordinary curve but not a simple curve. The following is the most general definition for curves. [1]

Definition 4.2 Let D_p be a very small neighborhood of p in a curve C, where the boundary of D_p meets C a finite number of times. This number is called the order of $p \in C$. We define, (1) if the order is one then p is called an end point, (2) if the order is two then p is called an ordinary point, and (3) if the order is greater than or equal to 3 then p is called a branch point.

We can see that point p in Fig. 4.1 has three branches. The following definition for 1D manifolds is equivalent to the definition we presented in Chap. 3 for k-manifold when $k = 1$.

Definition 4.3 A 1D manifold is the only curve that contains end points and ordinary points.

Therefore, a simple curve is a 1D manifold. In this book, we mainly consider simple curves.

4.2 Curves in Discrete Forms

Generally, a simple curve defined in the above section can be approximated by a set of line segments. An easy method is to sample the continuous curve with $n + 1$ points: $f_0, f_1, ..., f_n$ where $f_i = f(i/n)$. We can use these points vertices, then make

---

[1] Beginning readers can skip the rest of the materials in this section.

Fig. 4.2 A curve and its
piecewise linear
approximation

n edges as we assign $e_i = (f_{i-1}, f_i)$. Therefore, this graph will be a path and this path is the approximation of the original curve. When $n \to \infty$, the path is the curve.

For simplicity, let us look at a function as a curve shown in Fig. 4.2. We select a number of sequential points on the curve (or function) $f(x)$, linking all point using line segments that are linear functions. This is a type of decomposition of curves called the piecewise linear approximation of the curve.

Conversely, we can make a continuous curve based on a discrete curve (a path of discrete point), which is called the curve interpolation or approximation. We discuss this further in Chap. 11.

In general, a path in a graph can be viewed as a general discrete curve. Let us present the formal definition of discrete curves as the vertex paths of graphs below.

For a graph $G = (V, E)$, we can (naturally) define 0-cells and 1-cells as vertices and edges as we did in Chap. 3. Here, we give a more formal definition.

Definition 4.4 In $G = (V, E)$, a path of vertices $p_0, \cdots, p_n$ (where p_i and p_{i+1} are adjacent) is called a discrete curve. If $p_0 = p_n$, such a curve is called a closed curve.

Definition 4.5 A path $p_0, \cdots, p_n$ is called a simple discrete curve if p_i is not p_j, $i \neq j$ for any i and j except for p_0 and p_n.

Definition 4.6 The (discrete) length of a curve $p_0, \cdots, p_n$ is defined as the length of the path, which is n. The distance between two points (or vertices) is defined as the length of the shortest path between these two points.

Proposition 4.1 *The minimum distance between two vertices $d(u, v)$ naturally induces a metric.*

Proof According to the definition of metrics in Definition 3.1, let $d(u, v)$ be the distance between two vertices u and v in G. We can see: (1) $d(u, v) \geq 0$, since the length of every path is 0 if $u = v$ or greater than 0 if $u \neq v$. (2) $d(u, v) = d(v, u)$. This is because if there is a path $u = p_0, \cdots, p_n = v$ from u to v, then the path $p_n, \cdots, p_1$ from v to u has the same length. (3) Assume we have two shortest paths $u = p_0, \cdots, p_n = v$ and $v = q_0, \cdots, q_m = r$ i.e. $d(u, v) = n$ and $d(v, r) = m$. Then, we will have a path from u to r, $u = p_0, \cdots, p_n = q_0, \cdots, q_m = r$. However, this path may not be the shortest from u to r, so $d(u, v) + d(v, r) = m + n \geq d(v, r)$. $\quad\square$

We can put a discrete curve with edges on the (corresponding) path in 2D Euclidean space (E_2). This is called embedding. When we put a simple closed curve in E_2, we get a polygon (or a polygon-like shape if the edge is not straight). This polygon with its inside area is called a face-element. In a planar graph, we use F to denote the set of faces. An important result about planar graphs is called the Euler theorem of planar graphs: If $G = (V, E)$ is a connected planar graph, then

$$|F| + |V| - |E| = 2 \tag{4.2}$$

This theorem can be proven easily using mathematical induction [11]. Let us assume that there is only one polygon inside a plane. The number of vertices is equal to the number of edges in this polygon. Plus we only have two faces: the polygon and the rest of the plane. Therefore, this theorem holds and we can then attach another polygon to the original polygon. When we check each case, we can see that the intersection between these two polygons is just a vertex, and the intersection is the number of edges. This theorem is fundamental for proving theorems in digital topology (Chap. 9).

4.3   Digital Curves in Σ_2

Even though a digital space is a graph, there are several ways of representing a general digital curve through its connectivities. This type of characteristic shows the difference between digital geometry and classical discrete geometry. In other words, the differences between digital curves and discrete curves are not only that the digital curve is made using grid points, but also that several types of connectivities exist for digital curves. In this section, we present some forms of digital curves.

Σ_2 is the two dimensional grid space. Σ_2 could refer to direct (4-) adjacency or indirect (8-) adjacency graph. In image processing, a digital curve is a path of pixels. Each consecutive pair of pixels in the path is adjacent. Sometimes, it is hard to decide which adjacency, 4-adjacency or 8-adjacency, should be used. We may need to use different adjacencies in the same curve. When such a condition is met, our mathematics is turned into artificial intelligence. However, in this book, we are concerned with the concrete mathematics resolution. We will explicitly state when we are referring to the artificial intelligence resolution.

4.3.1   Digital Curve Representations

For a simple digital curve, which is made of digital pixels, all pixels are distinct, except for the first and the last.

Let us look at the similar example in Chap. 2, Fig. 2.11, which is a digitization. However, to make a digital curve, we must consider the connectivities based on 8-adjacency or 4-adjacency. In Fig. 4.3a, we show a curve that is in the space of 8-adjacency, the curve $P = P(p, q)$ can move diagonally. In Fig. 4.3b, we use the same sampled points in Fig. 2.11b. When we only allow 4-adjacency, some extra points will be added to keep the curve continuous. We call this curve $Q = Q(p', q')$. We have three types of representation.

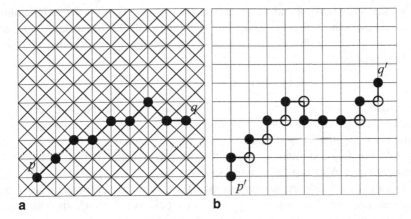

Fig. 4.3 Digital curves in Σ_2: **a** A curve in 8-adjacency, **b** a curve in 4-adjacency where some extra points are added to keep the curve continuous

Fig. 4.4 The direction of the next point in a curve when X is the current point

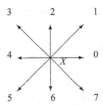

(1) Vector Point-Path The curve can be represented by a vector point-path. The following is the easiest example. If the bottom-left corner is the origin, we know the first point is $p = (1, 1)$. The curve is:

$P = P(p, q) = \{(1, 1), (2, 2,), (3, 3), (4, 3), (5, 4), (6, 4), (7, 5), (8, 4), (9, 4)\}.$

$Q \doteq Q(p', q') = \{(1, 1), (1, 2,), (2, 2), (2, 3), (3, 3), (3, 4), (4, 4), (4, 5), (5, 5), (5, 4), (6, 4), (7, 4), (8, 4), (8, 5), (9, 5), (9, 6)\}.$

(2) Parametric Representation: The second method uses the parametric representation. Just like a curve in continuous space, a curve in 2D can be represented by a pair of continuous functions $x(t)$ and $y(t)$. We use curve P as an example in Fig. 4.3a. For the curve in Fig. 4.3b, the method of representation is the same.

In digital spaces, $x(t)$ and $y(t)$ can be represented by two arrays $X[t]$ and $Y[t]$, $t = 0, 1, 2, ..., n$.

$X_t = \{1, 2, 3, 4, 5, 6, 7, 8, 9, \}. \; Y_t = \{1, 2, 3, 3, 4, 4, 5, 4, 4\}.$

(3) Chain-Code Representation The most effective representation is called the chain code created by Freeman [10]. It saves the most memory when we represent a digital curve in computers.

In Σ_2, a point has eight adjacent points in 8 directions. See Fig. 4.4. (In Σ_3, there are 26 adjacent points in 26-directions.)

If we know the first point and all of the moving directions of the following points, we can represent the curve. For example, the curve in Fig. 4.3 began at $(1, 1)$ then is followed by a point in direction 1, then in direction 1 again. So on so forth. We can code the curve as:

$$(1, 1) \text{ followed by } 1, 1, 0, 1, 0, 1, 7, 0.$$

How much space has the chain code method saved? More than half, and much more! The reason why is because we only have 8 directions that can be coded after the first point. A number N in real computer memory needs $\log_2 N$ memory space to store N. For data communication, chain code is an essential method. In algorithm analysis, we treat N as just using a unit space.

Originally, chain code was used to represent the contour of a 2D object. Chain code follows the contour of an object in the counter clockwise (or clockwise) direction. Therefore, chain code is also an edge detection method.

4.3.2 What is a Digital Point: A Vertex or a Pixel?

A TV monitor has many gridded unit squares. A unit square is called a pixel, or a picture element. One can use a point at the center of a pixel to represent this pixel, and put the color of the pixel as the value of this point. Meanwhile, a pixel is a unit square that has four edges and four corner points.

Another problem arises: what is a digital point? A digital point can be one of the following representations: (1) A digital point used to represent a pixel is called point-space representation, and (2) A digital point just as an "abstract" corner point, is called raster-space representation. Of these two representations, one is related to the Delaunay representation and the other is related to the Voronoi representation, both of which we discuss in this book.

Let us look at a simple example in Fig. 4.5. Fig. 4.5a shows an original image and Fig. 4.5b shows its digitization. The boundary pixels of the image are shown in Fig. 4.5c. The boundary curve is shown in Fig. 4.5d.

How do we save the boundary curve? We can either save the shed pixels (point representation) as individual array elements (Fig. 4.5c), or, using raster representation, we save the bold edge in Fig. 4.5c. Note that to save an edge element here, we must save a pixel pair (just like an element in E in graph $G = (V, E)$). By the way, a "corner point" will be represented by four pixels in raster representation, and this point is an abstract point.

4.3.3 A Property of Parametric Digital Curves

It is impossible to make a parametric representation of a discrete curve directly. This is because moving from one vertex to another has a numerous possibilities in terms

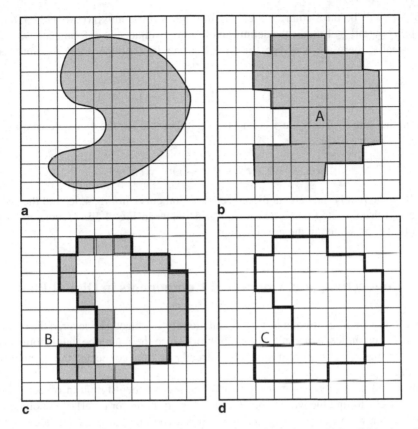

Fig. 4.5 Example of an image and its digitization: **a** The original image, **b** the digitization of the image, **c** the boundary pixels, and **d** the boundary by human interpretation

of different distances between the two vertices u and v. However, for digital space, each vertex is located at a grid point. It gives us the opportunity to have a parametric representation for digital curves.

Let $P(t) = (x(t)y(t))$ be a digital curve. If $x(t)$ and $y(t)$ are digital curves that hold the "continuity" condition, meaning that $|x(t+1)-x(t)| \leq 1$ and $|y(t+1)-y(t)| \leq 1$, then these properties are called gradual variation (Chap. 11) [5].

Formally, let $f(t)$ be a digital function, i.e. $f : \Sigma_1 \rightarrow \{1,2,\cdots,m\}$. Rosenfeld first defined $f(t)$ as digitally continuous if for all t, $|f(t+1) - f(t)| \leq 1$, a special case of gradual variation.

Proposition 4.2 *$p(t)$ is an 8-connected curve if each $x(t)$ and $y(t)$ is digitally continuous.*

Proof Since $|x(t+1)-x(t)| \leq 1$ and $|y(t+1)-y(t)| \leq 1$, $p(t+1)$ can only be located at 9 locations surrounding $p(t)$ and including $p(t)$. If we do not allow $p(t+1) = p(t)$, then $p(t)$ will be an 8-connected curve. In other words, $\max\{|x(t + 1) - x(t)|, |y(t + 1) - y(t)|\} \leq 1$ that is the definition of indirect adjacency or 8-adjacency. $\quad\square$

Basically, there are two types of digital curves in 2D. One is the 8-connected curve, i.e.

$$\max\{|x(t + 1) - x(t)|, |y(t + 1) - y(t)|\} \leq 1$$

and another one is strictly the 4-connected digital curve. For the 4-connected digital curve, we require:

$$|x(t + 1) - x(t)| + |y(t + 1) - y(t)| \leq 1$$

for all t. The pixels shed are an 8-connected digital curve in Fig. 4.5c, and the bold curve is a 4-connected curve due to the grid points in Fig. 4.5d. A more precise definition is presented in the next chapter.

4.4 Connectivity and Connected Components in Digital Plane

We know that finding the boundary cycle can identify an object. However, if the object contains many holes, this method will not work since there may be many boundary cycles. In this case, we need a method to directly find the connected component of a set in 2D.

Finding a connected component is another way of identifying an object. This method seeks to find the whole object, such as every point (or pixel) of the object.

As we know, in Euclidean space, two points are connected if there is a curve to link them. In digital or discrete space, we say that the points are connected if there is a (point)-path to link them. We have discussed the two common types of connectivity in Σ_2: 4-connectivity and 8-connectivity. The connectivity must be predefined before the actual task for finding the component. In image processing, finding a digital component and its boundary are essential tasks. This process is related to image segmentation.

A procedure (also called an algorithm) designed for finding a connected component is called breadth-first-search [7].

Example 4.1 We use an example to present the ideas of this algorithm. In Fig. 4.6, there are two (connected) components in 4-adjacency.

Let us present the procedure of finding the first component as follows: Start at point A, take its two neighbors B and D, and put these two neighbors in an array called Q. (In fact, Q is a queue, a first in first out data structure. But for now, we only consider Q to be an array for easy understanding.) $Q = \{B, D\}$. Pick up B from Q, get B's neighbors C and E, and put them into Q. Removing B from Q, we have $Q = \{D, C, E\}$. Repeat the process by picking up D, adding neighbor G to Q, (E was also a neighbor of D, but it is already in Q.), and remove D from Q. We now

Fig. 4.6 Finding connected components in 2D digital space

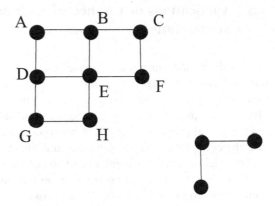

have $Q = \{C, E, G\}$, and so on and so forth. We will eventually visit all members in the component that contains A. We can use a simple mark to mark all of the points removed from Q, and these marked points plus A will be the component. To clarify, Q changes during each step of the process. We list all of the stages of Q and revisit the steps of the algorithm once more:

(Time 0) A was picked and Q is empty.
(Time 1) $Q = \{B, D\}$, A is marked.
(Time 2) $Q = \{D, C, E\}$, B is marked.
(Time 3) $Q = \{C, E, G\}$, D is marked.
(Time 4) $Q = \{E, F, G\}$, C is marked.
(Time 5) $Q = \{F, G, H\}$, E is marked.
(Time 6) $Q = \{G, H\}$, F is marked.
(Time 7) $Q = \{H\}$, G is marked.
(Time 8) $Q = \emptyset$, H is marked.

After these steps, we can move to another unmarked point and start the procedure over to get the second connected component.   □

This example not only presents the procedure of finding a connected component, but also gives an idea of how an algorithm would run. We will specifically deal with the algorithmic issue in later chapters.

There might be a question of this process: why do we not build a graph first to start the standard algorithm? This is because we do not want extra space to be used in storing the edges. The good thing about digital space is that the edges are assumed by its connectivity: 4-connectivity or 8-connectivity. Not spending extra time and space to store a graph is another difference and advantage of this method compared to the discrete method. In other words, in real time geometric data processing, we like to have a default setting so we can save time to do the process and skip the preprocessing. This is also the main difference between digital geometry and discrete geometry.

4.5   Applications of Connected Components: Image Segmentation

One of the main motivations of establishing the research area of digital geometry was the need for image processing, especially the need for image segmentation. Image segmentation is the basic approach in image processing and computer vision [10, 16]. It is used to locate specific regions and extract information from them. This is an essential procedure for image preprocessing, object detection and extraction, and object tracking. Image segmentation is also related to edge detection.

The goal of image segmentation is to partition an image into different components or objects. Can we say that image segmentation is used just to find connected components from the last section? The answer is both yes and no.

We say yes if the image is a binary image, meaning that this image only contains two values $\{0, 1\}$. We can also say no, since the image now has the property of each pixel having millions of possible values.

Segmentation partitions an image into connected subsets called segments (components). Each segment is uniform and no union of adjacent segments is uniform.

The formal description of segmentation is as follows: In a digital image F, if there exists a non-empty segmentation $F_1, F_2, ..., F_m$ satisfying:

(1) $F_i \cap F_j = \emptyset$, if $i = 1, .., m, j = 1, ..m; i \neq j$,
(2) $\cup_{i=1,...,m} F_i = F$,
(3) Each F_i is connected,
(4) Each F_i is "uniform," and
(5) If F_i and F_j are adjacent, then $F_i \cup F_j$ is not uniform,

then, $\{F_1, F_2, ..., F_m\}$ is called a segmentation of F.

The only thing we have not yet defined in this definition is the word "uniform." In essence, different people observe "uniform" differently. In common sense, uniform means only containing small variations, which could be colors, values, or patterns.

Let's introduce two simple methods for image segmentation in this section.
(1) The simplest method of image segmentation is called the thresholding method. This method is based on a clip-level (or a threshold value) to turn a gray-scale image into a binary image. The key for this method is selecting a threshold value. When a pixel value is greater than the threshold, we re-assign "1" to this pixel; otherwise, we assign "0" to this pixel. After that, we obtained a binary image. Then, we can use the algorithm to find a connected component in order to get the segmentation.

Here is a question of finding the threshold. A method called balanced histogram thresholding calculates a histogram, and the threshold will then be chosen to cut the image into a 0-value part and 1-value part of almost the same size (we can consider not counting the background of the image if it is appropriate to do so). Several other popular methods are used in industry including the maximum entropy method and Otsu's method (maximum variance) [10]. The following picture is the result of using the thresholding method through the maximum entropy method that calculated the best threshold is 23 [3, 14]. See Fig. 4.7.

Fig. 4.7 Example of
thresholding image
segmentation: **a** Original
image; **b** Segmented image
with the threshold at 23

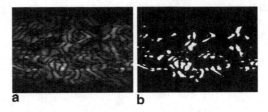

a   b

(2) Another simple method is to first detect the possible boundary of a segment, then use Chain-code to link the boundary. Finding the possible boundary of a segment is also called edge detection. For an image F, we calculate the derivative F_x and F_y, get the average value F', and normalize the biggest value to be 1. The ideal uniform region will be zero in F', except on the boundary. So, the bigger the value indicates the higher possibility of being the edge.

The Chain-code algorithm starts at a point on a boundary. Then the algorithm searchs for its neighbor using a counter clockwise angle to see if there is a neighbor whose value is almost 1. There are only eight directions to search. After the algorithm finds a point already visited, the algorithm stops to report a boundary curve or edge curve.

Image segmentation is a huge research area. In this book, we use many examples of it to explain the usage of digital and discrete geometry. There are many kinds of segmentation methods including region growing, split-and-merge segmentation, edge detection, graph cut, and variational principle based methods[10].

4.6 Constructing Digital Lines: Bresenham's Line Algorithm

In computer graphics [8], there are many occasions where we need to draw a line on the screen, such as in computer games. These lines are digital lines because the screen is formed by pixels. Bresenham's algorithm is a very elegant algorithm for computing all pixel locations for digital display purposes in computer graphics. It saves fractional computation time in the computers and uses a technique based on integer manipulation.

Giving two digital points $p = (x_1, y_1)$, $q = (x2, y_2)$ in Σ_2, how do we construct a digital line? The easiest way is to use line equations to determine each integer point.

$$y = \frac{y_2 - y_2}{x_2 - x_1}(x - x_1) + y_1. \tag{4.3}$$

When we assume $x1 \leq x2$, we calculate each y based on $x_1 < x < x_2$, where x is an integer. We also round x_1 and x_2 to the nearest integers. If we use this equation, the computer will need to multiply slope $k = \frac{y_2 - y_2}{x_2 - x_1}$ with an integer $(x - x_1)$ each time.

If we assume that $slope = \frac{y_2 - y_2}{x_2 - x_1} \leq 1$, when x is increased by 1, y may or may not change. In other words, put a digital line when x changes by 1 while y stays the same or also changes by 1.

Fig. 4.8 Bresenham's line
algorithm A: no
multiplication operations

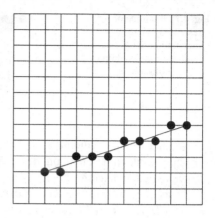

For simplicity, let us assume slope $k > 0$. Therefore, we only have two choices: Plot (mark) the point $(x + 1, y)$ or point $(x + 1, y + 1)$ on the screen that is $N \times M$ array.

If we could determine when y will change, we would not need to use the line equation explicitly in each time calculation as $x \leftarrow x + 1$. If we add 0.5 to the rounded error value, then getting a floor function would be more accurate in industry.

For example, if $k = 0.5$, we know that for every two changes of the x value, $x + 1$ and $x + 2$, we would need to increase the y value by 1.

Let *error* be a real number, the following algorithm will eliminate the multiplication operation. We can also design an algorithm that only uses integer operations. We assume $0 \leq k \leq 1$; the angle is from 0 to 45°, called the first octant. The other case is similar due to which variable of x or y should be the domain variable.

Algorithm 4.1 : (Bresenham's Line Algorithm A) No multiplication operation for real numbers.

Step 1   Let $x = x_1$ and calculate k. Let *error* $= 0$.
Step 2 $x := x + 1$, *error* $=$ *error* $+ k$, if *error* < 0.5, then $y := y$ else $y := y + 1$; *error* $:=$ *error* $- 1$.
Step 3   Repeat Step 2 until x equals x_2.

Assume $p = (2, 2)$ and $q = (11, 5)$. We have slope $k = 3/9 = 1/3$. See Fig. 4.8. The next point after p will be $(3, 2)$ since *error* $= 1/3 < 0.5$. The following point is $(4, 3)$, but the accumulation error would be *error* $= 2/3 - 1 = -1/3$. We can continue the calculations to get the following sequence of points plotted:

$$p = (2, 2), (3, 2), (4, 3), (5, 3), (6, 3), (7, 4), (8, 4), (9, 4), (10, 5), (11, 5) = q.$$

In *Bresenham's Line Algorithm A*, there is no fractional multiplication. The algorithm would be very fast, but it still has to add the fractional numbers. To avoid the floating point calculation, we can improve this algorithm on how many x changes are made in order to limit the number of y changes. If $\delta y = y2 - y1$ and $\delta x = x2 - x1$, when we accumulate δy until it passes half of δx, we increase y by 1 and let *error* $:=$ *error* $- \delta x$

Algorithm 4.2 : (Bresenham's Line Algorithm B) The algorithm only uses integer manipulations.

Step 1   Let $dx = x2 - x1$; $dy = y2 - y1$; $y = y1$, $error = 0$.
Step 2   For x from x_1 to x_2, plot or mark pixel (x, y).
Step 3 $error = error + dy$.
Step 4 $(error + error \geq dx)$ (meaning we pass the middle 0.5 line), then $y = y + 1$ and $error = error - dx$; and
Step 5 $x = x + 1$ repeat from *Step2*.

The idea presented in Bresenham's Line Algorithm has been used in circle drawing and other fast digital drawing algorithms [8, 13].

4.7 Hole Counting of Images

The number of holes (hole Counting) in an image is a topological feature for the image. In this section, we deal with a connected binary image. This topological property is important to determine if two images are similar or completely different. If they have a different number of holes, we can say that these two images are different without further calculation.

In this section, we present a simple formula for hole counting. Let us think about the following example: A person *Alice* is walking on a circle. When *Alice* finishes the circle, *Alice* would have walked a complete 360°. The second person *Bob* is walking on a simple cycle, a closed path. A closed path is not a perfect circle, but if *Bob* finishes the path back to the starting point, *Bob* would have also completed 360°. He may have put in more effort and walked a longer distance if the path twists and turns, but *Bob* has still completed 360°.

A 2D component M on a plane may have no holes or h holes. If there is no hole, then there is just a closed curve on the outside boundary. If there are h holes in M, we would have h small closed curves inside the outside closed curve. Therefore, we have a total of $h + 1$ closed curves in M. (The difference is that the outside curve "faces in" and the inside curve "faces out" to hold the component.)

Let us look at the digital case of this problem. In Fig. 4.9, we define *InWardTurns* as the total number of corner points, each of which points to the inside of the object. Likewise, the *OutWardTurns* is the number of total corner points, each of which points to the outside of the object.

The outside boundary curve always has 4 more outward points than inward points. However, each of the inside cycle has 4 more inward (corner) points than outward points. We assume that M has h holes. Then, we will have $h + 1$ cycles, one of which would be outside boundary cycle B. The rest of the cycles will be denoted as H_i, $i = 1, \cdots, h$. *InWardTurns*(C) and *OutWardTurns*(C) are the numbers of inward points and outward points of a cycle C, respectively. Therefore, we have

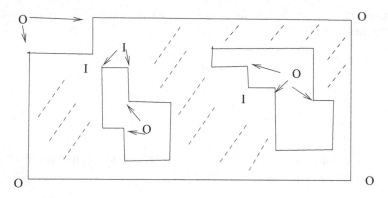

Fig. 4.9 Hole counting example

$InWardTurns(B) - OutWardTurns(B) = -4$ and $InWardTurns(H_i) - OutWardTurns(H_i) = 4$

So

$InWardTurns = InWardTurns(B) + \Sigma_i InWardTurns(H_i)$ and $OutWardTurns = OutWardTurns(B) + \Sigma_i OutWardTurns(H_i)$.

Therefore,

$InWardTurns - OutWardTurns = InWardTurns(B) - OutWardTurns(B) + \Sigma_{i=1}^{h}(InWardTurns(H_i) - OutWardTurns(H_i))$

Thus,

$$InWardTurns - OutWardTurns = -4 + 4 \cdot h$$

What is the total number of holes in a component in 2D digital space? We can get the following formula: the total number of holes in a connected component in 2D digital space in 4-adjacency is

$$h = 1 + (InWardTurns - OutWardTurns)/4. \tag{4.4}$$

The formal and topological proof requires more sophisticated knowledge in topology (see Chaps. 9 and 14).

To examine the correctness of the formula, we can count the total *InWardTurns* as $1 + 6 + 7 = 14$ and the total *OutWardTurns* as $5 + 2 + 3 = 10$. $h = 1 + (InWardTurns - OutWardTurns)/4 = 1 + (14 - 10)/4 = 2$. Therefore, if there is no hole, h will always be 0 based on the fact that the outside boundary curve always has 4 more outward points than inward points [4]. This theorem will be proved using Euler formula in Chap. 9.

Fig. 4.10 Examples of the area of triangles on grids

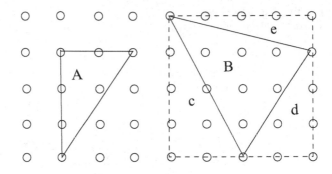

4.8 Pick's Theorem and Minkowski's Theorem*

Pick's theorem is used to obtain the area of a polygon whose vertices are at the grid points [9]. It is one of the most amazing theorems in geometry. Pick's theorem is also a bridge between grid space and continuous space in 2D. The vertices are located at the grid points but their measurements are done in Euclidean space.

Given a simple polygon in a 2D grid such that all the polygon's vertices are grid points, Pick's theorem says that the area A of this polygon equals $i + \frac{b}{2} - 1$, where i is the number of interior points and b is the number of boundary points.

This theorem is quite interesting and is hardly believable. Let us first examine some examples in Fig. 4.10. We can see that for the first triangle A, the area is $(3 \cdot 2)/2 = 3$. We know that there are six points on the boundary of A and one point inside A. Then $i + \frac{b}{2} - 1 = 1 + 6/2 - 1 = 3$, so the formula is verified for A.

For triangle B, we know $area(B) = 4 \cdot 4 - area(c) - area(d) - area(e) = 16 - 4 \cdot 2/2 - 2 \cdot 3/2 - 1 \cdot 4/2 = 7$ B contains six grid points inside and four points on the boundary. Therefore, $i + \frac{b}{2} - 1 = 6 + 4/2 - 1 = 7$. It follows that $area(B) = i + \frac{b}{2} - 1$.

We can prove first that Pick's theorem is true for the right triangle with two edges parallel to the grid lines. Then, we can prove the theorem for arbitrary triangles such as B. We leave this proof as a practice problem.

Theorem 4.1 *(Pick's theorem) Let A be a polygon on 2D grid space where each of the vertices of A is at a grid point. Then, the area of A is*

$$area\ (A) = i + \frac{b}{2} - 1. \tag{4.5}$$

where i is the number of interior points and b is the number of boundary points.

Proof We provide a brief proof here. We assume that all triangles satisfy Pick's theorem. Then, we can use mathematical induction to prove that a polygon can be split into two smaller polygons. If the smaller polygons satisfy the theorem, then the original polygon also satisfies Pick's theorem.

Looking at Fig. 4.11, we split the original polygon M into A and B. And we assume

Fig. 4.11 Examples of the
area of triangles on grids

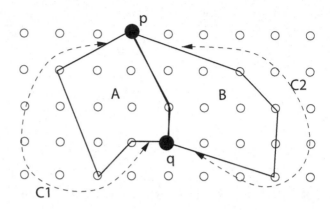

$area(A) = i_A + b_A/2 - 1$ and $B = i_B + b_B/2 - 1$
$area(M) = area(A) + area(B)$

Let us assume that on the partition curve p to q, there are x points in the middle of the path, excluding the two end points.

$b_A = C1 + 2 + x$ and $b_B = C2 + 2 + x$

The number 2 comes from counting points p and q on the boundary.

$i_M = i_A + i_B + x$ and $b_M = C1 + C2 + 2$

Therefore,

$area(M) = area(A) + area(B) = i_A + b_A/2 - 1 + i_B + b_B/2 - 1$
$= i_A + b_A/2 - 1 + i_B + b_B/2 - 1$
$= i_A + i_B + (b_A + b_B)/2 - 2$
$= i_A + i_B + (C1 + 2 + x + C2 + 2 + x)/2 - 2$
$= i_A + i_B + x + (C1 + 2 + C2)/2 + 1 - 2$
$= (i_A + i_B + x) + (C1 + 2 + C2)/2 - 1$
$= i_M + b_M/2 - 1$

We have proven the theorem.   □

However, there is no polyhedron generalization of Pick's theorem in 3D or higher dimensions.

In 2D space, Minkowski's Theorem is about the integer point existence in a convex region T in 2D where T is symmetric with respect to the origin, e.g. points $(-x,-y)$ and (x,y) are both in T. If the area of T is greater than 2^2, T must contain an integer point other than the origin [15].

Minkowski's Theorem is the foundation of the geometry of numbers, a branch of number theory for Diophantine approximation. We present Minkowski's Theorem in R^n as follows:

Theorem 4.2 *(Minkowski's Theorem) Let T be a centrally symmetric convex in R^n. If the volume of T is greater than or equal to 2^n, then there exist integers $x_1, \cdots, x_n$, not all zero, such that point $(x_1, \cdots, x_n) \in T$.*

In fact, this theorem can be extended to other lattices.

4.9 Remark

In Euclid's Elements, a curve does not have a width. However, Hilbert's curve can fill an entire Euclidean plane[12]. This somehow generates a contradiction. This contradiction was caused by the foundations of mathematics. We point out this issue here to express a concern that even mathematics is not perfect.

A non-trivial curve must lay on a 2D or higher dimensional space. It means that to define a curve, one must first determine its ambient (containing) space. In order to understand a lower dimensional object, we need a higher dimensional space, which may not yet be defined. In terms of the length and straightness of a digital line, without a discrete line, we can only talk about digital curves. Bresenham's lines are the digitization of digital lines. To define a digital line and plane may not be very fruitful. We may want to limit this concept within Euclidean space in order to better deal with it [13]. The next chapter will discuss digital surfaces and manifolds [2, 6]. More constructive algorithms will be discussed in Chap. 6. A good reference dealing algorithms is the book written by Cormen et al. [7].

Connected components are related are topological concepts [1, 2]. So, image processing especially image segmentation is highly related to topological methods. For discrete method for smooth data, see [5].

References

1. P. S. Alexandrov, Combinatorial Topology, New York: Dover, 1998.
2. L. Chen, *Discrete Surfaces and Manifolds: A theory of digital-discrete geometry and topology*, 2004. SP Computing.
3. L. Chen, λ-connectedness determination for image segmentation, Applied Imagery Pattern Recognition Workshop, 2007. AIPR 2007. 36th IEEE Volume, Issue, 2007, pp 71–79.
4. L. Chen, Determining the number of holes of a 2D digital component is easy, *http://arxiv.org/abs/1211.3812*, Nov. 2012.
5. L. Chen, Digital Functions and Data Reconstruction, Springer, NY, 2013.
6. L. Chen and J. Zhang, Digital manifolds: A Intuitive Definition and Some Properties, Proceedings of the Second ACM/SIGGRAPH Symposium on Solid Modeling and Applications, Montreal, 1993, 459–460.
7. T. H. Cormen, C.E. Leiserson, and R. L. Rivest, Introduction to Algorithms, MIT Press, 1993.
8. J. D. Foley, Andries Van Dam, Steven K. Feiner and John F. Hughes, Computer Graphics: Principles and Practice. Addison-Wesley. 1995.
9. J. E. Goodman, J. O'Rourke, Handbook of discrete and computational geometry, CRC Press, Inc., Boca Raton, FL, 1997.
10. R. C. Gonzalez, and R. Wood, *Digital Image Processing*, Addison-Wesley, Reading, MA, 1993.
11. F. Harary, Graph theory, Addison-Wesley, Reading, Mass., 1969.
12. Ito, K. and Ugakkai, N.S., eds., Encyclopedic Dictionary of Mathematics, MIT, Cambridge, MA, 1987, 2nd edition.
13. R. Klette and A. Rosenfeld, Digital Geometry, Geometric Methods for Digital Picture Analysis, series in computer graphics and geometric modeling. Morgan Kaufmann, 2004.
14. C.T. Lu, Y. Kou, J. Zhao, and L. Chen, Detecting and tracking region outliers in meteorological data, Information Sciences, Vol 177, pp 1609–1632, 2007.

15. H. Minkowski, Geometrie der Zahlen, Leipzig and Berlin: R. G. Teubner, 1910.
16. A. Rosenfeld and A.C. Kak, *Digital Picture Processing*, 2nd ed., Academic Press, New York, 1982
17. J. Stewart, Calculus, Brooks/Cole Publishing Company, Pacific Grove, CA, 4th ed, 1999.

Chapter 5
Surfaces and Manifolds in Digital Space

Abstract The digital surface is one of the main topics of this book. We know that digital curves are digital paths and discrete surfaces can be described as triangulations. Therefore, is a digital surface a simple digitization of a continuous surface? The answer is no. This is because the basic 2D cell of digital surface in direct adjacency is a unit square and is not flexible enough to stick perfectly onto a continuous surface in order to compare with triangles of different sizes. Their 2-cells are only perpendicular to coordinates. In indirect adjacency, there are may be multiple choices for choosing a triangle to fit the original surface. In addition to that, the digital point might not on the surface which is not the same as a traditional triangulation. Digital surfaces are much more difficult to deal with. In this chapter, we mainly focus on the Morgenthaler and Rosenfeld definition of digital surface and the Chen and Zhang definition for direct adjacency. The Chen–Zhang definition has led to the classification of simple surface points in 3D, which is regarded as a significant result in digital geometry.

We will also define digital manifolds in direct adjacency. We cover more profound topics such as the general discrete manifold in Chap. 7 and digital topology in Chap. 9.

Keywords Surface · Digital surface · Digital manifold · Definition · Classification of surface points

5.1 Introduction to Surfaces and Digital Surfaces

The simplest example of a surface is a continuous function on a plane: $z = f(x, y)$. But this surface is not general enough to define a sphere (as we explained in Chap. 3). This is because a sphere has two values at many points on the x, y-plane.

A more general definition for a surface is called the parametric surface: Like the general form of curves, we use t as a parameter in [0,1] to define a curve. Now we can use $(u, v) \in A$, a rectangular region in $R \times R$, to define a surface in 3D Euclidean space:

$$s(u, v) = (x(u, v), y(u, v), z(u, v)). \tag{5.1}$$

Example 5.1 A sphere with radius r centered at $(0, 0)$ has the parametric form as follows:

© Springer International Publishing Switzerland 2014
L. M. Chen, *Digital and Discrete Geometry,* DOI 10.1007/978-3-319-12099-7_5

$$\begin{cases} x = r \sin(v) \cos(u) \\ y = r \sin(v) \sin(u) \\ z = r \cos(v) \end{cases} \qquad (5.2)$$

where $u \in [0, 2\pi]$ and $v \in [0, \pi]$. We can easily verify that $x^2 + y^2 + z^2 = r^2$. For more about the parametric surface, see [28].

A sphere is a special case of surfaces. Its radius is fixed. With regard to an arbitrary surface, it is difficult to determine the rectangular area $A = [0, 2\pi] \times [0, \pi]$. Such A and the function on A may change from a point to a different point on the same surface. Thus, differential geometry is needed for defining a general surface. In the differential form, "A" would become some what of a moving reference frame along with each point on the surface. We discuss this in Chap. 13.

In any case, we assume that a surface can be decomposed by triangles, or more generally, polygons. This is called the piecewise linear representation of the surface. See Fig. 1.2 [6].

The question now becomes: In order to make a digital surface or a digital surface decomposition of a continuous surface, can we just use the unit square and attach it to a continuous surface just as we did using triangles?

The answer is no. This is because even though we can use a unit square, we have to use them in different rotations or angles. Digital space 2-cells only allow the unit square to be in upright, right angle directions. In other words, if U is a digital 2-cell, then U must be in the XY, XZ, or YZ-plane of a 3D Euclidean space. (See Fig. 5.1a).

Another question can be asked: Can one just digitize a continuous surface? Meaning, if we obtain all digital/integer vector points, then it must be a digital surface. The answer depends on various factors. Even though one can digitize a continuous surface[1], there is no guarantee that the digitized points form a "surface." It would depend on the resolution of digitization since one may get a unit cube where each point of the cube is on the original surface. In this case, we have a 3D cell in the "digital surface."

5.2   Definitions of Digital Surfaces

The digital surface can be viewed as the polygonal decomposition of continuous surfaces, where each polygon is a unit square. However, this is a property, not a definition. We cannot use this to identify a digital surface when a set of points is given. If one does want to try using this as a definition, every set of digital points would be a digital surface since we can always make a continuous surface that can go through all digital points of the set. Rosenfeld was probably the first person to realize that digital surfaces must have a separate definition mathematically.

---

[1] In most cases, one cannot even store a continuous surface into a computer if there is no formula to describe the surface.

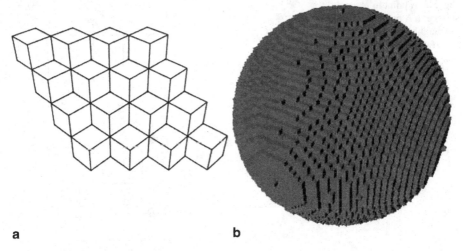

Fig. 5.1 Examples of digital surfaces: **a** A digital plane, and **b** a digital sphere made by J. Lachaud and the $DGtal$ Group

A digital surface in 3D grid space does not resemble a surface in 3D Euclidean space. A digital surface is the collection of the faces of unit cubes. In other words, a digital surface consists of unit squares parallel to $XY-$, $XZ-$, and $YZ-$ planes in 3D. Figure 5.1 shows two examples of digital surfaces: a digital plane and a digital sphere, respectively.

5.2.1 Morgenthaler-Rosenfeld Definition of Digital Surfaces

Let us first look at a neighborhood of a digital point p in 3D digital space, denoted as N_p, In Fig. 5.2 a. N_p contains 27 points including p meaning that p has 26 points in its neighborhood. For p, a is a directly adjacent point; b and c are indirectly adjacent points. b is closer to p when comparing the distance between p and c.

In fact, $p = (x, y, z)$ in Σ_3 has six direct neighbors, namely, $(x \pm 1, y, z)$, $(x, y \pm 1, z)$, and $(x, y, z \pm 1)$. This is also called 6-adjacency in 3D. p also has 12 2D-diagonal neighbors, namely, $(x \pm 1, y \pm 1, z)$, $(x, y \pm 1, z \pm 1)$, and $(x \pm 1, y, z \pm 1)$; and 8 3D-diagonal neighbors, namely, $(x \pm 1, y \pm 1, z \pm 1)$. So p has $26 = 6 + 12 + 8$ adjacent points total and is called 26-adjacency. If we only allow 2D-diagonal neighbors, then p has $18 = 6 + 12$ adjacent points, called 18-adjacency.

Therefore, we have a total of 3 types of adjacencies, 6-, 18-, and 26-adjacencies. A path of digital points with α adjacency, $\alpha = 6, 18, 26$, is called an α connected path or α-path [26, 24]. A set of points are called α-connected if any two points in the set are α-connected, meaning that there is an α-path linking them inside this set.

When a surface S passes through a point p and p's neighborhood is on S, then $S(p)$ is a subset of N_p. In addition, $S(p)$ should look like a "plate," and $S(p) - \{p\}$ should

Fig. 5.2 A point in N_p and surface: **a** A point p and its 3D neighborhood N_p, and **b** A surface S passes point p in N_p

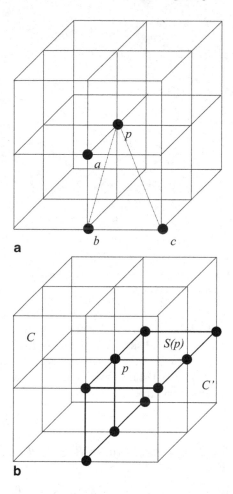

look like a "circle" or a closed digital curve. Finally, $S(p)$ should cut N_p into two components, i.e. $N_p - S(p)$ should contain two parts that are not connected. These observations will lead to the Morgenthaler–Rosenfeld definition of digital surfaces. See Fig. 5.2.

In Fig. 5.2b, p is surrounded by 8-points on S. These 8 points form a closed curve. $N_p - S(p)$ is the union of two disconnected components C and C'. In 1982, Morgenthaler and Rosenfeld gave a mathematical definition of digital surfaces. In order to understand the definition, we first present a simple form of it as follows.

Definition 5.1 (Morgenthaler and Rosenfeld) A point p in S is an $(6, 26)$-(simple) surface point if

(1) $S(p)$ is a 6-component,
(2) $N_p - S(p)$ has exactly two 26-components and p is 26-adjacent to both of these 26-components,
(3) Each of the 6-neighbors of p is 26-adjacent to both 26-components of $N_p - S(p)$.

Fig. 5.3 A corner point p of S in N_p. a is 26-adjacent to p

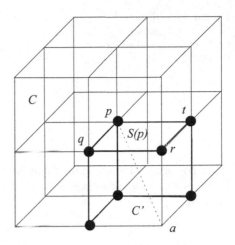

The simplest connectivity in 3D digital space is 6-connectivity. In Definition 5.1, it is understandable that we want a surface that is 6-connected so that each $S(p)$ is a 6-component and the first condition in Definition 5.1 is met. We discussed that $N_p - S(p)$ should have exactly two components, see C and C' in Fig. 5.2. Let's look at the example in Fig. 5.3. It is very common for a surface to contain a corner point, such as p in Fig. 5.3. The component $C' \subset N_p - S(p)$ only contains a point a. a is 26-adjacent to p, which is why condition 2 of the definition must be satisfied.

We now give the complete Morgenthaler–Rosenfeld definition of digital surfaces below:

Definition 5.2 (Morgenthaler and Rosenfeld) Let $\alpha, \beta \in \{6, 18, 26\}$. A point p in S is an (α, β)-(simple) surface point if

(1) $S(p)$ is an α-component,
(2) $N_p - S(p)$ has exactly two β-components and p is β-adjacent to both β-components,
(3) each of the α-neighbors is β-adjacent to both β-components of $N_p - S(p)$.

Again, why did Morgenthaler and Rosenfeld need to use β-components for the complement of $S(p)$, $N_p - S(p)$? This is because a surface point p should also be adjacent to some point in both components of the complement. However, in Figs. 5.2 and 5.3, the component C' does not have a point that is 6-adjacent to p. So, the Morgenthaler–Rosenfeld definition is very elegant in avoiding this problem.

The Morgenthaler–Rosenfeld definition provides a total of nine types of digital surfaces. However, most of these do not really exist in terms of usefulness by Kong and Roscoe [5, 19]. The Morgenthaler-Rosenfeld definition seems to be a complete definition for digital surfaces in Σ_3. However, Chen found a counter-example, which contains a visually true surface-point but not an (α, β)-surface point [4, 5]. See Fig. 5.4. We discuss the definition of digital surfaces further in the historical remark of this chapter.

The Morgenthaler–Rosenfeld definition is used to define a surface through its simple surface points. If every point in S is an (α, β)-surface point, then S is called

Fig. 5.4 Visually true
surface-points but not
(α, β)-surface points for any
α and β

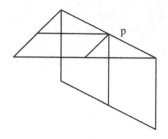

an (α, β)-surface. Therefore, this definition is about the definition for closed surfaces. It cannot deal with surfaces on boundaries. Next, we provide a definition to deal with this case.

5.2.2 Parallel-Moves and Chen-Zhang Definition of Digital Surfaces

The Morgenthaler–Rosenfeld definition of digital surfaces was a milestone in digital geometry. It showed the necessity and feasibility of digital geometry as a new research area, a topic that continues to inspire rigorous study today. However, the definition was based on set-theoretical methods, which lacked certain geometric intuition. Plus, it was difficult to design an algorithm to actually recognize a digital surface. Chen and Zhang found another definition of digital surfaces using so called parallel-moves of line-cells. This definition was based on the observation that surfaces are made from moving curves [8, 9].

First, we give a recursive definition for generating an i-cell. Let Σ_3 be a three-dimensional digital space. A pair of points p, p' in Σ_3 can be a line-cell if they are adjacent. A surface-cell, or a closed path, consists of 4 points p_1, p_2, p_3, and p_4 which satisfy

$$\begin{cases} d_D(p_i, p_{i+1}) = 1 \text{ for } i = 1, 2, 3. \\ d_D(p_1, p_4) = 1 \\ d_D(p_1, p_3) \neq 1 \text{ and } d_D(p_2, p_4) \neq 1 \end{cases} \qquad (5.3)$$

where $d_D(p, q)$ is the distance of the direct adjacency. For example, in Fig. 5.3, points p, q, r, and t form a surface-cell. They satisfy the Eq. (5.3).

Definition 5.3 (Parallel-move) A line-cell $\{q, q'\}$ is called a parallel-move of $\{p, p'\}$ if p and p' are adjacent to q and q' respectively; but neither p and q' nor p' and q are adjacent.

We can see that in Fig. 5.3, line-cell (t, r) is a parallel-move from (p, q). Note that a line-cell $\{p, p'\}$ might not have an order, but when we associate this line-cell with its parallel, it is better to indicate the order, such as (p, p') or (p', p).

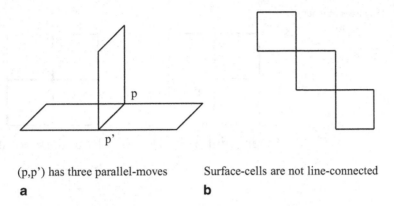

(p,p') has three parallel-moves   Surface-cells are not line-connected

a   **b**

Fig. 5.5 Two instances which are not considered to be surfaces

It is also easy to see that line-cell (p, q) in Fig. 5.3 has four parallel-moves in N_p: left, right, up, and down. Therefore,

Lemma 5.1 *Each line-cell has four parallel-moves in Σ_3.*

We also have,

Lemma 5.2 *A line-cell and each of its four parallel-moves form a surface-cell in Σ_3.*

Now, we can give our definition of digital surface based on parallel-moves:

Definition 5.4 (Chen–Zhang) A connected subset S of Σ_m is a digital surface if any point $p \in S$ is included in some 2-cell of S, and (1) Any two 2-cells are line-connected in S, (2) Every line-cell in S has only one or two parallel-moves in S, and (3) S does not contain any 3-cells.

We can explain that all of the three conditions in Definition 5.4 are necessary.

If we assume there is a line-cell that has three parallel-moves, then it must not be in any surface. See Fig. 5.5a. It goes against the principle of a local neighborhood that a simple surface point must be similar to a 2D plate. Therefore, condition (2) must be valid in this definition.

For condition (1) of the definition, if two surface-cells in S are not line-connected, then S must contain something like Fig. 5.5b. The last condition is easy to see.

Another advantage of Definition 5.4 is that it can be used to describe the boundary points. p is an inner point in surface S if every line-cell containing p has exactly 2 parallel-moves in S. p is a boundary point of S if there exists a line-cell containing p that has exactly one parallel-move. The set of all boundary points is denoted by ∂S. See Fig. 5.6.

The concept of parallel-moves cannot be directly used in general discrete spaces such as indirect adjacency. Some principle of Definition 5.4 can be extended to general cases of discrete manifolds. We discuss general discrete manifolds in Chap. 7.

We can conclude that Definition 5.4 is intuitive and reasonable. But why is this definition better than the Morgenthaler–Rosenfeld definition in direct adjacency? It

Fig. 5.6 Examples of inner and boundary points

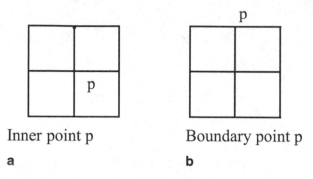

Inner point p   Boundary point p

a   b

is better because it can be directly used in designing fast algorithms for digital surface tracking and recognition. We discuss these topics in Chap. 6. In addition, it can be used to discover the classification of simple digital points in the next section, a basic result presented in this chapter.

5.3   The Classification of Digital Surface Points

After defining the digital surface, we can ask the important question: How many different types of simple surface points are there in 3D digital space? We show two types in Figs. 5.2b and 5.3. In 1993, Chen and Zhang proved that there are only six types of simple digital surface points in 3D that are in direct adjacency [10]. This is called the classification theorem for digital surfaces in direct adjacency in three-dimensional space. Classification of simple surface points become one of the most important results in digital geometry and topology. It lead directly to the discovery of the digital Gauss–Bonnet Theorem for calculating the genus of 3D objects. Many researchers have used or observed the some of these shapes in their research.

In this section, we use the parallel-move based definition of digital surfaces to prove this theorem. In fact, there is an equivalence between the two definitions of digital surfaces given by Morgenthaler and Rosenfeld [24] and Chen and Zhang [9] in direct adjacency. We give the proof in Chap. 15 due to its complexity [11].

The classification theorem deals with the categorization of simple surface points. It states that there are exactly six different types of simple surface points [10].

5.3.1   Simple Surface Points and Regular Inner Surface Points

The definition of parallel-move based surfaces is simple and intuitive. The question is: What is the relationship between such a digital surface and a Morgenthaler–Rosenfeld surfaces? We can show that a closed regular surface is precisely a Morgenthaler–Rosenfeld simple surface.

In order to establish the relationship between Morgenthaler–Rosenfeld's simple surfaces and our parallel-move based surfaces, we need to introduce the concept of regular surface points.

If p is a point of a parallel-move based surface S, then p is regular if all of S's surface-cells, including p, are line-connected in S. If p is both inner and regular, then p is called a regular inner surface point.

To deal with general cases, we can expand the meaning of a regular surface point to any (point-)connected set as follows.

Definition 5.5 Let S be a connected subset of Σ_3. Assume $p \in S$ and $S(p) = S \cap (N_{27}(p) \cup \{p\})$. p is called a regular surface point of S if:

(1) Each line-cell in S containing p has at least 1 and at most 2 parallel-moves in $S(p)$.

(2) Any two surface-cells containing p in S are line-connected in $S(p)$.

(3) $S(p)$ does not contain any 3D-cell.

We say p is a regular inner surface point if p is a regular surface point and each line-cell containing p has exactly two parallel-moves in $S(p)$.

The following theorem establish the relationship between the two definitions of digital surfaces given by Morgenthaler and Rosenfeld [24] and Chen and Zhang [9].

Theorem 5.1 *A Morgenthaler-Rosenfeld simple surface point is a regular inner surface point in direct adjacency. That is, a regular closed surface is a Morgenthaler-Rosenfeld surface.*

Theorem 5.1. can be used to prove the classification theorem, the stand alone proof is given in Chap. 15.

5.3.2 Isometric and Geometric Equivalence in 3D

Isometric transformation is a mapping that preserves the distance of two points in an object to be the same as it is transferred to another space. Isometric is a type of geometric equivalence. In order to explore the structure of simple closed surfaces in Σ_3, we like to define the digital geometric equivalence below.

Definition 5.6 Let S and S containing p be two subsets in N_p, a neighborhood having p at the center. S and S are geometric equivalent if and only if there is a one-to-one mapping $f : S \to R$ which satisfies:

(1) $f(p) = p$,
(2) $d(x, y) = d(f(x), f(y))$, where $x, y \in S$ and d is the distance for 6-connectivity, and
(3) $D(x, y) = D(f(x), f(y))$, where D is the distance for 26-connectivity.

In this definition, the distance refers to the length of the shortest path.

Lemma 5.3 *The geometric equivalence relation is a mathematical equivalence relation.*

Fig. 5.7 Simple surface points, each of which has four adjacent points

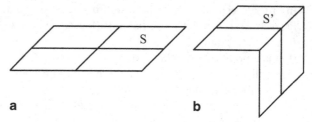

a   b

Proof It is easy to see that conditions (1) and (2) are necessary in Definition 5.6 for this equivalence. It is not obvious that condition (3) is needed. However, we can see that S and S in Fig. 5.7 are equivalent without condition (3). □

5.3.3   The Theorem of the Classification

The classification theorem presented in this section states that there are exactly six different types of simple surface points. A simple surface point p concerns with the point p and its surrounding points. The geometric equivalence relation described in Definition 5.6. can classify all N_p's subsets with point p into a number of geometric equivalence classes. Among these classes, only a few of them make p a simple surface point.

Only direct adjacency (6-adjacency) is considered for simple surface points here. Based on the geometric equivalence, all simple surface points will be classified.

According to Theorem 5.1, we can easily get the following lemma.

Lemma 5.4 *If p is a simple surface point, then each line-cell containing p in N_p has exactly two parallel-moves; any two surface-cells are line-connected in N_p, and there is no 3D-cell in N_p.*

Theorem 5.2 *There are exactly 6 types of simple surface points showed in Fig. 5.8 that are not geometrically equivalent to each other.*

Proof Start with a point p and one of its directly adjacent points p'. (This is because an isolated point cannot be a simple surface point.) According to Lemma 5.4 and geometric equivalence, the 1-cell (p, p') must have two parallel-moves. So, we can only derive two cases in Fig. 5.9, and nothing else.

Thus, a simple surface point with three surface-cells in $N(27, p)$ can be derived from case (b) shown in Fig. 5.9. The result is case (1) shown in Fig. 5.8.

Because each line-cell must have exactly two parallel-moves, from Fig. 5.9, we can develop three cases with 3 surface-cells in Fig. 5.10. These are the only possibilities for having simple surface points without duplication under the geometric equivalence relation.

Continuing the derivation, we can develop 6 cases with 4 surface-cells from Fig. 5.10, see Fig. 5.11.

Fig. 5.8 All types of simple
surface points

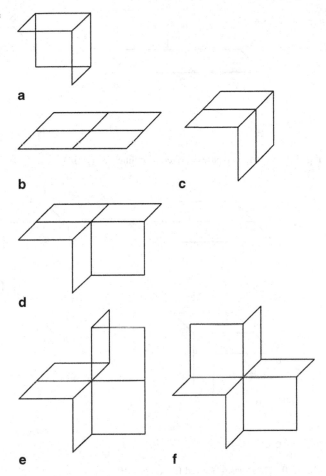

Therefore, we get case (2) and case (3) in Fig. 5.8 from case (a) and case (c) in Fig. 5.11. Again, we can develop the 6 cases with 5 surface-cells in Fig. 5.12 from Fig. 5.11.

Next, we can get case (4) in Fig. 5.8 from case (e) in Fig. 5.12. We can see that point p in case (d) cannot generate a simple surface point. Cases (a), (b), (c), and (f) in Fig. 5.12 have only one option of being a simple surface point. When we add a surface-cell to cases (a), (c), or (f), we get the simple surface point as case (5) in Fig. 5.8. When we add a surface-cell to case (b) in Fig. 5.12, it becomes case (6) in Fig. 5.8. Thus, we have completed the proof.   □

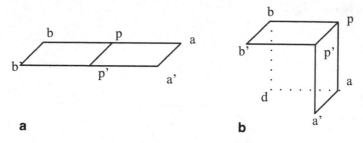

Fig. 5.9 Exactly two cases derived from line-cell (p, p')

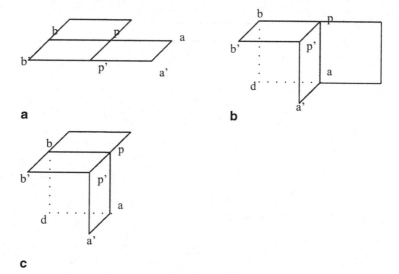

Fig. 5.10 Three cases derived from Fig. 5.10

5.4 Digital Manifolds

In this section, we extend Chen–Zhang's definition of digital surfaces to digital manifolds. This definition unifies all the definitions of curves, surfaces, and manifolds in digital spaces in direct adjacency. In other words, it provides a simple, formal, and uniform definition for digital curves, surfaces, and manifolds.

5.4.1 k-Cells and Connectivity

Two k-cells are said to be k'-dimensional adjacent (k'-adjacent), $k > k' \geq 0$, if they share a k'- cell. A (simple) k-cell path with k'-adjacency is a sequence of k-cells

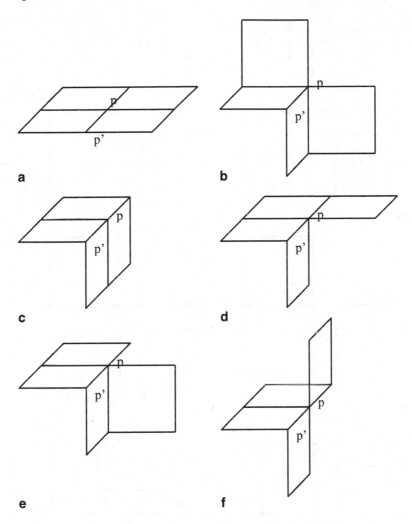

Fig. 5.11 Six cases with four surface-cells derived from Fig. 5.10

$v_0, v_1, ..., v_n$, where v_i and v_{i+1} are k'-adjacent and $v_0, v_1, ..., v_n$ are different. Two k-cells u and v are called k'-dimensional connected if they are two end elements of a (simple) k-cells path with k'-adjacency.

Let S be a subset of Σ_m. We denote $\Gamma^{(0)}(S)$ as the set of all points in S, and $\Gamma^{(1)}(S)$ as the line-cells set in S. Inductively, we use $\Gamma^{(k)}(S)$ be the set of k-cells of S.

Therefore, two elements p and q in $\Gamma^{(k)}(S)$ are k'-adjacent, then $p \cap q \in \Gamma^{(k')}(S)$, $k' < k$.

In Chap. 9, we will discuss simplicial complexes that is a collection of simplexes. Let S is a n-dimensional object. $\Gamma^{(k)}(S)$, $k = 0, ..., n$ represent the similar concepts of simplicial complexes. It is called finite topology [20]. We discuss this topology of digital manifolds in Chap. 9.

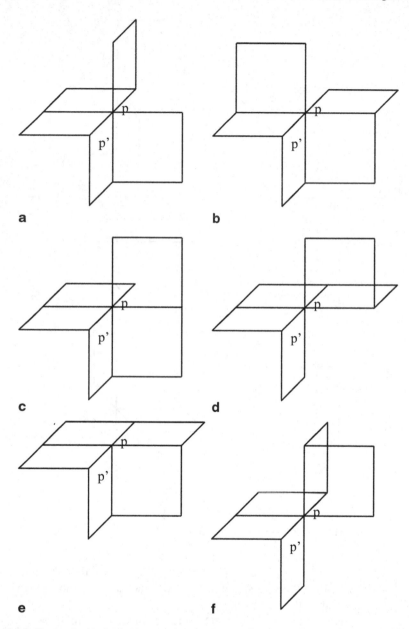

Fig. 5.12 Six cases with five surface-cells derived from Fig. 5.11 b, d, e, and f

The difference between the Chen–Zhang's definition and the simplicial (or cell)-complexes based definition of surfaces is that simplicial type of complexes need to store all $\Gamma^{(k)}(S)$, $k = 0, \ldots, n$ of S. The digital manifolds we define next will not store $\Gamma^{(k)}(S)$, $k = 0, \ldots, n$. Instead of that, we generate a data structure for calculation. We will discuss the algorithm issues in Chap. 6.

5.4.2   Definition of Digital Manifolds

In the discussion of digital manifolds, we first extend the concept of parallel-moves to high dimensions. Then, we give a recursive definition of the m-(dimensional-) cells based on parallel-moves. Afterwards, we can define n-dimensional digital manifolds in Σ_m.

A more general definition of parallel-moves is given as follows:

Definition 5.7   Let A be a subset of Σ_m. A' is called a parallel-move of A if

(1) $|A| = |A'|$,
(2) $A \cap A' = \emptyset$, and
(3) there exists a biconjunction mapping $f : A \to A'$ such that $d(a, f(a)) = 1$ for all $a \in A$.

In addition, a and b are adjacent in A if and only if $f(a)$ and $f(b)$ are adjacent in A'.

Based on the general definition of parallel-moves of a set, we have:

Lemma 5.5   *Let V be a parallel-move of V', then V' is the parallel-move of V.*

Parallel-moves are mainly created for constructing k-cells. We can get a k-cells using the following procedure:

Let $\{p_1, ..., p_{2^k}\}$ be a k-cell of Σ_m, $k \leq m$. Suppose vector $p_i = (x_1^{(i)}, ..., x_m^{(i)})$, $i = 1, ..., 2^k$, then there are $(m - k)$-components so that vectors $p_1,...,$ and p_{2^k} have the same value in each of the $(m - k)$-components, e.g., if the jth component is such a component. Then, $x_j^{(1)} = x_j^{(2)} = ... = x_j^{(2^k)}$. Any parallel-move of $\{p_1, ..., p_{2^k}\}$ can be obtained by adding or subtracting 1 at any of these $(m - k)$-components. Therefore,

Lemma 5.6   *Let V be a k-cell, then V has exactly $(2 \cdot (m - n))$ parallel-moves in Σ_m.*

Let p be a point and p' be a parallel-move of p. We can see that $\{p, p'\}$ is a line-cell. If an i-cell A' is a parallel-move of an i-cell A, then $\{A, A'\}$ is an $(i + 1)$-cell. Formally,

Lemma 5.7   *Let V and V' be k-cells, where V is a parallel-move of V'. Then $V \cup V'$ forms a $(k + 1)$-cell.*

Definition 5.8   A connected subset M of Σ_m is an n-dimensional digital manifold if any point $p \in M$ is included in some n-cell of M and

(1) Any two n-cell are $(n - 1)$-connected,
(2) Every $(n - 1)$-cell in M has only one or two parallel-moves, and
(3) M does not contain any $(n + 1)$-cell.

We can give an equivalent definition of Definition 5.8 in the next. Theis one is good for us to understand a digital manifold.

Definition 5.9   A connected subset M of Σ_m is an n-dimensional digital manifold if $M = \cup\{a | a \in \Gamma^{(n)}(M)\}$ and

(1) $\Gamma^{(n)}(M)$ is $(n-1)$-D connected,
(2) Each element in $\Gamma^{(n-1)}(M)$ has only one or two parallel-moves, and
(3) $\Gamma^{(n+1)}(M)$ is empty.

Now we define the boundary of a digital manifold as exactly having one parallel-move in M, the set we considered in Σ_m.

Definition 5.10 Let M be an n-dimensional digital manifold. The boundary of M, denoted by ∂M, is the set of all $(n-1)$-cells in M, which only have one parallel-move in M. M is called an n-dimensional closed digital manifold if ∂M is empty.

We also want to define the regular (ordinary) point for a digital manifold that is also the extended definition in that of in digital surfaces.

Definition 5.11 Let M be an n-dimensional digital manifold. We say $p \in M$ is an n-regular (ordinary) point if all the n-cells containing p are $(n-1)$-connected among (i.e. inside of) these cells (sometimes referred to as local-connected). A manifold M is n-regular if any point $p \in M$ is n-regular.

Obviously, 1-D digital manifolds are curves, and 2-dimensional digital manifolds are surfaces. Rosenfeld gave a strict definition of digital curves [25]: C is a simple (closed) curve if and only if each point p in C has exactly two adjacent points in C. We can see that Definition 5.8 is equivalent to Rosenfeld's curve definition except we do not allow the boundary of a 2-cell to be a curve. We have also proven that a Morgenthaler–Rosenfeld surface is just a regular closed surface in Σ_3 [11].

5.4.3   Properties of Digital Manifolds*

In this subsection, we will discuss more profound properties of digital manifolds, especially the boundary of a digital manifold. In common sense, the boundary of an nD-manifold is a closed $(n-1)$ D-manifold. However, in digital Case, we need more efforts to reach the same goal. We will next present a concept for bridges that is similar to the cuts in graph theory. This concept is related to genus in topology, but we like to keep the contents of this chapter to be in geometry not topology.

Definition 5.12 Let M be an n-dimensional digital manifold. We say an $(n-1)$-cell c in M is a bridge if $M-c$ is not connected or point-connected. M is said to be bridge-free if M does not contain any bridge.

Definition 5.13 Let M be an n-regular digital manifold, c be an $(n-1)$-cell in M, and $M_c = \cup\{a|a \in \Gamma^{(n)}(M) \& a \cap c \neq \phi\}$, i.e., the union of all n-cells containing c or part of c. We say c in M is a local-bridge if $M_c - c$ is not (point-) connected in $M_c - c$. M is said to be local-bridge-free if M does not contain any local-bridge.

The surface shown in Fig. 5.13a is bridge-free, the surface in (b) has bridges, and the surface in (c) is bridge-free but has local-bridges.

It seems an irregular situation that a surface has a bridge while a closed regular surface has no bridge.

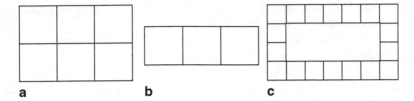

Fig. 5.13 **a** A 2-regular surface that has no bridge; **b** A 2-regular surface that has bridges; **c** A 2-regular surface that has only local-bridges but no bridges

Theorem 5.3 *If surface S is regular and local-bridge-free in Σ_3, then ∂S is the union of some closed digital curves and is regular.*

Proof First, we prove that each point in ∂S has at least two adjacent points (two parallel-moves). If ∂S is empty, then, this theorem is true. Suppose that $\partial S \neq \phi$. Assume $\{p, q\} \in \partial S$, then $\{p, q\}$ has only one parallel-move, namely $\{p', q'\}$ (p' is adjacent to p). Thus, $\{p, q, p', q'\}$ is a surface-cell.

Because S is local-bridge-free, there must be a point in $\{p', q'\}$ not in ∂S.

(1) If $p' \in \partial S$, then p has two adjacent points in ∂S. In addition, $\{q, q'\}$ has two parallel-moves because q' is not in ∂S. One of these two parallel-moves is $\{p, p'\}$, and another one can be denoted by $\{q_1, q_1'\}$, where q_1 is adjacent to q. If q_1 is in ∂S, then q has two adjacent points (parallel-moves). If q_1 is not in ∂S, then line-cell $\{q, q_1\}$ has two parallel-moves, one of which is $\{q', q_1'\}$. The other can be denoted by $\{q_2, q_2'\}$ in S, where q_2 is adjacent to q, and so forth. Since q has only a finite number of adjacent points (at most 6 in Σ_3), there must be an i such that $\{q_i = p, q_i'\}$, and each $q_1, ..., q_{i-1}$ is adjacent to q but not in ∂S. Similarly, $\{q, q_{i-1}\}$ has two parallel-moves: $\{q_{i-2}', q_{i-1}'\}$ and $\{q_i = p, q_i'\}$ in S. Therefore, $\{q, q_i = p\}$ has a new parallel-move $\{q_{i-1}', q_i'\}$. However, $\{q, q_i = p\}$ has only one parallel-move $\{p', q'\}$. Therefore, q has two adjacent points (parallel-moves).
(2) If $q' \in \partial S$, then using the same procedure as above, we can prove that both p and q have two adjacent points in ∂S.

Now, we discuss the case where both p' and q' are not in ∂S. We still use the same method as above. For p, we generate the sequence $\{p_1, p_1'\}, ... \{p_i, p_i'\}$; for q, we generate the sequence $\{q_1, q_1'\}, ... \{q_i, q_i'\}$. As a result, for any line-cell in ∂S, each point of the line-cell has two adjacent points in ∂S. Thus, each point in ∂S has two adjacent points in ∂S.

Second, we need to prove that any p in ∂S has only two adjacent points in ∂S. If there is only one surface-cell containing p, then p has only two adjacent points. If there are exactly two surface-cells containing p, then p has three adjacent points in ∂S implying that there is a local-bridge. Therefore, p has only two adjacent points in ∂S.

If there are three or more surface-cells, $S_1, ..., S_k$ containing p, $k \geq 3$, then we know $S_1, ..., S_k$ are line-connected among these surface-cells. Also, p has two adjacent points, namely q and r, in ∂S. Both $\{p, q\}$ and $\{p, r\}$ has one parallel-move.

If p, q, and r are in a surface-cell S_i, then there is a S_j, $i \neq j$ such that $S_i \cap S_j = \{t, t'\}$ is a line-cell. Note: $p \in S_j$. If $t = p$, then $t' = q$ or $t' = r$. Thus, $\{p = t, q = t'\}$ or $\{p = t, r = t'\}$ has two parallel-moves. So, p, q, and r cannot be in a surface-cell S_i when each of $S_1, ..., S_k$ contains p.

If $q \in S_i$ and $r \in S_j$, there are two possibilities: $S_i \cap S_j = \{p, t\}$ or $S_i \cap S_j = \{p\}$. If $S_i \cap S_j = \{p, t\}$, for the third surface-cell S_m, there is a line-connected surface-unit path in $S_1, ..., S_k$ from S_i (or S_j) to S_m. Let the path be $S_i, S_n, ..., S_m$. Then, $S_i \cap S_n = \{p, q\}$ or $\{p, t\}$. If $S_i \cap S_n = \{p, q\}$, $\{p, q\}$ has two parallel-moves. If $S_i \cap S_n = \{p, t\}$, then $\{p, t\}$ has three parallel-moves, S_i, S_j, and S_n. This is impossible because each line-cell has at most two parallel-moves.

For the second possibility, $S_i \cap S_j = \{p\}$, we will prove that p has only two adjacent points q and r in ∂S.

Let us assume that there is a third point $t \in \partial S$ and t, adjacent to p. Thus, two cases can be derived.

(1) If $t \in S_i$ or $t \in S_j$, then this case is the same as the first possibility.
(2) If q, r, and t are in three different surface-cells, then let $t \in S_m$. We have already discussed the cases where $S_i \cap S_m$ or $S_j \cap S_m$ is a line-cell. So, we can assume $S_i \cap S_m = \{p\}$ and $S_j \cap S_m = \{p\}$. Because there are only six adjacent points for p in Σ_3, $S_i \cup S_j \cup S_m$ contains all these six points. Suppose that these points are $\{q, r, t, q', r', t'\}$. There are only six line-cells containing p in Σ_3, and each surface-cell in $\{S_1, S_2, ..., S_k\} - \{S_i \cup S_j \cup S_m\}$ contains two of the six line-cells. However, these line-cells cannot be $\{p, q\}$, $\{p, r\}$, or $\{p, t\}$. It is impossible for a surface-cell to be line-adjacent to all three surface cells S_i, S_j, and S_m in this case. On the other hand, if there are two surface-cells in $\{S_1, S_2, ..., S_k\} - S_i \cup S_j \cup S_m$, then there is an element in $\{\{p, q'\}, \{p, r'\}, \{p, t'\}\}$ contained by these two surface-cells. This element is already in each S_i, S_j, and S_m. Thus, this line-cell has three parallel-moves, which is impossible. Therefore, we have completed the proof of Theorem 5.4.   □

We now generalize Theorem 5.4 to digital k-manifolds. The key is to define the regular i-cell as more than just a regular point. The concept of local-bridge-free will be contained by the new definition of regular manifolds.

Let M be a digital k-manifold. The boundary of M, ∂M, is the set of all $(k - 1)$-cells in M, each of which is contained by only one k-cell in M. M is said to be closed if $\partial M = \emptyset$. A point p is called k-inner if $p \notin \partial M$. A point p in a k-manifold M is called regular if all k-cells containing p are $(k - 1)$-connected. Moreover, we may state:

Definition 5.14 An i-cell $A_i \in M$, $i = 0, .., k - 2$, is regular if all k-cells containing A_i are $(k - 1)$-connected (in these k-cells). M is regular if all i-cells in M are regular for all $i = 0, ..., k - 2$.

For $A_i \in M$, the set containing these k-cells is denoted by $S_k^{(i)}(A_i)$. An inner point p in a k-manifold M is called simple if $S_k(p) - \{p\}$ is a simple $(k - 1)$-manifold.

Theorem 5.4 *The boundary of a regular n-dimensional digital manifold is regular and is the union of several $(n - 1)$-dimensional closed digital manifolds.*

Proof We just present a simple proof here. We know that ∂S is the union of the collection of $(n-1)$-cells, each of which belongs to only one n-cell in S. We might as well assume here that the ∂S is the collection of $(n-1)$-cells.

First, we need to prove that each $(n-2)$-cell is included in two $(n-1)$-cells in ∂S. Suppose $A \in \partial S$ is an $(n-1)$-cell, we want to prove that each $(n-2)$-cell in A is contained by two $(n-1)$-cells including A. Let a be an $(n-2)$-cell in $A \in \Delta$, where $\Delta \in S$ is an n-cell. There is a unique $A' \in \Delta$ that contains a. If A' is in ∂S, then there are two $(n-1)$-cells containing a. Otherwise, A' is contained only by another Δ_1, and Δ_1 contains another A'' containing a. If $A'' \subset \partial S$, then a is contained by two $(n-1)$-cells; Because all n-cells containing a are $(n-1)$-connected, there must exist a $A^{(m)} \in \partial S$ because we have a finite numbers of cells. (Another explanation is that an $(n-1)$-cell is impossible to be contained by three n-cells according to the definition of the digital manifold.)

Second, if there are three $(n-1)$-cells containing a, then all n-cells containing a will not be $(n-1)$-connected based on the above process. So, ∂S is a closed $(n-1)$-pseudo-manifold or several closed $(n-1)$-pseudo-manifolds.

We can also prove that every i-cell is regular in ∂S as well. □

5.5 Historical Remarks: Analysis on General Digital Surfaces in 3D

Before the Morgenthaler–Rosenfeld definition of digital surfaces, people understood digital surfaces through classic discrete surfaces such as the triangular surface or the imperial understanding and experience of digital surfaces.

The Morgenthaler–Rosenfeld definition gave the first mathematical definition of digital surfaces. It was a necessary step in digital geometry. The definition contains a total of nine types of digital surfaces. Kong and Roscoe presented a detailed work to analyze Morgenthaler–Rosenfeld's surfaces [19] and Chen made some additional analyses [3, 4, 5]. The results show that most of the nine types do not really exist.

In fact, to require that a surface point $S(p)$ separates N_p into two completely disconnected components is not really reasonable. An 8-curve in the digital plane cannot separate its neighborhood into two components, so why should we require that of 18- or 26-connected surfaces?

Chen suggested another general surface definition in [5], where 6-connectivity is used for $N_p - S(p)$.

On the other hand, indirect adjacency may cause ambiguous interpretations for a set S. For example, suppose that we allow 18-adjacency for a set S containing points as shown in the following figure, Fig. 5.14a. From this, we can see that for 6-adjacency, we have a unique interpretation. However, for 18-adjacency, we could have an interpretation such as (b), which is a surface. If we interpret (a) to be (c), then it is not a surface. If we interpret (a) as (d), then it is a different surface. All of these interpretations are reasonable under 18-adjacency.

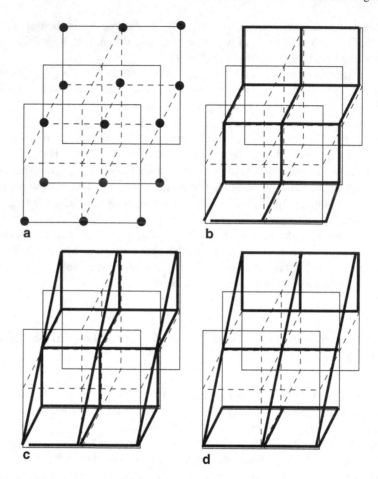

Fig. 5.14 A digital point set in S and some of the interpretations: **a** Original data points, **b** and **d** Two different surfaces, and **c** Not a surface

Therefore, 18- or 26- adjacency introduces ambiguities for the interpretation of surfaces. That is why a general solution allowing indirect adjacency is very difficult. As we know, there are only six types of simple surface points for direct adjacency [10, 22], but it is really hard to get all simple surface points to allow indirect adjacency. The ambiguity is real, and it is impossible to reduce these cases to be unique. On the contrary, we are able to see the power of indirect adjacency.

Researchers are still interested in identifying how many possible interpretations of digital surface points there are in just N_p. The number is very large, 10,580 cases, as indicated in [12].

A unified definition may not be very practical in defining digital surfaces. However, for the purpose of surface recognition, we have to have a unified definition for surfaces, or at least a few possible definitions. This problem is still open, i.e. finding the most reasonable surface if there is more than one interpretation of possible digital surfaces [14, 27].

The classification of digital surfaces is very important [10]. The six classes of the digital surface points in 6-adjacency play an important role in digital geometry. Using this result, the digital form of the Gauss-Bonnet theorem was found in [7]. The detailed proof will be provided in Chap. 14.

For digital manifolds, the general definition of a m-D digital manifold contains many philosophical meanings. The definition we provided in this chapter has algorithmic advantages [13]. It is not just the digitization of a continuous manifold in digital space.

The digital manifold definition of this note can also represent non-orientable cases. For instance, a Mobius band is a surface that fits under the definition of digital surfaces [1]. We will discuss it in Chap. 6. Herman and Webster discussed the surface of some 3-complexes generated by voxels, and they proved that the boundary (surface) of a 3-complex is the union of several line-connected components of the 2-cells [2, 16].

Theorem 5.5 is a general form for some results discovered by Chen and Zhang [8], Letecki [21], and Letecki, Eckhardt, and Rosenfeld [23]. A regular set is called a well-composed set in [21, 23]. Letecki then expanded this concept to 3D [21]. We present the general theorem, Theorem 5.5 here. Some information about this development can be found in [5, 18] The more general definition for discrete manifolds is presented in Chap. 7.

References

1. P. S. Alexandrov, Combinatorial Topology, New York: Dover, 1998.
2. E. Artzy, G. Frieder and G.T. Herman, The theory, design, implementation and evaluation of a three-dimensional surface detection algorithm, Comput. Vision Graphics Image Process. 15, 1981, 1–24.
3. L. Chen, Generalized discrete object tracking algorithms and implementations, Melter, Wu, and Latecki ed, Vision Geometry VI, SPIE 3168, pp 184–195, 1997.
4. L. Chen, A note on (alpha, beta)-type digital surfaces and general digital surfaces, in Melter, Wu, and Latecki ed, *Vision Geometry VII*, SPIE Proc. 3454, pp 28–39,1998.
5. L. Chen, *Discrete Surfaces and Manifolds: A theory of digital-discrete geometry and topology*, 2004. SP Computing.
6. L. Chen, Digital Functions and Data Reconstruction, Springer, NY, 2013.
7. L. Chen, and Y. Rong, Digital topological method for computing genus and the Betti numbers, Topology and its Applications, Volume 157, Issue 12, 2010, Pages 1931–1936.
8. L. Chen and J. Zhang, Digital manifolds: A Intuitive Definition and Some Properties, Proceedings of the Second ACM/SIGGRAPH Symposium on Solid Modeling and Applications, Montreal, 1993, 459–460.
9. L. Chen and J. Zhang, A digital surface definition and fast algorithms for surface decision and tracking, in R.A. Melter and A.Y. Wu, Vision Geometry II, SPIE Proceedings 2060, 169–178. (Also see Mathematical Review 95g:68119.)
10. L. Chen and J. Zhang, Classification of simple digital surface points and a global theorem for simple closed surfaces, in R.A. Melter and A.Y. Wu, Vision Geometry II, SPIE Proceedings 2060. (Also see Mathematical Review 95g:68119.)

11. L. Chen, H. Cooley and J. Zhang, The equivalence between two definitions of digital surfaces, *Information Sciences*, Vol~115, pp 201–220, 1999.
12. Ciria, J.C., Domnguez, E., Francis, A.R., Quintero, A.: A plate-based definition of discrete surfaces. Pattern Recogn. Lett. 33(11), 1485–1494 (2012)
13. T. H. Cormen, C.E. Leiserson, and R. L. Rivest, Introduction to Algorithms, MIT Press, 1993.
14. R. C. Gonzalez, and R. Wood, *Digital Image Processing*, Addison-Wesley, Reading, MA, 1993.
15. F. Harary, Graph theory, Addison-Wesley, Reading, Mass., 1969.
16. G.T. Herman and D. Webster, A topological proof of a surface tracking algorithm, Comput. Vision Graphics Image Process. 23, 1983, 162–177.
17. E. Khalimsky, R. D. Kopperman and P. R. Meyer, Computer graphics and connected topologies on finite ordered sets, Topology and its Applications, Vol. 36, 1990, 1–17.
18. R. Klette and A. Rosenfeld, Digital Geometry, Geometric Methods for Digital Picture Analysis, series in computer graphics and geometric modeling. Morgan Kaufmann, 2004.
19. T.Y. Kong and A.W. Roscoe, Continuous analogs of axiomatized digital surfaces, Computer Vision, Graphics, and Image Processing, Vol 29, 60–86, 1985.
20. V. A. Kovalevsky, Finite topology as applied to image analysis, Computer Vision, Graphics and Image Processing, Vol. 46, 1989, pp. 141–161.
21. L. Latecki, "3D well-composed pictures," *The Proc. of SPIE on Vision Geometry IV,* Vol 2573, pp. 196–203, 1995. also see Graphical Models and Image Processing Vol 59, 164–172, 1997.
22. L.J. Latecki, Discrete Representation of Spatial Objects in Computer Vision, Kluwer Academic Publishers, 1998.
23. L. Latecki, U. Eckhardt, and A. Rosenfeld, "3D well-composedness of digital sets," *The Proc. of SPIE on Vision Geometry II,* Vol 2060, pp. 61–68, 1993.
24. D.G. Morgenthaler and A. Rosenfeld, Surfaces in three-dimensional images, *Imform. and Control* 51, 1981, 227–247.
25. A. Rosenfeld, Connectivity in digital pictures, Journal of the ACM, Vol. 17, 1970, pp. 146–160.
26. A. Rosenfeld, Three-dimensional digital topology, Inform. and Control 50, 1981, 119–127.
27. A. Rosenfeld and A.C. Kak, *Digital Picture Processing*, 2nd ed., Academic Press, New York, 1982
28. J. Stewart, Calculus, Brooks/Cole Publishing Company, Pacific Grove, CA, 4th ed, 1999.

Chapter 6
Algorithms for Digital Surfaces and Manifolds

Abstract In Chap. 2, we introduced some algorithms for graphs. In this chapter, we specifically design algorithms for digital object recognition and tracking. These algorithms are mainly for digital surfaces and manifolds. There are two types of questions to solve in this chapter: (1) Given a set of data M, decide or recognize whether the data represents a geometric shape, specifically a curve, surface, or solid object, and (2) Extract the curve or surface components of the data set. The main task is to extract the boundary of a surface or a 3D manifold. We also design algorithms for these problems for higher dimensional manifolds. In this chapter, we deal with various important tasks in digital and discrete geometry in an ideal situation such as no noise with perfect data formats. We then design algorithms to find solutions for these problems. In Chap. 11, we specifically discuss the data in the format of randomly collected points, called cloud data or scattered data sets that usually do not form a specific geometric shape. In such a case, the researcher needs to estimate the best possible shape for the data. These types of problems are usually related to geometric processing.

Keywords Algorithm · Time complexity · Digital surface · Digital manifold · Recognition algorithm · Tracking algorithm · Orientability testing

6.1 What is an Algorithm?

An algorithm is a sequence of instructional steps for solving a problem. The algorithm can be designed to be very complicated if we are not able to find a simple formula or equation to fit the problem.

In mathematics, algorithms are always secondary choice for solving a problem. Along with the fast development of digital computers, people today are very comfortable with using algorithms to solve problems. Instead of spending intensive time to find simple formulas, listing algorithmic steps would bring us to the same results. In addition, we let the computer do the job for us.

Two types of measurements, time and space costs, are used to evaluate whether an algorithm is a good algorithm. These are called time complexity and space complexity, respectively. The third measurement is related to the "length" of the algorithm and is a measure of the total instructions programmed in memory. In other words, it is

© Springer International Publishing Switzerland 2014

L. M. Chen, *Digital and Discrete Geometry*, DOI 10.1007/978-3-319-12099-7_6

the size of the algorithm, called Chaitin–Kolmogorov complexity. This is somehow related to the simplicity of the number of instructions and "formulas " used for a problem. The most widely used complexity measure is time complexity.

We say that the bubble-sort sorting algorithm is $O(n^2)$ meaning that this algorithm requires $c \times n^2$ steps to solve the problem where c is a constant. This algorithm is in polynomial time.

The first algorithm that is not trivial is probably the Euclid's algorithm, an algorithm for finding the greatest common divisor (GCD) of two integer numbers. An algorithm is needed since there is no perfect formula for getting $GCD(a, b)$ as a function of a and b. Therefore, we write a procedure to represent this. In computer programs, we also use the term function, which is a name borrowed from mathematics.

In computer science, the most basic algorithms are sorting and searching, for example bubble-sort and binary-search. Binary-search uses the divide-and-conquer technology that is very popular in algorithm design. Other techniques such as dynamic programming and greedy algorithms are also useful.

In additional to good algorithmic technology, data structures, the way of holding the input data, is also important to efficient algorithms, especially in geometric problems.

6.1.1   Easy Problems and NP-hard Problems

Some problems are easy to solve such as sorting and searching problems. However, for other problems, even though we can find algorithms to solve the problem, there is no quick or fast algorithm to complete the task. We may need exponential time to solve them, meaning that the program would take years to run for a large input size. A problem is said to be an NP-problem if we can check the answer (if there is any that we know of) quickly, e.g. within $O(n)$ time. Very many problems are found to have such a property. For instance, the traveling salesmen problem has this property: Finding a route for a traveler who plans to visit one city in each of the 50 U.S. states exactly once, is there a route such that the total distance is smaller than a given number K?

If we know the sequence of the list of 50 cites, then we can easily check to see if the answer is correct. If we do not know the answer, then we may need to spend 10 years to find it. [1, 11]

It was proven that the traveling salesmen problem is the hardest of the NP-problems, called an NP-complete problem. It means that if the traveling salesmen problem can be solved in polynomial time, then all NP-problems can be solved in polynomial time with regular computers. This is a famous unsolved problem in mathematics and computer science called the $P =?NP$ problem. Here, P means that the problem can be solved in polynomial time by regular computers (for example, a Turing machine), and NP means that the problem can be solved in polynomial time by a non-deterministic Turing machine.

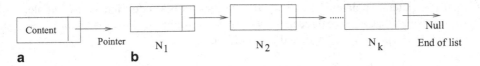

Fig. 6.1 A linked list: **a** A node configuration, **b** An example of a linked list

6.1.2 Geometric Algorithms

Geometric algorithms are very board. It can refer to a graph algorithm with or without a geometric metric. It can also mean algorithmic geometry, sometimes called computational Geometry, that deals with algorithms for Euclidean geometric shapes. Geometric algorithms can also deal with geometric processing with inputs that are sampled points or cloud data points. It can even mean geometric curves and surface fitting algorithms.

In this chapter, we mainly deal with the algorithm for digital surfaces and manifolds. We discuss other types of geometric algorithms in Chaps. 10–12.

6.2 Data Structures for Digital Data Sets

Data structures are the way to store input or processed data. When an algorithm is running, it needs to obtain the input data or to save temporary data for processing. It also needs to output the final results. The format of storing data, the data structure, is important to the algorithm.

For example, a searching algorithm is used to find whether or not a number is in an array. If the number is randomly inserted in the array, then we have to search for the number one by one, making the time complexity $O(n)$ (meaning that there is a constant c such that the time spending is at most $c \cdot n$). However, if we have already arranged (sorted) the data in ascending order, a data structure called a priority queue, then we can use the binary search algorithm, which only needs $O(\log n)$ time.

A. Arrays The simplest data structure for digital data is the array. A surface S in Σ_3 can be represented as an $N \times N \times N$ array, which is a three dimensional array. A point in S will be assigned a "1" at the location in this array. Otherwise, we assign "0" to the rest of the locations.

This array data structure is not very effective since there may be too many zeroes.

B. Linked Lists A linked list is a data structure that is constructed by "nodes." A node contains two parts: the content and the pointer. The content part holds data and the pointer links to the next node. See Fig. 6.1. In Chap. 2, we discussed two methods of representing graphs, one of which is called an adjacency list. It is similar to linked lists.

In computers, a pointer is a memory address that holds a node. Here is an example: Let us assume that we have three nodes $n1 = < 2, pointer1 >$, $n2 = < 5, pointer2 >$, and $n3 = < 8, pointer3 >$. A linked list could be

$$n1 = < 2, n2 >, n2 = < 5, n3 >, n3 = < 8, null >,$$

where *null* indicates the end of the list, or

$$n1 = < 2, n3 >, n3 = < 8, n2 >, n2 = < 5, null > .$$

For a graph $G = (V, E)$, we can use linked lists to represent G. Let us use the example in Fig. 4.6, where the first component is a subgraph with vertices $\{A, B, C, D, E, F.G, H\}$. We can represent all edges as follows (using n_A to represent the node holding A):

$$< A, n_B >, < B, n_D >, < D, null >;$$
$$< B, n_A >, < A, n_C >, < C, null >;$$
$$< C, n_B >, < B, n_f >, < F, null >;$$
$$\cdots\cdots\cdots$$

The above representation is clear for computer programmers, but for mathematicians, we simplify the data structure as

$$A \rightarrow B, D;$$
$$B \rightarrow A, C;$$
$$C \rightarrow B, F;$$
$$\cdots\cdots\cdots$$

We do this because we cannot represent the meaning of "$\rightarrow$" easily in computers. This is one reason why computer scientists created data structures.

Let $S = \{p_1, \cdots, p_n\}$ be a subset of Σ_3. Then the adjacency-list of S is given as follows:

$$p_1 \rightarrow p_1(0), ..., p_1(k_1)$$
$$p_2 \rightarrow p_2(0), ..., p_2(k_2)$$
$$\cdots\cdots\cdots\cdots\cdots$$
$$p_n \rightarrow p_n(0), ..., s_n(k_n)$$

From p_1 to p_n, we list all points in S. $p_i(0)$ to $p_i(k_i)$ are all directly adjacent points of point p_i in S, and they are denoted by $AL_P(p_i)$.

$$AL_P(p_i) = \{p_i(0), ..., p_i(k_i)\}, i = 1, ..., n.$$

C. List of All k-cells The data structure for computer graphics usually uses the list of k-cells, where the points are represented as vectors. Then, we get the index of the vectors to represent the 1-cell that is formed by two points before we list the set of 1-cells. Therefore, this data structure is usually saved in a file as follows:

1) The number of points, list of points;
2) The number of 1-cells, list of 1-cells;
3) The number of 2-cells, list of 2-cells. $\cdots$

This data structure is nice and easy to understand, but not good in terms of efficiency. It not only requires extra space, but it also does not create an effective connection among specific cells. In algorithm design, which is not only for display purpose, for instance in 3D image processing, this data structure would make the algorithm very slow.

D. Stacks and Queues A *Stack* is a special data structure that holds a number of elements. It manages the data in the fist-in and last-out mechanism. Just like a stack of books, one can only access the top book all the time. The first book that was pushed in the stack will be the last to be able to take it out. Two operations related to stacks are *Push* and *Pop*.

A *Queue*, on the other hand, is in the fist-in and first-out order. Two operations related to queues are *Insert* and *Remove*. We always insert an element to the end of a queue and remove an element from the front of the queue [11].

6.3 Algorithms for Decision and Tracking of Digital Surfaces

In this section, we start to design basic algorithms for digital surfaces. Our algorithms include those for the decision algorithm and the tracking algorithm [4, 5, 9].

We assume that none of the objects considered reaches the border of Σ_m, the ambient space that holds the objects. We first want to present some of the basic problems for digital surfaces.

Problem 6. 1 The surface decision problem: Given a subset S of Σ_3, we want to determine if S is a digital surface.

Problem 6. 2 The surface tracking problem: Given a solid object, we want to determine (and extract) its boundary surface.

6.3.1 Algorithms for the Surface Decision Problem

How can we tell if a set S is a digital surface? This can be decided in accordance with Definition 5.4. Three procedures are shown in order to accomplish the following three tasks:

1. Find all parallel-moves of every 1-cell (line-cell), $(s_i, s_i(j))$ in S, and decide whether or not each $(s_i, s_i(j))$ has one or two parallel-moves.
2. Search for all 2-cells (surface-cells) by the line-adjacency in S and decide whether or not all surface-cells are line-connected and every point in S will be in one of the 2-cells. (We want also check if the set of 2-cells covers S.)
3. Find all parallel-moves of each surface-cell in S and decide whether any surface-cell has parallel-moves in S. We want to know if S contains a 3-cell or not.

First, we want to obtain all the parallel-moves of each line-cell in S and decide whether or not it satisfies the first condition of Definition 5.4. The following lemma will help design the algorithms.

Lemma 6.1 $|\Gamma^{(1)}(S)| = (1/2)\Sigma_{i=1}^{|S|}(n_i + 1).$ $|\Gamma^{(1)}(S)| \leq 3|S|.$

Proof All 1-cells in S can be represented as follows:

$$\Gamma^{(1)}(S) = \{< s_i, s_i(j) >: i = 1, |S|; j = 1, n_{|S|}\}.$$

where $[a, b] = [b, a]$,so $|\Gamma^{(1)}(S)| = (1/2)\Sigma_{i=1}^{|S|}(n_i + 1)$. On the other hand, each point in Σ_3 has at most 6 directly adjacent points, so $|\Gamma^{(1)}(S)| \leq 3|S|$ holds.   □

We also know that: If $p, q \in \Sigma_3$ and $d(p,q) = 1$, then there is no point $r \in \Sigma_3$ with $d(p,r) = 1$ and $d(r,q) = 1$. If $d(p,q) = 2$, there are at most two paths whose lengths are 2 between p and q.

Algorithm 6.1   : Find all parallel-moves of every line-cell and decide whether or not each line-cell has one or two parallel-moves.

Step 1   Take every point s_i in the point-adjacency-list sequentially.

Step 2   For each point $s_i(j)$ adjacent to s_i, take all the adjacent points $AL_P(s_i(j))$ of $s_i(j)$, except s_i.

Step 3   For each point p in $\{AL_P(s_i(j)) - s_i\}$, find all q's in $\{AL_P(p) - s_i(j)\}$. If q in $\{AL_P(s_i)\}$, then (p,q) are the parallel-moves of $(s_i, s_i(j))$. According to above Lemma and discussions, for a certain p, there is at most one such q $(\neq s_i(j))$ in $\{AL_P(s_i)\}$.

Step 4   For each line-cell $u_k = (s_i, s_i(j))$, set its parallel-moves to the set $AL_L(u_k)$. Thereafter, we have the line-cells' parallel-move-list:

$$u_k \rightarrow AL_L(u_k).$$

Observe the number of elements in $AL_L(u_k)$ to decide whether or not each line-cell has one or two parallel-moves.

We know that each point has six direct adjacent points in Σ_3. So each $AL_P(s_i)$ has at most six elements in 3D, we have the following:

Lemma 6.2 *The time complexity of Algorithm 6.1 is* $O(|\Gamma^{(1)}(S)|)$. *This can also be expressed as* $O(|S|)$.

Second, we design an algorithm to search all surface-cells in S and to decide their line-connectedness based on the parallel-move-list of the line-cells, where the parallel-move-list is obtained using Algorithm 6.1. A surface-cell has four line-cells, since one line-cell and its one parallel-move construct a surface. Two surface-cells are line-adjacent if and only if there exists a line-cell that is the intersection of these two surface-cells. In other words, the line-cell has two parallel-moves to be contained by these two surface-cells, respectively. We already know that (u_k, v) is a surface-cell in S if $v \in AL_L(u_k)$, so we use the breadth-first-search technique on the line-cells' parallel-move-list $u_k \rightarrow AL_L(u_k)$.

Algorithm 6.2 Search for all line-adjacent surface-cells and decide whether or not all surface-cells are line-adjacent and cover the set S.

Step 1   Let Q be a queue. Choose a line-cell u_k in S and mark it.

Step 2   For each unmarked element v of $AL_L(u_k)$, put v into Q. Also, put the other two line-cells in the surface-cell (u_k, v) and do not put u_k or v into Q.

Step 3   Take an element from Q and mark it, then repeat step 2 until Q turns out to be empty.

Step 4   If the marked line-cells cover the set S, that is, all line-cells are marked in their parallel-moves' adjacency-list, then all surface-cells are line-adjacent and cover set S.

Because Algorithm 6.2 is the standard breadth-first-search algorithm,

Lemma 6.3 *The time complexity of Algorithm 6.2 is $O(|\Gamma^{(1)}(S)|)$, which is also $O(|S|)$.*

The last thing for deciding whether a surface is line-adjacent is to check whether each surface-cell has a parallel-move. In the same manner described in Algorithm 6.1 and based on Algorithm 6.3, we could get all the parallel-moves of each surface-cell in S and test whether or not each surface-cell has a parallel-move in S. The following lemma is helpful in the design of the algorithm.

Lemma 6.4   (1) $\Gamma^{(2)}(S) = (a, b) \mid a$ is i^{th} $1-cell, b \in AL_L(i), i = 1, ..., |\Gamma^{(1)}(S)|\}$.

2) $|\Gamma^{(2)}(S)| = (1/4)\Sigma_{i=1}^{\gamma_1}(m_i + 1), \gamma_1 = |\Gamma^{(1)}(S)|$.

3) $|\Gamma^{(2)}(S)| \leq |S|$.

Algorithm 6.3 : Find out the parallel-moves of all surface-cells in S, and decide whether any of the surface-cells have parallel-moves in S, we take the following steps. (If S is a surface, then there is no such a parallel-move. This is because a surface-cell and its parallel move will be a 3-cell.)

Step 1   Take every line-cell u_k in the line-cell's parallel-move-list sequentially.

Step 2   For each line-cell $u_k(j)$, which is a parallel-move of u_k, i.e., $u_k(j) \in AL_L(u_k)$, take all the elements in $AL_L(u_k(j)) - u_k$.

Step 3   For each element p in $\{AL_L(u_k(j)) - u_k\}$, find all q's in $\{AL_L(p) - u_k(j)\}$. If q in $\{AL_L(u_k)\}$, then (p, q) is a parallel-move of $(u_k, u_k(j))$. According to Lemma 4.2.4, for a certain p, there is at most one such q such that $(\neq u_k(j))$ in $\{AL_L(u_k)\}$.

Step 4   For each surface-cell, $z_t = (u_k, u_k(j))$, put its parallel-moves into the set $AL_S(z_t)$. Thereafter, we have a surface-cell's parallel-move-list:

$$z_t \to AL_S(z_t).$$

Observe the number of elements in $AL_S(z_t)$ to decide whether any surface-cell has a parallel-move.

Based on Lemma 6.3 and Lemma 6.4, we have:

Lemma 6.5 *The time complexity of Algorithm 6.3 is $O(|\Gamma^{(2)}(S)|)$, which can also be expressed as $O(|S|)$.*

To summarize, deciding whether a subset S of Σ_3 is a digital surface can be done by using the three above algorithms. If we start with an adjacency-list of S, then we have,

Theorem 6.1 *There is an algorithm to decide a digital surface in $O(|S|)$, where S is a subset of the three dimensional space Σ_3.*

6.3.2 Surface Tracking

Now, we discuss the surface's tracking problem. First we need to define the solid objects in Σ_3. A solid object is a point-connected subset of Σ_3; and it consists of $3D$-cells, one of which can be made by combining a surface-cell and its one parallel-move. Thus,

Lemma 6.6 *Let V be a subset of Σ_3. V is a solid object if and only if $\cup\{z|z \in \Gamma^3(V)\} = V$ and for every $u \in \Gamma^2(V)$, $|AL_S(u)| = 1$ or 2. Also, ∂V consists of all u with $|AL_S(u)| = 1$.*

Lemma 6.7 *There is an $O(|V|)$ time algorithm to track the surface of a solid object V.*

Proof First of all, we need to decide whether V is a solid object. Using Algorithms 6.1 and 6.3, along with Lemma 6.7, we can find all the $3D$-cells. V is the solid object when these $3D$-cells cover the whole set V. Herman and Webster indicated that all the surface-cells of the face surface in a solid object are line-connected [16]. We knew that the face of a solid object may not be a digital-surface, but we can find out all the surface-cells, each of which only has one parallel-move to obtain the face surface of V in Algorithm 6.3. In order to track the surface of V, we made a little change in functionality (of Algorithm 6.3). Hence, the complexity of the algorithm after the change still satisfies $O(|V|)$.   □

6.4 Algorithms for Digital k-Manifolds

We defined digital k-manifolds in Chap. 5. In this section we present the algorithms to recognize and extract the digital k-manifolds. First, we discuss the data structure that stores digital manifolds in each dimension and also the construction of the i-cells using the principle of parallel moves found in Chap. 5. Second, we give the general algorithms for finding or recognizing the digital k-manifolds.

6.4.1 Data Structures of Digital k-Manifolds

Unlike triangulated manifolds, digital manifolds do not need predefined k-cells. Digital k-cells can be obtained as needed (on-line or dynamically). In 1993, Chen and Zhang defined a digital manifold using parallel-moves. In Chap. 5, we defined $3D$ digital manifolds and proved that the face surface of a regular $3D$ digital manifold is a digital surface.

In this subsection, we attempt to solve the following problem: Given a set of points, M, in finite dimensional Euclidean space, we want to decide if M is a k-dimensional digital manifold and get the boundary of M, which is a $(k-1)$-manifold.

We also want to know when M contains the majority of k-cells and a few $(k+1)$-cells, can we still extract the k-dimensional components? It is possible for the $(k+1)$-cells to be caused by some sampling errors.

Let the Euclidean space be m-dimensional. All data points are sampled at integer grid points.

Given a set of points in Σ_m, in order to design a generalized algorithm, we need to design a special data structure. This data structure will not only hold $\Gamma^{n)}(M)$, the n-cells, but it can also indicate the relationship and connection among $\Gamma^{i)}(M)$ for all i.

This data structure also gives a guideline for the future implementation of these algorithms. The terminology we use here are mostly compatible with $C++$ and $Java$.

This data structure is represented as a vector V, where each component is a location that saves a structure.

$$V = (C_0, ..., C_i, ..., C_m)$$

C_i is a structure (or class-object in $C++$ and $JAVA$) for i-cells. In order to get C_i, the set of $i-cells$, our input is the set C_{i-1}. An element of C_i is the pair of two elements in C_{i-1}. These two elements are parallel-moves to each other.

We also need to use the adjacency linked-lists in graph theory to represent the data connections for later algorithms.

Let S be a subset of Σ_3. The adjacency-list (AL) of S is given as follows:

$$s_1 \rightarrow s_1(0), ..., s_1(n_1)$$
$$s_2 \rightarrow s_2(0), ..., s_2(n_2)$$
$$..................$$
$$s_{|S|} \rightarrow s_{|S|}(0), ..., s_{|S|}(n_{|S|})$$

s_1 to $s_{|S|}$ are all points in S; $s_i(0)$ to $s_i(n_i)$ are all directly adjacent points in S, denoted by $AL_P(s_i)$.

$$AL_P(s_i) = \{s_i(0), ..., s_i(n_i)\}, i = 1, ..., |S|. \tag{6.1}$$

According to the definition of parallel-move of points, the adjacency-list could be viewed as a point's parallel-move-list.

The adjacency-list not only can represent a graph of vertices, but it can also represent a graph of cubes where adjacency is two cubes sharing a square.

A new discovery is to regroup sharing adjacency and parallel-move adjacency. We usually use adjacent points as two points that have a distance of "1." Since this case only applies to graphs and to connected components, we also noticed that point parallel-move generates graph edges. Therefore we put the edge connected components into the parallel-move category. In this paper, two points u and v are said to be adjacent if $|u - v| = 1$. This is direct adjacency.

Algorithm 6.4, The General Algorithm For i-cells : Obtain all i-cells based on parallel-moves of all surface-cells in S in each dimension. The output of this algorithm is the vector that contains all i-cells.

C_0:   (a) Contain all points of $M = \{v_1, v_2, ..., v_{|}M|\}$. Each element is treated as a 0-cell.

(b) Get the parallel-move (adjacency) list for the 0-cell, which is the (graph) adjacency-list.

C_1:   (a) The Input: C_0.

(b) Get all 1-cells based on the parallel-move list in C_0,

(c) Determine all (shared) 0-adjacency: Use 1-cells as the vertex set, and if two 1-cells share a 0-cell from C_0, then put an edge in between these two 1-cells.

(d) Get all connected components of (shared) 0-adjacency.

(e) Get the parallel-move (adjacency) list for 1-cells.

C_2:   (a) The Input: C_1.

(b) Get all 2-cells based on the parallel-move (PM) list for 1-cells in C_1.

(c) Determine all (shared) 0-adjacency and 1-adjacency: Use 2-cells as the vertex set,

and if two 2-cells share an i-cell from C_i ($i = 0, 1$), then

put an edge between these two 2-cells.

(d) Get all connected components of (shared) 0-adjacency and 1-adjacency, respectively.

(e) Get the parallel-move (adjacency) list for 2-cells.

. . . . . .

C_k: (a) The Input: C_{k-1}.

(b) Get all k-cell based on the parallel-move list for $(k - 1)$-cells in C_{k-1}. If there are no k-cells, then empty all C_k, $C_{k+1},...,$ C_m.

(c) Determine all (shared) i-adjacency: Use k-cells as the vertex set, and if two k-cells share an i-cell from C_i ($i = 0, 1, ..., k - 1$), then put an edge between these two i-cells.

(d) Get all connected components of (shared) i-adjacency ($i = 0, 1, ..., k - 1$), respectively.

(e) Get the parallel-move (adjacency) list for all i-cells.

We should note that i-cells are only constructed by parallel-moves in our algorithm. If an i-cell c is the union of two $(i - 1)$-cells a and b in digital space, then a is parallel to b, $a \cap b = \emptyset$, and $c = a \cup b$. In Chap. 5, we discussed how c is a partial graph or partial structure that contains all edges (faces) of its subsets.

6.4.2   Decision and Recognition Algorithms

A unified method is studied in this section for recognizing and tracking digital manifolds and their boundaries for any dimension in a given digital space. In the last subsection, a general purposed data structure was proposed and the algorithm is designed. The algorithm can accept any data set from m-dimensional digital space and can recognize and extract 1D objects, 2D objects, and k-dimensional objects.

Digital manifolds are the digital forms of curves, surfaces, solid-objects, and multi-dimensional objects. The digital manifold recognition and tracking is an important research topic in 3D image processing and computer vision. It can be used for medical image data extraction and seismic image object recognition.

Since most of these 3D images are obtained by 3D mathematical inversion and reconstruction, the boundary of an object is generally not very clear. In such a situation, researchers need an interactive system to recognize and display specific objects on the screen. By adjusting parameters, such as thresholds, they can then decide if the object viewed is the desired object.

A general definition for digital manifolds in direct adjacency was first proposed in [8]. We discussed this intensively in Chap. 5.

For a set M in digital space Σ_m, we can now discuss the algorithm for C_k. We can track all $(k-1)$-connected components. If they are each $(k-1)$-cells in C_k and are included in one or two k-cells in C_k, then C_k is a union if several semi k-manifolds. We also want to know whether there are $(k+1)$-cells in C_k. If there are none, then we would have k-manifolds.

The boundary of a k-manifold is a collection of $(k-1)$-cells where each cell is adjacent to a point that is in $\Sigma_m - M$.

Algorithm 6.5   (Finding a k-manifold in C_k) This algorithm is designed for all k.

Input: C_{k-1}, C_k, and C_{k+1} of a data set M. They are all in linked-list format meaning that for an element a in C_i, we can get the $(i-1)$-adjacent neighbor in C_i for $i = k-1, k, k+1$.

Output: The connected components of M, each of which is a k-manifold.

Step 1:   Select an element a from C_k. Get all $(k-1)$-connected components.

Step 2:   For each $(k-1)$-connected component A in C_k, check if each point is a regular point. Then, check if a $(k-1)$-cell is included in just one or two k-cells. If any of the two conditions is not satisfied, then this component is not a regular k-manifold.

Step 3:   For each element a in A, check its parallel-moves in C_{k+1}. In other words, for each element b_{k+1} in C_{k+1} that contains a and b_{k+1}, check if $b_{k+1} - a$ is an element in A. If so, this means that A contains a parallel-move of a. That is to say that A contains a $(k+1)$-cell. Then, A is not a k-manifold. Otherwise, A is a (regular) k-manifold.

The following algorithm extracts the boundary of the k-manifold A. The primary characteristic of a boundary of a k-manifold is the following: (1) The boundary is a collection of $(k-1)$-cells, (2) Each of these $(k-1)$-cells has only one k-cell in

M that contains this $k - 1$-cell, and (3) These $k - 1$-cells are $(k - 2)$ connected components.

Algorithm 6.6 This algorithm gets B_k, which is a boundary of the k-manifold in C_k. We usually say that we track the boundary of a manifold.

Input: (a) C_k, C_{k-1}, and C_{k-2}

Output: (a) A list containing all $(k - 1)$-cells that are $(k - 2)$-connected. This boundary is called a semi-boundary manifold of a k-component. (b) Decide if the semi-boundary manifold is a manifold.

Step 1:   From the output of C_k, we can get all $(k-1)$-connected components that are k-manifolds (Algorithm 6.5). Find a $(k - 1)$-cell that only has one parallel-move in C_{k-1}. That means this $(k - 1)$-cell is on the boundary. Track all of its neighbors that share $(k - 2)$-adjacency along with the neighbor whose $(k - 1)$-cell is on the boundary (there is only one in C_{k-1}). We will get all connected sets on boundary. We can search for all elements in the connected set C_k.

Step 2:   Deciding if a boundary is a manifold is easy. We just need to check if each point is regular and whether each $(k - 2)$-cell can be contained by exactly two $(k - 2)$-cells. The boundary of a k-manifold should be closed.

In summary, this section uses a unified data structure for algorithm design. The algorithms are based on the definitions presented in Chap. 5.

6.5   Algorithms for the Orientability of Digital Surfaces

A surface is not orientable if it contains a subsurface, which is topologically equivalent to a Mobius band. In this section, we want to solve the surface orientation problem in digital space. This problem can be stated as follows: Given a surface S, we want to determine if S is orientable?

A unsolved problem is determining the number of Mobius bands on a closed surfaces in high dimensional digital space. As an example, we show a digital Mobius band below (Fig. 6.2). See [9, 17, 18]

How do we decide whether a surface is a non-orientable surface? In pure mathematics, we decide whether the surface contains a subsurface, which is topologically equivalent to a Mobius band. However, to simulate the mathematical method for the discrete decision problem here is impossible. In fact, the digital world has its own advantages since a wonderful algorithm for deciding whether a digital surface is non-orientable was finally found [5, 9].

Let p_1, p_2, p_3, p_4, $d_D(p_i, p_{(i+j)mod(4)}) = 1$ and be a surface-cell; then, its two normal lines can be represented as (p_1, p_2, p_3, p_4) and (p_4, p_3, p_2, p_1) (see Fig. 6.3a). Meanwhile, all of the loop-shifts within are equivalent. For example, $(p_1, p_2, p_3, p_4) = (p_3, p_4, p_1, p_2)$.

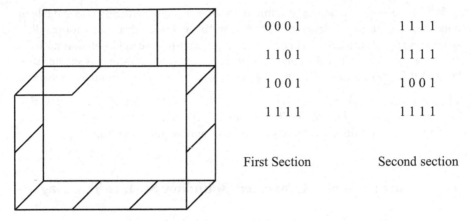

0 0 0 1	1 1 1 1
1 1 0 1	1 1 1 1
1 0 0 1	1 0 0 1
1 1 1 1	1 1 1 1

First Section   Second section

Fig. 6.2 The Mobius band in digital space

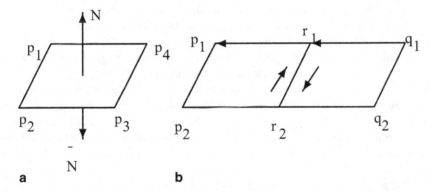

Fig. 6.3 Surface-cell orientations

Let A and A' be two surface-cells that are line-adjacent (see Fig. 6.3b). There are 4 normal lines (p_1, p_2, r_2, r_1), (r_1, r_2, p_2, p_1), r_1, r_2, q_2, q_1, and (q_1, q_2, r_2, r_1). The pairs, (p_1, p_2, r_2, r_1) with r_1, r_2, q_2, q_1 and (r_1, r_2, p_2, p_1) with (q_1, q_2, r_2, r_1), are called adjacent normal-lines.

Generally, if N and N' are two normal-lines of two adjacent surface-cells A and A', respectively, then N and M are called adjacent when the orders of r_1, r_2 appearing in N and N' are different, where $A \cap A' = \{r_1, r_2\}$. We can determine the connectedness of normal-lines based on their adjacency. We can see that a surface S is nonorientable if and only if all normal-lines of S are connected.

Lemma 6.8 *Let S be a surface, where N and $\bar{N}$ are two normal-lines of some surface-cell of S. S is nonorientable if N and $\bar{N}$ are connected.*

Proof It is easy to see that all normal-lines of S can be separated into two connected components, and N and $\bar{N}$ are in different connected components. Therefore, the two connected components are connected when N and $\bar{N}$ are connected.   □

We can design an optimal algorithm to decide whether a surface S is orientable or nonorientable. In fact, after Algorithm 6.2, we have a line-adjacency (not parallel-adjacency) list of surface-cells of S. Using the breadth-first-search in all normal-lines of S, S is nonorintable if the amount of the connected normal-lines is greater than $|\Gamma^{(2)}(S)|$ for a continuous search; otherwise, S is orientable. In summary, we have:

Theorem 6.2 *There exist an $O(|S|)$ time algorithm for the surface orientable problem: Given a surface S, decide whether S is orientable.*

An detailed algorithm for more general case will be given in Chap. 7.

6.6   Isosurface and λ-Connected Boundary Surface Tracking

In a 3D image, an isosurface is a surface where the value of each point on the surface is the same. In other word, if there is a continuous function on a data volume, the points with the constant value forms an isosurface. There are many applications for extracting this type of surfaces for finding the constant valued pressure, temperature, or velocity for real world problems.

A digital image is usually a gray scale image or colored image. There is no easy way to accurately scale an image into a binary image [13, 15, 16, 21].

Using the upper-lower thresholds, we can clip a gray scale image into a binary image. However, we can also just use the thresholds, the lower bound and upper bound, to track the surface of a solid object. This is called an isosurface. Therefore, the algorithm should be the same as the one we discussed above where we are looking for the value to fall into a range and not just specifically for a value of "1" in the binary image.

Here is another case where the value not in a bound.

Chen developed a so-called λ-connectedness method that can be applied to describe such a problem [3, 5, 6]. A corresponding segmentation technique has been studied intensively.

To involve λ-connectedness in surface tracking is not the same as using this method in segmentation.

First we need to define what a λ-connected boundary point is based on the target element. For example, our objective could be to find an element (point) with a specific value or to find a certain location and then obtain the value. Let us assume that we start at a point with the value $f(p)$. We know that a λ-connected set starting at p always exists. However, the boundary of the connected set may not be λ-connected on its own. This would cause a serious problem for tracking because we do not know the boundary unless we determine the whole connected set. However, this idea is against the purpose of tracking the boundary of an object since it avoids searching the whole component.

λ-connectedness allows a small change on the values of neighboring points. If it is gray-scale, then we can allow the value to change gradually. For instance, if we are currently at a point with value 100, then we would look for the adjacent point with values 100, 101, or 99 to first move to. The formal definition of λ-connectedness is included in Chap. 12, where we discuss this further.

Isosurfaces and λ-connected surfaces are two types of surfaces. One is in the "vertical" value clip level and the other is in the "horizontal" connectivity. There are two ways of solving this problem of λ-connected surfaces:

(1) Assume that the image can be divided into several "continuous" components, each of which satisfies the condition of normal λ-connected segmentation.

(2) The boundary part can be repeated. In this way, we can still find the boundary of λ-connected sets.

The solutions will be found in Chap. 12.

6.7 Remarks

In this Chapter, several algorithms for solving problems regarding surface decision, surface tracking, and surface orientation are presented. These algorithms are developed on the basis of the digital surface definition given in Sect. 4.4.1 [9, 10]. In the past, some research works were related to find the boundary surfaces [2, 15]. Rosenfeld and Reed showed an algorithm for surface decision [19, 20].

This chapter provided a systematic technique for topological boundary tracking algorithms based on the work presented in [4, 5]. We also accomplished these algorithms with λ-connectedness. See Chap. 9 for more details of the definition of λ-connectedness. We developed a λ-connected tracking algorithm that preserves the boundary of a given digital surface or the union of digital surfaces. We apply these techniques in real image processing such as human brain image extraction.

The algorithm that extracts 1D and 2D components can be found in [4]. The generalization of the algorithm that is able to find cycles in each i-skeleton for homology groups will be discussed in Chapter 13 and 14. The smallest i-cycle, that is not an $(i + 1)$-cell, plays an important role in homology groups related to the generators of the groups [7].

References

1. A. Aho, J.E. Hopcroft, and J.D. Ullman, The Design and Anlysis of Computer Algorithms, Addison-Wesley, Reading, MA, 1974.
2. E. Artzy, G. Frieder and G.T. Herman, The theory, design, implementation and evaluation of a three-dimensional surface detection algorithm, Comput. Vision Graphics Image Process. 15, 1981, 1–24.
3. L. Chen, The lambda-connected segmentation and the optimal algorithm for split-and-merge segmentation, *Chinese J. Computers* Vol 14, pp 321–331, 1991.
4. L. Chen, Generalized digital object tracking algorithms and implementations The Proc. of SPIE on Vision Geometry VI, Vol 3168, 1997.
5. L. Chen, *Discrete Surfaces and Manifolds: A theory of digital-discrete geometry and topology*, 2004. SP Computing.

6. L. Chen, Digital Functions and Data Reconstruction, Springer, NY, 2013.
7. L. Chen, and Y. Rong, Digital topological method for computing genus and the Betti numbers, Topology and its Applications, Volume 157, Issue 12, 2010, Pages 1931–1936.
8. L. Chen and J. Zhang, Digital manifolds: A Intuitive Definition and Some Properties, Proceedings of the Second ACM/SIGGRAPH Symposium on Solid Modeling and Applications, Montreal, 1993, 459–460.
9. L. Chen and J. Zhang, A digital surface definition and fast algorithms for surface decision and tracking, in R.A. Melter and A.Y. Wu, Vision Geometry II, SPIE Proceedings 2060, 169–178. (Also see Mathematical Review 95g:68119.)
10. L. Chen and J. Zhang, Classification of simple digital surface points and a global theorem for simple closed surfaces, in R.A. Melter and A.Y. Wu, Vision Geometry II, SPIE Proceedings 2060. (Also see Mathematical Review 95g:68119.)
11. T. H. Cormen, C.E. Leiserson, and R. L. Rivest, Introduction to Algorithms, MIT Press, 1993.
12. M.R. Gary, D.S. Johnson, Computers and Intractability. H.Freeman Press, 1979.
13. R. C. Gonzalez, and R. Wood, *Digital Image Processing*, Addison-Wesley, Reading, MA, 1993.
14. F. Harary, Graph theory, Addison-Wesley, Reading, Mass., 1969.
15. G.T. Herman, "Geometry of Digital Spaces," Birkhauser, Boston, 1998.
16. G.T. Herman and D. Webster, A topological proof of a surface tracking algorithm, Comput. Vision Graphics Image Process. 23, 1983, 162–177.
17. R. Klette and A. Rosenfeld, Digital Geometry, Geometric Methods for Digital Picture Analysis, series in computer graphics and geometric modeling. Morgan Kaufmann, 2004.
18. C.-N. Lee and A. Rosenfeld. Simple connectivity is not locally computable for connected 3D images. Computer Vision, Graphics, Image Processing, 51:87– 95, 1990.
19. D.G. Morgenthaler and A. Rosenfeld, Surfaces in three-dimensional images, *Imform. and Control* 51, 1981, 227–247.
20. M. Reed and A. Rosenfeld, Recognition of surfaces in three-dimensional digital images, Information and Control, Vol 54, 507–515, 1982.
21. A. Rosenfeld and A.C. Kak, *Digital Picture Processing*, 2nd ed., Academic Press, New York, 1982

Part III
Discretely Represented Objects: Geometry and Topology

Chapter 7
Discrete Manifolds: The Graph-Based Theory

Abstract This chapter presents a general-purpose definition of discrete curves, surfaces, and manifolds. This definition only refers to a simple graph, $G = (V, E)$ and its topological structure. Similar to digital manifolds defined in Chap. 5, the ideas presented in this chapter still use recursive definitions for discrete curves, surfaces, solid objects, and so on. Specifically, a vertex is a point-cell, and an edge is a line-cell. A surface-cell (2-cell) is defined as a special simple closed "curve"—a closed semi-curve with the minimum cycle property. In general, a k-cell will be a closed semi $(k - 1)$-manifold with minimum "cycle" property. For a graph G, all i-cells, $i = 0, ..., n + 1$, will provide topological structure to the discrete space G. An n dimensional discrete manifold M is defined as: (1) M consists of n-cells and any two n-cells are $(n - 1)$-dimensionally connected, (2) each $(n - 1)$-cell in M is contained by one or two n-cells, and (3) there is no $(n + 1)$-cell in M. We also consider the definition of orientable and non-orientable surfaces with a corresponding decision procedure. Finally, some unconventional examples of the definitions, such as quadtree surface-cell representation, an octree solid-cell representation, Voronoi decomposition, and Delaunay simplifications are presented.

Keywords Discrete curve · Discrete surface · Recursive definition · Discrete manifold · Topological structure · Graph · Algorithm

7.1 What Should be a Discrete Manifold

Basically, discrete manifold usually refers to the piece-wise linear approximation of a continuous manifold. For instance, a triangulation of a 2D manifold is a 2D discrete manifold. But this is only a description and not a definition.

Why do we need to define a discrete manifold? We need to define a discrete manifold because there is a need to know the purpose of computerized data storage, or computing purpose. It is simple for someone to mathematically define a manifold, but we could not store such a manifold in a computer in digital form.

For instance, we can define a single variable as a continuous or smooth function, but we cannot store a general continuous function in a computer, we can only store an approximation of the continuous function. This is because we cannot store every point of a function in the domain of [0, 1]. One can also say that we can store a

© Springer International Publishing Switzerland 2014   107
L. M. Chen, *Digital and Discrete Geometry,* DOI 10.1007/978-3-319-12099-7_7

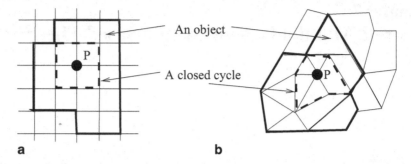

Fig. 7.1 Examples of discrete plane points: **a** An inner point of a digital plane, and **b** an inner point of a triangulated plane

function if it is a polynomial such as $f(x) = a_0 + a_2x^2 + a_3x^3$ by storing a_0, a_2, and a_3. However, not every function can be represented as a polynomial. We can only store a finite number of polynomials. To summarize, we can only store a finite number of lines or polynomial curves that are an approximation of the original function.

On the other hand, when a set of points is collected from MRI or CT machines, how can we tell whether this data set is a manifold or not? We must first have a definition for discrete manifolds.

Researchers found certain characteristics of discrete manifolds and what these manifolds had in common. First, all data points of a manifold are collected discretely. Second, the distance between two sample points may indicate the existence of a line between the points if they are nearby. Therefore, we will first have a graph $G = (V, E)$ where only V is real and E is just an interpretation.

The next question is: Can we define a discrete k-manifold based on a graph? In other words, can we build a topological structure on a graph? Similar to partitioning a continuous manifold into triangles or squares to get a discrete manifold, called the top-down method, we now want to define discrete 1-cells, 2-cells, and k-cells, which is called the bottom-up method.

In Chap. 5, we defined the digital manifold based on k cubic cells. This is because we can have a unique way of defining k-cells in digital space Σ_n. How we define discrete manifolds on a graph is the challenge in this chapter.

This chapter generalizes the definition of digital manifolds to discrete manifolds. To begin, let us look at the common properties of a triangulated manifold and its corresponding digital manifold. Since a plane is a simplest surface, we examine the an inner point of a digital plane and its counterpart in a triangulated plane in Fig. 7.1.

We can see the following in the bounded areas of Fig. 7.1a and b: (1) Each 1-cell (edge) is included in two 2-cells, which is true for every inner (internal) point of the 2D manifold. (2) Point p and its neighboring points form a disk. In other words, its analog is homeomorphic to a disk in 2D Euclidean space. (3) The neighborhood of p forms a closed cycle that is "homemorphic" to a circle. In addition to those three requirements, a surface should not contain any 3D cells.

These observations become the basis of our definitions of discrete manifolds. In Chap. 9, we discuss the close relationship between discrete manifolds and k-simplicial complexes. However, a k-discrete manifold is not exactly equivalent to a k-simplicial complex.

7.2   Discrete Curves on Graphs

For a graph $G = (V, E)$, a discrete curve is a path P. A simple curve is just a simple path. There is only one small caveat: P should not contain any 2-cells. This is a little difficult problem for us since we must define a 2D topological structure on G in order to appropriately define a discrete curve, or 1D manifold.

Before the formal definition, we need some terminologies for sets. Let S be a set. Assume S' is a subset of S, denoted by $S' \subseteq S$. If S is not a subset of S', then S' is called a proper-subset of S, denoted by $S' \subset S$.

Let $G' = (V', E')$ be a graph where $V' \subset V$ and $E' \subset E$ for graph $G = (V, E)$. $G' = (V', E')$ is called a sub-graph of G. If E' consists of all edges in G whose two joining vertices are in V', then this sub-graph $G' = (V', E')$ is called a partial-graph of G, denoted by $G' \preceq G$. If V' is a proper-subset of V, then we denote $G' \prec G$.

We note that for a certain subset, V' of V, the partial-graph G' with vertices V' is uniquely defined.[1]

Definition 7.1 Each element of V is called a point-cell, 0-cell, or point. Each element of E is called a line-cell or 1-cell.

A path is a subgraph, and a simple closed path can be viewed as a 2-cell (surface-cell). However, we want a 2-cell that only contains the minimum number of vertices, if possible. We call this the minimal cycle, meaning that it does not contain any other 2-cells. The idea comes from Σ_2 where every 2-cell contains only four points. For general graph G, we cannot require that every 2-cell have a unique shape or the same number of vertices, but we can restrict the graph to not contain any other 2-cell. Therefore,

Definition 7.2 If a partial- graph D of G is a simple cycle, then D has no proper sub-graph that is a closed path. Such a D is also called a minimal-cycle.

A 2-cell in this book must be a minimal cycle, but a minimal cycle may or may not be a surface-cell. Let's take a look at the G shown in Fig. 7.2.

In Fig. 7.2a and b, every surface-cell is clear. However, G in (c) and (d) has the same V and E, but each could have a different interpretation.

The simple cycle $C = \{a, b, c, f, i, h, g, d\}$ (without point e in the center) is a minimal cycle. If $C = \{a, b, c, f, i, h, g, d\}$ is not a surface-cell, then G looks like a

---

[1] The concepts of partial-graphs and subgraphs are important in this chapter. In our earlier publications, such as [3, 6, 7], partial-graphs were defined as subgraphs, which we correct here. However, the name itself is not a huge problem in understanding the context.

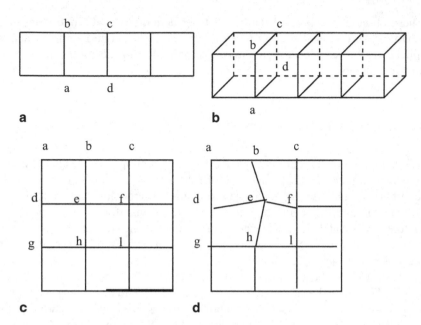

Fig. 7.2 Examples of discrete spaces **a** A minimal cycle in $2D$; **b** A minimal cycle in $3D$; **c** $\{a, b, c, f, l, h, g, d, a\}$ not considered to be a 2-cell; **d** $\{a, b, c, f, l, h, g, d, a\}$ considered to be a 2-cell

plane (in Fig. 7.2c). If C is a surface-cell, then G would have a 3-dimensional-cell (3-cell) as shown in Fig. 7.2d. This becomes a difficult thing to think, a minimum cycle, even though it looks like a 2-cell, we do not have to define it as a 2-cell. On the other hand, a minimum cycle having many vertices, we still can define it as a 2-cell.

Thus, it is important to know: (1) A geometrical interpretation of G is needed to give a geometrical frame (a topological structure in [21]) to G, (2) Defining a topological structure requires determining a class of cellular complexes [25], and (3) The set of 2-cells is a subset of the set of minimal cycles in G.

Let C be the set of all minimal cycles in G. Defining a subset of C as the set of surface-cells is a way of generating a geometrical frame for G.

Definition 7.3 Let C be the set of all minimal cycles in G. A subset of C, U_2, is called a 2-cell set if for any two different minimal cycles in U_2, u and v, $u \cap v$ is connected in $u \cap v$.

In Definition 7.3, $u \cap v$ is a subset of a path. That $u \cap v$ is connected in $u \cap v$ means that the intersection is "point-connected" or "0-connected." We want to avoid the case shown in Fig. 7.3. Even though B can be a 2-cell in continuous space, we would still like to keep the simplicity here because we can always decompose B into two 2-cells in continuous space. There is no need to keep such a case as in the definition.

Fig. 7.3 The intersection of
two cells should not be
disconnected as shown in this
figure

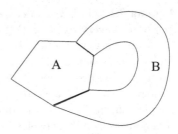

Each element of U_2 is called a surface-cell, with respect to the pair $< G, U_2 >$. If $u \cap v$ is empty, then u and v are not adjacent. We want $u \cap v$ to be a connected path because we need the intersection of two surface-cells to be on the "edges" of these two surface-cells. If $u \cap v$ is just a node (point), we say u and v are point-connected. If $u \cap v$ contains line-cells, then u and v are said to be line-connected.

A partial-graph is unique for a certain subset of the vertices of G. If G' is a subgraph, then $G(G')$ denotes the partial-graph with all vertices in G'. For a subset of the vertices of G, called V', $G(V')$ will be the partial-graph of G with all vertices in V'.

Now we are going to define two more concepts before defining a discrete curve. First, a simple path will be also called a pseudo-curve . Second, define a semi-curve.

Definition 7.4 A semi-curve D is defined as having one of the two following properties: (1) D is a simple path $P = \{p_0, ..., p_n\}$ such that $G(P) = P$ if (p_0, p_n) is not an edge, or (2) D is $G(P) = P \cup \{(p_0, p_n)\}$ if (p_0, p_n) is an edge in G.

We can view a semi-curve as a subset of vertices of G since $G(P) = P$. A semi-curve is a partial-graph D of G where each vertex has one or two adjacent vertices in D. In other words, a semi-curve D is a simple path $p_0, ..., p_n$ such that p_i and p_j are not adjacent in G if $i \neq j \pm 1$, except where $i = 0$ and $j = n$.

If C is a closed semi-curve, then $G(C)$ does not contain any other 2-cells except that C itself could be a 2-cell. In other words, if C is a closed semi-curve and $C \notin U_2$, then C will separate a 2D plane into two components. See Fig. 7.2c. This property is called the Jordan curve property that holds true in an Euclidean plane.

Definition 7.5 For a graph $G = (V, E)$ and a set of 2-cells in $< G, U_2 >$, $D \subset V$ is called a curve if D is a semi-curve and D does not contain any 2-cells in U_2 (meaning that D does not contain any subset that are the vertices of a 2-cell in U_2).

As a matter of fact, we intentionally did not use $G(D)$ in Definition 7.5 because $G(D)$ and D are in fact equivalent to each other. We can summarize the above definitions to provide a concise definition for discrete curves:

For a graph $G = (V, E)$ and a U_2 of G, $D \subset V$ is said to be a discrete curve with respect to $<G, U_2>$ if: (1) D is connected (0-connected or point-connected), (2) Each point (0-cell) is contained by one or two line-cells, and (3) D does not contain any surface-cells in U_2.

It follows easily that:

Lemma 7.1   (1) *If a semi-curve C is not a curve, then C is a 2-cell. (2) If U_2 is empty, then every simple path (pseudo-curve) is a curve.*

In summary, we would like to define a 2-cell as a semi-curve. However, we simultaneously want a discrete curve to have properties of the Jordan curve theorem (in most cases): A simple closed curve separates a surface into two disconnected nonempty sets. The definition that "a simple discrete (or digital) curve is just a simple path in G" [24] is a commonly accepted, intuitive definition of curves, or pseudo-curves in this chapter. The reason we do not use it as a mathematical definition of discrete curves is because it breaks too easily this fundamental theorem in mathematics. There is a simple path, such as $\{e, f, l, h\}$ in Fig. 7.2c, that does not separate the surface into two disconnected nonempty sets. Even though the analog of a discrete path in Euclidean space would be good for this property, maintaining the property of the Jordan curve theorem is important in discrete space as well.

7.3   Discrete Surfaces

After extensive discussion on discrete curves, now it is much easier to define discrete surfaces. We cover three types of discrete surfaces here: (1) discrete pseudo-surfaces, (2) discrete semi-surfaces, and (3) discrete surfaces. We treat pseudo-surfaces as naturally defined triangulated surfaces or mesh surfaces. The semi-surface is almost a discrete surface except that a closed semi-surface can be a 3-cell. Finally, a discrete surface holds most of the mathematical properties in discrete space just as a (continuous) surface does in Euclidean space.

The philosophy of building the concepts of discrete surfaces has the same hierarchy as the philosophy behind discrete curves: A pseudo-surface is a subgraph that is "soft" and may have multiple interpretations on its supporting vertex set. A semi-surface is a partial-graph that is "solid" and has one unique interpretation. A discrete surface is a semi-surface, if it is not a 3-cell.

Simplicial complexes or cellular complexes of combinatorial topology or algebraic topology can be based on a similar concept such as the pseudo-curve/surface. It is clear that we like to build more specific and sound discrete surfaces or manifolds in real computations. The best examples of these, digital surfaces and manifolds, are presented in Chaps. 5 and 6. (these examples are like the convex-hull since they are unique when the points are fixed).

7.3.1   A Special Set of 2-cells

Pseudo-surfaces are the same as "naive discrete surfaces" that can be defined as a collection of 2-cells, 1-cells, and 0-cells, as we do in Chap. 9 (as cell complexes). For a continuous space, it is fine since we can choose whether or not we want to put a 3-cell into the complex. If we do not, it is a surface. In other words, we select what we want to form a surface. However, in discrete case, since we have to consider the computational cost, there are so many options when the set of 0-cells and 1-cells are fixed that selecting U_2 has an exponential number of options and no computing machine can handle it.

Consequently, in the computing sciences, the selection of U_2 is a critical issue. We must tell the machine what the best way to get a discrete surface is. In order to determine the geometrical and the topological structure of graph G in a unique manner, we give a default definition of U_2 here. Unless otherwise specified, this default definition applies to all of our cases.

Definition 7.6 (Default Definition) Let G be a simple graph, the default definition of U_2 of G is:

(1) Assume m is the minimum value of the lengths of all simple cycles. Then, all simple cycles with length m are surface-cells, i.e. in U_2.

(2) If a point p is not a point of any surface-cell in U_2 by (1), then a minimum value of the simple cycle containing p will be included in U_2.

We at least have one U_2 defined and will skip the explanation on how to obtain a U_2 in further sections.

7.3.2   Discrete Semi-surfaces, 3-cells, and Discrete Surfaces

We now consider discrete surfaces. Based on prior knowledge of defining a discrete curve, the definition of discrete surface would also be dependent on the definition of 3-cells. In this case, a 3-cell would be a minimal closed semi-surface. Of course, our definitions are not dependent on how we select U_2, the set of surface-cells or 2-cells. For G, we use $\mathcal{G}_2 = <G, U_2>$ to represent the set $G = (V, E)$ with adding 2-cells U_2. If D is a partial-graph of G, $U_2(D) = \{s | s \in U_2 \& V(s) \in D\}$, where $V(f)$ is the set of vertices of 2-cell s. $<D, U_2(D)>$ is a partial structure of $<G, U_2>$. (By the way, since $V = U_0$ and $E = U_1$, $<G, U_2> = <U_0, U_1, U_2>$.)

Definition 7.7 Let D be a partial-graph of G and $S = <D, U_2(D)>$. S is a semi-surface if and only if each 1-cell of D is included in one or two 2-cells of S. S is called a closed semi-surface if and only if each 1-cell in S is included in exactly two surface-cells of S.

We also can view S or D in Definition 7.7 as the vertex set of D (or S) because D is uniquely defined by its vertex set. In other words, we can use the "vertex subset of D" as a substitute for the "partial-graph of D" in the definition. A 3D-cell (solid-cell or 3-cell) is always a closed semi-surface, but a closed semi-surface may not be a 3-cell.

Similarly, we can define:

Definition 7.8 A closed semi-surface is said to be a minimal closed semi-surface if it does not contain any proper subsets that are closed semi-surfaces.

Any minimal closed semi-surface can be a 3-cell. A 3-cell in this chapter must be a minimal unit of the 3D structure. A special set of minimal closed semi-surfaces forms a topological structure on G, denoted by U_3, a set of 3-cells. Similar to Definition 7.3, such a special set is defined as follows:

Definition 7.9 Let S be the set of all minimal closed semi-surfaces in G. A subset of S, U_3, is a 3-cell set if for any two different elements in U_3, u and v, $u \cap v$ is empty, a vertex, point-connected if it contains line-cells, or line-connected if it contains surface-cells (in $u \cap v$).

In most cases, $u \cap v$ just contains a point (0-cell), a line-cell, or a 2-cell. Each of the elements of U_3 is a 3-cell with respect to $< G, U_2, U_3 >$, in other words, the discrete space.

Definition 7.10 Subset S of G is a discrete surface with respect to $< G, U_2, U_3 >$ if and only if S is a semi-surface and S does not contain any subset that is a 3-cell in U_3.

In fact, S is a point set (or a vertex set). S has a unique interpretation in $\mathcal{G}_3 =< G, U_2, U_3 >$. That is $< G(S), U_2(S), U_3(S) >$, a partial structure of $\mathcal{G}_3$. The follow definition is about the boundary of a discrete surface:

Definition 7.11 The boundary of surface S, denoted by ∂S, is a subset of S such that for any point $b \in \partial S$, there is a line-cell e containing b where e is contained by exactly one 2-cell in S.

Corollary 7.1 *A discrete surface is closed if and only if $\partial S = \emptyset$.*

7.4 Properties of Discrete Surfaces

In this section, we examine some basic properties of discrete surfaces. As we know, the discrete surface defined in this chapter is a "sound" or "hard" discrete surface. This means that there is only one interpretation for this surface. On the contrary, when we are able to generate multiple surfaces on a vertex set (the supporting set), we refer such type of discrete surface to a pseudo-surface, which is similar to triangulated or meshed surfaces. The foundation related to simplicial complexes that will be discussed in later chapters.

In this chapter, the "hard" discrete surface is a special case of 2D cell complexes. This special case is very helpful in algorithm design for real world calculations related to geometric problems.

In Chap. 2, we presented the complete graph K_5 and the bicomplete graph $K_{3,3}$. These are two exclusive cases of planar graphs [19]. The following lemma will confirm that K_5 and $K_{3,3}$ are not discrete surfaces.

Lemma 7.2 *K_5 and $K_{3,3}$ are not discrete surfaces under the default definition of 2-cells.*

Proof Let us redraw K_5 and $K_{3,3}$ below.

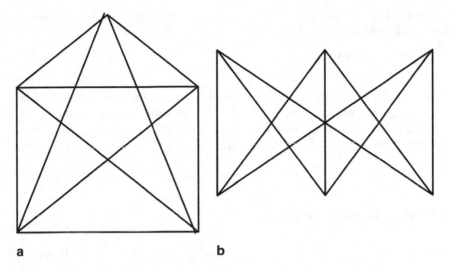

Fig. 7.4 The complete graph K_5 and the bicomplete graph $K_{3,3}$

Fig. 7.5 A planar graph that
is not a discrete surface

According to the default 2-cell definition (Definition 7.6), every 3-point-cycle is a surface-cell in K_5 and every 4-point-cycle is a surface-cell in $K_{3,3}$. We can see that every edge in (a) is contained by more than three 2-cells. In (b), every edge is also contained by at least three 2-cells (Fig. 7.4).   □

A discrete surface can be viewed as a planar graph. However, a planar graph sometimes is not a discrete surface. The following result is interesting and it shows the relationship of discrete surfaces and planar graphs.

Lemma 7.3 *A discrete surface is a planar graph, but a planar graph may not be a discrete surface.*

Proof We know that G is a planar graph if and only if G does not contain K_5 or $K_{3,3}$. According to Lemma 7.2, a discrete surface is a planar graph. However, looking at the following example, we can see that this planar graph is not a surface (Fig. 7.5).   □

The reason a planar graph may or may not be a discrete surface is that it is dependent on how the 2-cell set U_2 is formed. The next lemma shows that if we accept the simplex as a member of U_2 as we see it on a triangulated surface and we have a unique interpretation of the surface, then this surface is a discrete surface as defined in this chapter. This leaves a huge problem in finding U_2, which is far to unique, refer to the Catalan number in Sect. 1.2.1.

Lemma 7.4 *The simplex (or convex polygon) decomposition of a continuous surface forms a discrete surface if U_2 only contains surface triangles.*

Proof We define U_2 to be

$$U_2 = \cup\{\partial \Delta\}$$

where Δ is a simplex on the decomposed surface. Any line segment is contained by at most two simplexes. On the other hand, the intersection of two simplexes is empty, a point, or a line segment. There is no 3-cell. (In fact, in computer graphics, we do not include any 3-cells in the data set.) This matches the definition of discrete surfaces.

7.5   Regular Surface Points

In Chap. 4, we introduced the concept of ordinary points on curves. A 1D manifold is a curve in which every point is an ordinary point. An ordinary point means that there are no branches for any point on the curve. For the same reason, we do not want "branches" on a surface, i.e. the neighborhood of each point on a discrete surface must be similar to a 2D disk. We call such a point a regular (or ordinary) surface point for discrete surfaces.

Let S be a subset of G, $S(p)$ denotes all points in the set of all 2-cells in S containing p. In other words, if q is in a 2-cell that contains p, then $q \in S$.

Definition 7.12 A point p in a discrete surface S is a regular (or ordinary) point if the set of all surface-cells containing p are line-connected among these surface-cells. If a point in S is not regular, then it is called irregular.

This definition will reject the case of that the intersection of the neighborhoods of two surface points is only a point. We give an example shown in Chap. 5.

We may generalize the above definition. For any $< G = (V, E), U_2, U_3 >$, a point $p \in V$ is said to be a regular surface point if $S(p)$ (meaning the partial-graph generated by $S(p)$ with all line and surface-cells) is a surface and all surface-cells in $S(p)$ are line-connected.

Lemma 7.5 *Assume S is a discrete surface and there is a point $p \in S$ that is not on the boundary of S. Let p have only two adjacent points p', p'' in S. Then: (1) There are two 2-cells A, B such that $A \cap B$ contains p', p, p'', and (2) If p', p'' are adjacent in S, then p', p, p'' form a surface-cell.*

Proof The first statement is easy to prove. Assume A and B contain p', p. Because A is a simple cycle, A must contain two adjacent points of p. Therefore, A must contain p'' since p has only two adjacent points. For the same reason, B contains p''. See Fig. 7.6a

In the second statement, if $\{p', p, p''\}$ is not a 2-cell, then there are two surface-cells, C and D, where C contains 1-cell $\{p', p''\}$ and D does as well. Now $A \cap C$ contains p', p'', but p', p'' are not connected in $A \cap C$. This does not satisfy the

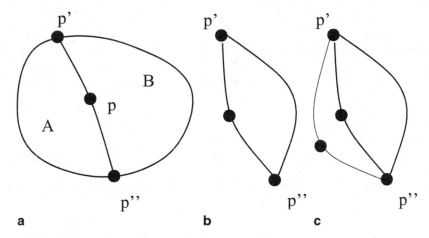

Fig. 7.6 Two adjacent points case of an inner point p

definition of 2-cells, i.e., $A \cap C$ must be connected in $A \cap C$. So there is a contradiction. See Fig. 7.6b and c. □

If p only has two adjacent points p', p'' in surface S, and p', p'' are not adjacent, then p is called an abundant point . This means we can delete p and make p', p'' adjacent without changing the structure of S, except for the two surface-cells containing p. (This is true except in the case where we are making non simple graphs. If we delete "p," then it would generate a cell that only has two 1-cell (edges) as its boundary. In Fig. 7.6c, this cell can be merged with other cells, i.e. we can delete both edges (p p') and (p,p')').)

We could assume that S does not contain any abundant points for the rest of this chapter. It is also true that if a surface S does not contain any abundant points, then the intersection of any two surface-cells in S can at most contain one line-cell. However, in general, we should not make such an assumption.

In order to know about the neighborhood of a point p inside of S, we want $S(p)$ to be a disk-like shape that is homemorphic to a 2D disk in 2D Euclidean space. We would like to have $S(p) - \{p\}$ be a simple cycle that is homemorphic to a circle in Euclidean space. We now can prove this important observation.

Lemma 7.6 *Let S be a discrete surface without abundant points. If p is an inner and regular point of S, then there exists a simple cycle containing all points in $S(p) - \{p\}$ in S.*

Proof Let p be an inner and regular point of S. Assume q is a point other than p in $S(p)$ and a surface-cell A contains q. Then, A is a semi-curve. Thus, q has two adjacent points a, b in A. If $p = a$ (or $p = b$), then $\{q, p\}$ is a line-cell in S. According to the definition of S and since p is an inner point, there are two surface-cells containing $\{q, p\}$ in S, one of which is A and we assume the other is B. B must be in $S(p)$ because B contains p. Then q must have an adjacent point r in B, and $r \neq p$.

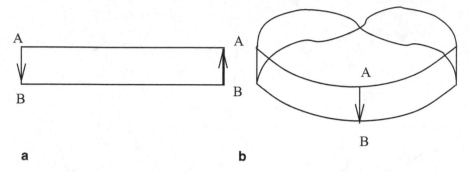

Fig. 7.7 **a** A paper strip with two ends; **b** A Mobius strip by attaching flipped two ends. (redraw needed)

If $b = r$, then $S(q) = A \cup B$. Otherwise, S is not a surface. Thus, q is an abundant point of S.

If $b \neq r$, we know $r \in S(p) - \{p\}$, then q has two adjacent points in $S(p) - \{p\}$.

If $p \neq a$ and $p \neq b$, then q has two adjacent points in $S(p) - \{p\}$. $S(p) - \{p\}$ is a finite set, so such a procedure generates a closed path $P = p_0, \ldots p_n, p_0$, which includes all points in $S(p) - \{p\}$.   □

In order to make the connection to digital surface in Σ_3, we want to define the following concept.

Definition 7.13 Let S be a discrete surface. Then, p is said to be a simple surface point if $S(p) - p$ is a closed curve.

A regular inner point is a simple surface point in digital space Σ_m. For a digital surface S in Σ_3, a simple surface point under Morgenthaler–Rosenfeld's definition is a regular inner surface point in the case of direct adjacency (Chap. 5; [14]).

We continue our discussions on special properties of regular discrete surfaces in the following two sections on orientable properties and separable properties.

7.6 Orientability of Discrete Surfaces

In 3D, a surface-cell has two sides, just like a coin has a face and a tail. However, in a Mobius strip, one can move from one side to another side smoothly. This type of surfaces is called none-orientable. See Fig. 7.7.

We can make a Mobius strip by taking a long, thin strip of paper and attaching the two short ends together so that the top of one end is at the bottom of the other (Fig. 7.7a).

The purpose of orientation is simple in that it is to determine the outside and inside of a closed curve in a 2D or a closed surface in 3D ($(m - 1)$D hyper surface in mD space). We have presented a surface that is not orientable in 3D digital space in Chap. 6.

Let C be a simple cycle in $G = (V, E)$. C is $p_0, p_1, \cdots, p_n, p_0$, so it has two orientations: (a) Beginning from p_0 to $p_1,..., p_n$ and returning to p_0, and (2) Beginning from p_0 to $p_n,..., p_1$ and returning to p_0. These two orientations are called two normals of C. If $p_0, p_1, ..., p_n, p_0$ is clockwise, then $p_0, p_n, ..., p_1, p_0$ is counter-clockwise, and vice versa.

Two orientations of a simple cycle C are denoted by $N(C)$ and $\bar{N}(C)$. That is to say, if $N(C) = p_0 p_1 ... p_n p_0$, then $\bar{N}(C) = p_0 p_n ... p_1 p_0$, and vice versa. These two orientations are called natural orientations based on cycle C's looping direction.

Consider a partial-graph $G(C)$ with all vertices in C. $G(C)$ may contain more edges than C. In such a case, it is possible to have another orientation for $G(C)$, i.e., a permutation of $p_0, p_1, \cdots, p_n$, say, $q_0, q_1, \cdots, q_n$, where $q_0, q_1, ..., q_n$ is a simple cycle other than $N(C)$ and $\bar{N}(C)$ regardless of any shift of the points in a loop. Therefore, the number of orientations of $G(C)$ may be greater than two.

However, if $G(C) = C$, i.e., each point in C has exactly two adjacent points in $C \in G$, then, $G(C)$ has only two orientations.

This result is easy to prove because there is no permutation of $p_0, p_1 ... p_n$ that is a simple path other than $p_0, p_1, ..., p_n$ and $p_n, ... p_1, p_0$. Any surface-cell A has exactly two orientations since a surface-cell is a closed semi-curve so $G(A) = A$. Thus, if A is a surface-cell, then an orientation of A (view A as a simple cycle) is an orientation of $G(A)$ (or $G(A)$ consists of all vertices).

Suppose S is a discrete surface on G, i.e. $<G, U_2, U_3>$ has been defined, and S is a discrete surface with respect to $<G, U_2, U_3>$. If A, B are line-adjacent (1-adjacent) two surface-cells in S, then A, B are two simple cycles. We have a total of four orientations to consider for A and B: $N(A)$, $\bar{N}(A)$, $N(B)$, and $\bar{N}(B)$. Let $A \cap B = \alpha$ where α is an arc (non-closed simple path). If $\alpha = a_1, .., a_k$, then we can denote $\alpha^{-1} = a_k, ..., a_1$.

Our purpose is to pass one normal to another normal from A to B or from B to A (through α or α^{-1}). This is a different method compared to the method in topology where A and B are combined as a bigger cell by dismissing the intersection to get two orientations of $A \cup B$. The method presented here is better for algorithmic solutions, meaning that we can easily decide if a surface is orientable or not. This is because we do not need to store a new 2-cell of $A \cup B$. (The union is not viewed as a 2-cell here.)

Definition 7.14   Let A, B be two line-adjacent surface-cells where $A \cap B = \alpha$. Then $n_A \in \{N(A), \bar{N}(A)\}$ and $n_B \in \{N(B), \bar{N}(B)\})$ are said to be normal-adjacent if α is a part of n_A and α^{-1} is a part of n_B, or α^{-1} is a part of n_A and α is a part of n_B.

For two 2-cells A and B, no matter how we name $N(A)$ or $N(B)$. Now, we can define the connectivity of normals between any two 2-cells in a surface S.

Definition 7.15   Two normals n_A and n_B of two 2-cells A and B are called normal-connected if there is a normal path $n_A = N_0, \cdots, N_m = n_B$ where N_i, N_{i+1}, $i = 0, \cdots, m - 1$, are normal-adjacent.

What we want to do is a little more clear now. We want to establish a connection through normals for all 2-cells in S. Think of S as a meshed sphere, the normals will be separated into two components: one indicating the inside of the sphere and the other indicating the outside of the sphere. If S is a Mobus strip, then we only have one component of those normals. In fact, we have established the following lemmas:

Lemma 7.7 *Normal-connectedness is an equivalence relation.*

Proof First, we define n_A and n_B as normal-adjacent. Second, it is easy to know that if n_A to n_B are normal-connected, then there is a path $n_A = N_0, \cdots, N_m = n_B$. So, n_B to n_A would be normal-connected by the reversed path $n_B = N_m, \cdots, N_0 = n_A$, and we have symmetry. Third, we prove the transitivity. If n_A to n_B and n_B to n_C are normal-connected, then we have $n_A = N_0, \cdots, N_m = n_B$ and $n_B = N'_0, \cdots, N'_{m'} = n_C$ that are normal-connected paths. So $n_A = N_0, \cdots, N_m = N'_0, \cdots, N'_{m'} = n_C$ is a normal-connected path. Then n_A to n_C is normal-connected.   □

Lemma 7.8 *Let S be a discrete surface, then all normals have at most two normal-connected components by normal-adjacency.*

Proof Let S be a discrete surface as defined in this chapter. We know that a 2-cell only contains two normals and any two 2-cells are line-connected (1-connected) in S. Let A and B be line-connected by $A = A_0, \cdots, A_m = B$. On the other hand, fix a cell A so that it has two normals N_A and $\bar{N}_A$. Let us assume there is a normal N, which belongs to an arbitrary cell B. Then, path $P = B, \cdots, A$ would induce a normal-connected path $N, \cdots, n_A$. Therefore, no matter what, any normal of B will be normal-connected either by N_A or $\bar{N}_A$. Thus, any normal will be either connected to N_A or $\bar{N}_A$. So there are at most two normal components.   □

We have the following definition for orientable surfaces:

Definition 7.16 Surface S is said to be orientable if its normal set has two normal-connected components.

A good example of a non-orientable surface is shown in Fig. 6.2. The following lemma is trivial based on the definition.

Lemma 7.9 *Let A be a surface of S. If $N(A)$ and $\bar{N}(A)$ are normal-connected, then S is not orientable.*

We can also prove that if S has m surface-cells and if S is orientable, then two normal-connected components of the normal set of S are equally sized by Lemma 7.8. This is because that we have total of $2m$ normals of S.

In Sect. 6.5, we presented an algorithm for deciding if a digital surface is orientable. This algorithm can be modified for general discrete surfaces.

In pure mathematics, to decide an orientable surface, we need to test whether the surface contains a subsurface that is topologically equivalent to a Mobius strip (or band) [26]. See Fig 7.7. We can also see that a Mobius strip has a single boundary curve, and this fact has fundamental significance to non-orientability. However, even though any non-orientable surface contains a Mobius strip, it is sometimes very hard to find out there is a one embedded in a bigger surface if the surface is formed in a complicated way.

It seems actually impossible using a constructive procedure in continuous space. However, in discrete space, we only have a finite number of 2-cells. An algorithm would stop after all cells have been checked in normal connectedness.

Fig. 7.8 Two normals in a
Mobius strip moving to the
same normal smoothly.
(Original figure was obtained
from Dick Palais
http://virtualmathmuseum.
org/Surface/gallery_m.html)

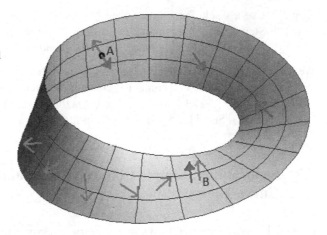

In practice, we want a procedure to decide if all normals are normal-connected
instead of finding a Mobius strip.

A discrete Mobius strip is shown in Fig. 7.8. The two normals of a 2-cell A will
be moving to the same normal at a 2-cell B.

We can design an optimal algorithm, $O(|S|)$ time, to decide whether a surface S
is orientable or non-orientable. We give the detailed steps of this algorithm in below.

Since the algorithm for deciding a discrete surface is similar to the algorithms
presented in Chap. 6 and the procedure is not very complicated, we just give the
algorithm for orientability here. The idea of the algorithm is simple: We use normal-
connectedness to link 2-cells. If the total number of normals of 2-cells exceeds the
number of 2-cells in the surface S, then there must be a 2-cell whose two normals
are connected by normal-connectivity, so S is not orientable. Otherwise it must be
orientable.

Algorithm 7.1 Decision Algorithm for Orientability.

Input: (1) A discrete surface $S = (G = (V, E), U_2, U_3)$. (2) Two orientations
(two normals) of each 2-cell A in the format $N_A = <a_1, ..., a_n, a_1>$ and $\bar{N}_A = <
a_n, ..., a_1, a_n>$, if $A = \{a_1, ..., a_n\}$ is a 2-cell and $a_1, ..., a_n$ is a simple cycle.

Output: S is orientable or not?

Step 1   Set the set of normals $\mathcal{N} \longleftarrow \emptyset$.

Step 2   Get a 2-cell A, and put N_A into $\mathcal{N}$. Because S is line-connected, there
must be an edge (1-cell) that is shared with another 2-cell, denoted as B.
If N_A has a common part of the sequence of N_B, then $<a_1, ..., a_n, a_1>$
$\cap <b_1, ..., b_n> \neq \emptyset$ and put $\bar{N}_B$ into $\mathcal{N}$. Otherwise, put N_B into $\mathcal{N}$. (The
normal-connectedness must be in reverse order in terms of direction in their
intersection.)

Step 3   Use a stack *Stack* or a queue *Queue* to save all 2-cells visited[2]. Repeat
          Step 2 until $\mathcal{N}$ contains more than $|U_2|$ members. In such a case, S is
          non-orientable. Otherwise, $|\mathcal{N}|$ is $|U_2|$ and S is orientable.

We leave the algorithm analysis to readers. This algorithm would be linear time considering that the largest cell contains a constant, limited number of points. However,
it is at most $O(|V| + |E| + |U_2|k^2)$, where k is the length of the longest cycle, which
is a 2-cell in U_2. See [1, 15] for methods of algorithm analysis.

7.7   Separability, Simple Connectedness, and Jordan Curve Theorem

One of the most important tasks in geometry is separating a geometric object into
smaller pieces. The methods of Voronoi and Delaunay, along with other partitioning
methods are all about the separation of a space. However, we never discussed weather
or not a space is separable.

For a small or local space, we can observe this property and determine whether or
not we can separate it. However, when a space is extremely large, we really do not
know whether it is possible. One of the most exciting theorems is called the Jordan
curve theorem. It answers a basic part of this question. (The theorem is exciting not
only because of the theorem itself, but also because of its proofs since people are
still debating whether we have a solid proof.)

In topology, the Jordan curve theorem states [26, 27]:

Theorem 7.1 *A simply closed curve J in a plane Π decomposes $\Pi - J$ into two
components.*

In fact, this theorem holds for any simply connected surface, and a plane is a
simply connected surface. So what is a simply connected space?

A connected topological space M is said to be simply connected if for any point
p in M, any simply closed curve containing p can contract to p. The contraction
means a continuous mapping among a series of closed curves, e.g. from the original
cycle to a point. See Fig. 7.9a. We have a boundary surface of a sphere. The simple
cycle a will separate the closed surface into two components. Also, the cycle b and
c are the contractions of a. When we continue to contract c, it would be becoming
the point p.

The Jordan curve theorem is not true for a general continuous surface. For example
in Fig. 7.9b, when M is the boundary surface of a donut, then it cannot always be
separated into two components by a closed curve such as cycles a and b. Meanwhile,
a cycle is not always contractible to a point on the surface. Neither a nor b can be
contracted into p. See Fig. 7.9b.

---

[2] Use depth-first search or breadth-first search to go though every cell in S like in Step 2. We omit
the details to make the algorithm simpler and easier to understand. For search details, see Chap. 6.

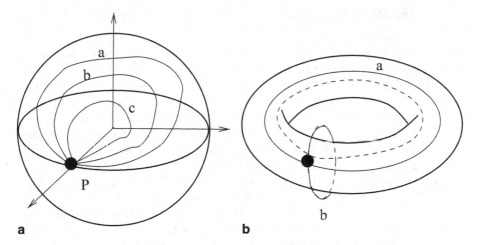

Fig. 7.9 Simply connected surfaces: **a** A sphere that is simply connected, and **b** A donut which is not simply connected

The first proof of the Jordan curve theorem believed to be correct was made by Veblen in 1905. Tutte proved this theorem in planar graphs in the 1970s.

To simulate the proof of the Jordan theorem in discrete space is not easy. First, we must define a "discretely" continuous mapping, i.e. discrete deformation. Second, we need the concept of "discrete contraction."

These concepts have a connection to the gradual variation introduced in Chap. 11. Due to the complexity of the proof and the technical change of the methodology from the recursive definition of cells and surfaces to discrete deformation, we must use a completely different aspect of the discrete geometry method. We give a proof in Chap. 15.

Digital spaces can be non-Jordan meaning that the Jordan curve theorem may not apply to the some digital spaces. For example, in Σ_2 with 8-adjacency, assume a, b, c, d are four points of a unit square. A path passes the diagonal points a and c, it seems partitioned b and d to be belong to two different parts. However, b and d are adjacent here. So the Jordan curve theorem is not true for 8-adjacency.

In addition,, $\{a, b, c, d\}$ is a surface-cell. Moreover, $\{a, b, c\}$, $\{a, b, d\}$, etc. are 8-adjacency surface-cells too. Such cases may generate ambiguities [4, 5]. It is not able to described by cell complexes neither.

7.8 Discrete k-Manifolds

A discrete k-manifold is the generalization of discrete surfaces. In addition, the definition of kD digital manifolds in Σ_m in Chap. 5 should be a special case of the discrete k-manifold. In this section, we introduce a general definition of a k-dimensional discrete manifold on a graph G.

7.8.1   The Recursive Definition of Discrete k-Manifolds

The methodology we use is still the same, we define a discrete k-manifold recursively in the following way (assuming that we defined k-cells on the closed minimal $(k-1)$D semi-manifold): (1) Define the (k)D semi-manifold based on k-cells, (2) Define $k+1$-cells as the closed minimal (k)D semi-manifold, and (3) Define k-manifold.

We have already defined $<G, U_2, U_3>$, along with 2-cells and 3-cells in the above sections. Let S be a set of k-cells, $k \geq 3$. We say that two k-cells, A and B, are $(k-1)$-connected if there is path from k-cells $A = A_1, \cdots , A_n = B$, such that $A_i \cap A_{i+1}$ is a $k - 1$-cell (or a set of $k - 1$-cells that are $k - 2$-connected). S is said to be $(k - 1)$-connected if any two k-cells, A and B in S are $(k - 1)$-connected.

Let M be a subset of V. The partial-structure (space) of M, with respect to $\mathcal{G} =< V = U_0, E = U_1, U_2, ..., U_k >$, $\mathcal{G}(M) = \{u | V(u) \in M \& u \in U_i\}$, suggests that each i-cell $u^{(i)}$ of the supporting set $(V(u^{(i)})$ is in M.

So we give the general k-dimensional semi-manifold S as follows:

Definition 7.17   A kD semi-manifold S is a set of k-cells in $<G = (V, E)$, $U_2, ..., U_k>$ satisfying $S = \{$all k-cells in $\mathcal{G}(V(S))$, i.e. S is a partial-space. This semi-manifold satisfies: (1) S is $(k - 1)$-connected in S, and (2) Each $(k - 1)$-cell in S is contained by one or two k-cells.

Definition 7.18   A closed kD semi-manifold (k-semi-manifold) S is called minimal if there is no proper subset of S that is a closed k-semi-manifold. A $(k + 1)$-cell set, U_{k+1}, is a subset of all closed minimal k-semi-manifolds of G, such that for any pair $u, v \in U_{k+1}$, $u \cap v$ is empty, a 0-cell, an unclosed 1-semi-manifold, an unclosed 2-semi-manifold, $\cdots$, or an unclosed k-semi-manifold.

As we mentioned before, in most cases, $u \cap v$ is empty, a 0-cell, a 1-cell,..., or a k-cell.

Definition 7.19   A k-semi-manifold S is said to be a k-manifold if S does not contain any $(k + 1)$-cells with respect to $<G, U_2, ..., U_{k+1}>$.

Thus, a k-manifold S is a set of k-cells in $<G = (V, E), U_2, ..., U_{k+1}>$ satisfying: (1) S is $(k - 1)$-connected in S, (2) each $(k - 1)$-cell in S is contained by one or two k-cells, and (3)S does not contain any $(k + 1)$-cells.

An n-cell set, U_n, can be defined using the existing $U_0 = V, U_1 = E, ...U_{n-1}$ of G. We also call $U_0, U_1, ...U_n$ an n-dimensional discrete topological structure (DTS).

We can see that if $U_i = \emptyset$ then $U_n = \emptyset$ for all $n > i$.

The dimension of $\mathcal{G}$ is defined as the largest index of $U_n \neq \emptyset$ for all possible selections of DTS.

Two k-cells u, v are i-adjacent if $u \cap v$ consists of several i-cells. In addition, u and v are i-connected if there is a path $u_1, ..., u_m$ where u_t, u_{t+1} are i-adjacent, $t = 1, ..., m - 1$, where $u = u_1$ and $v = u_m$.

We have discussed the structure of discrete k-manifolds, which have a bottom-up type of construction. We can also describe a manifold with a top-down construction. We always assume $\mathcal{G}(M)$ to be equivalent to M. A definition based on 0-cells (points) is presented next.

7.8.2   An Alternative Definition of Discrete k-Manifolds

Definition 7.20 Given an n-dimensional discrete-topological-structure of G, $\mathcal{G} =<$ $G = (V, E), U_2, ...U_n>$, a connected subset M of $V(G)$ is a discrete k-manifold, $k = 0, ..., n-1$, if for any $a \in M$, there exists a $u_i \in U_i, i = 1, ..., k+1, V(u_i) \subset M$, where $i = 0, ..., k+1$ and: (1) Any two k-cells are $(k-1)$-connected in M, (2) Every $(k-1)$-cell in M is contained by only one or two k-cells, and (3) M does not contain any $(k+1)$-cells.

A discrete n-manifold can be defined in $<G, ..., U_n>$ by ignoring $(n+1)$-cell testing. Let M be a discrete k-manifold. The boundary of M, ∂M, is the set of all $(k-1)$-cells in M, each of which is contained by only one k-cell in M. M is said to be closed if $\partial M = \emptyset$. A point p is called k-inner if $p \notin \partial M$. A point p in a k-manifold M is called regular if all k-cells containing p are $(k-1)$-connected [7, 11].

7.8.3   Boundary of Discrete k-Manifolds

The boundary of a discrete k-manifold is the key issue in actual practice. We first define the regular (ordinary) i-cell in a manifold.

Definition 7.21 A i-cell $A_i \in M, i = 0, .., k-2$, is regular (ordinary) if all k-cells containing A_i are $(k-1)$-connected (in these k-cells). M is regular if all i-cells in M are regular for all $i = 0, ..., k-2$.

For $A_i \in M$, the set containing these k-cells is denoted by $S_k^{(i)}(A_i)$. An inner point p in a k-manifold M is called simple if $S_k(p) - \{p\}$ is a simple $(k-1)$-manifold. A k-pseudo-manifold P is a collection of k-cells such that (1) any two k-cells in P are $(k-1)$-connected in P, and (2) any $(k-1)$-cell in $c \in P$ is contained by one or two k-cells in P. We treat P as a "sub-graph" for pseudo-manifolds.

Theorem 7.2 *For G, if $S \subset V(G)$ is a regular n-manifold, then ∂S is a closed regular $(n-1)$-pseudo-manifold or several closed regular $(n-1)$-pseudo-manifolds.*

Proof We present the idea of a proof here. We know that ∂S is the union of the collection of $(n-1)$-cells, each of which belongs to only one n-cell in S. We might as well assume here that ∂S is the collection of $(n-1)$-cells.

First, we want to prove that each $(n-2)$-cell is included in two $(n-1)$-cells in ∂S. Suppose an $(n-1)$-cell $A \in \partial S$, then we want to prove that all $(n-2)$-cell in A is contained in two $(n-1)$-cells, including A. Let a be a $(n-2)$-cell in $A \in \Delta$, where $\Delta \in S$ is an n-cell. There is a unique $A' \in \Delta$ that contains a based on the construction of an n-cell. If A' is in ∂S, then there are two $(n-1)$-cells containing a. Otherwise, A' can only be contained by another (the only) Δ. In addition, Δ contains another $A'' \neq A$ containing a. If $A'' \in \partial S$ then a is contained by two $(n-1)$-cells on the boundary of S; otherwise, we will have $A^{(3)}$ that contains a. And so on. Because all n-cells containing a are $(n-1)$-connected, there must be an $A^{(m)} \in \partial S$ containing

a because we either have a finite number of cells or an $(n - 1)$-cell is contained in three n-cells. It is impossible that an $(n - 1)$-cell is contained in three n-cells. So A and $A^{(m)}$ are two (n-1)-cells containing a.

Second, if there are three $(n - 1)$-cells containing a, then all n-cells containing a will not be $(n - 1)$-connected based on the above process. So, ∂S is a closed $(n - 1)$-pseudo-manifold or several closed $(n - 1)$-pseudo-manifolds.

We can also prove that every i-cell is regular in ∂S as well. $\square$

7.8.4   Examples of Discrete Manifolds

The best example of a discrete manifold is a digital manifold. This only has one single interpretation in a natural way, direct adjacency. There are other types of "natural" discrete manifolds.

Voronoi Diagrams and Delaunay Decompositions are two good examples of discrete manifolds in any dimensions [16, 18]. For a continuous manifold, we randomly select N points, the Voronoi diagrams and Delaunay Decomposition will generate a discrete Manifold in any dimension. The detailed algorithm will be presented in Chap. 10.

Some real images are stored by quadtree and octree schemes. Quadtree and octree techniques are powerful methods that represent a 2D or 3D image. They are also used in image compression and solid modeling. A leaf of quadtree or octree can be viewed as a 2-cell or a 3-cell. Their partition forms a topological structure of a manifold of an object. [3, 7, 9]

7.9   Remark

The relationship between the definition in this chapter and the definition of the simplicial complex [20] in topology is that a pseudo-manifold forms a simplicial complex. The simplicial complex representation of a manifold in Euclidean space forms a discrete manifold. On the other hand, a discrete manifold is only a special class of cell complexes. Its supporting set (V or U_0) only has one interpretation in the space. The space we define in this section $\mathcal{G} = <G = (V, E), U_2, ..., U_k>$ can hold multiple discrete manifolds in a unified manner, but a simplicial complex, or a celluar complex, is only specific to a partition. The example of such a manifold is the digital surface discussed in Chap. 5 [12–13].

The simplicial complex when represents an n-manifold must rely on a continuous manifold, which is not a general space that contains many objects. In other words, we must use a continuous space when talking about a simplicial complex that is n-manifold. Again, we cannot store a general continuous space into a computer [17]. Similarly, a vertex set has an exponential number of interpretations even in triangulation. It is computationally impossible to consider a set that contains all simplicial complexes for a supporting set [18].

In [7], we present a more general definition of discrete surfaces that ignores the requirements, including that a k-manifold should not contain a $(k+1)$-cell. This is the of pseudo-manifold method. One only cares if there is an interpretation of the discrete set that is a k-manifold. In this case, some 8-adjacency in 2D and 26-adjacency in 3D digital space will be covered. Some original thoughts about geometrical and topological structures can be found in [2, 21–23].

Two issues that relate to research philosophy in digital and discrete geometry are the following. First, we know that the boundary of 2-cell is a closed curve and the boundary of 3-cell is a closed closed surface. Generally, The boundary of a k-cell are closed except 1-cell. Why? We say the boundary of a k-cell is a closed semi $k-1$ manifold. Each pair of $(k-1)$-cells on the boundary is $(k-2)$-connected. For 1-cell, we have two 0-cells on the boundary, they are (-1)-connected. The (-1)-cell in fact is a empty cell. We can see that every pair is empty-connected. There is no need for their connectivity.

An observation is also interesting. There are two types of connections that related to distance measurement in discrete manifolds: edge-connection and inner-connection of cells. Let us look at an example: We have defined that a 2-cell A is a minimal cell meaning that there is no other 2-cell is a proper subset of A, A does not contain any 1-cell and 0-cell inside of A. The boundary of A will contain some 1-cells and 0-cells. From one point to another point in A, the distance is usually the shortest point-path between two points. These points are on the edge of a cell. However, on the other hand, a 2-cell or its boundary can contract to any of its boundary point using just one-unit time. It means that two points in the same 2-cell has an internal connection. Just like a unit square in 8-adjacency in 2D. Therefore, the cell-edge-connection in this chapter is the direct adjacency in digital space. The inner connection is the general or indirect adjacency. When allowing the inner connection to make a curve or path, we will have a non-Jordan space. In addition, using inner connection, we will not be able to make a simplicial complex that will be discussed in Chap. 9.

We can see that only triangle cells maintain the property that edge-connection is the same as inner-connection every pair of points in a triangle has the same graph-distance that is one.

In general, for a k-cell, there are also edge-connection and inne-connection. Here edge means the boundary of a cell. Only simplexes have the property that two such connections are the same. This might be another reason that triangles and simplexes have the simplest topological structure. However, it is hard to obtain simplexes computationally comparing to the digtal or cubic cells.

More discussion on the Jordan curve theorem can be found in Chap. 15 and in [6, 7, 30]. Topics on digital topology and discrete topology will be discussed in Chaps. 9 and 14 [6, 8, 10, 24, 28, 29].

References

1. A. Aho, J.E. Hopcroft, and J.D. Ullman, The Design and Anlysis of Computer Algorithms, Addison-Wesley, Reading, MA, 1974.
2. E. Artzy, G.Frieder and G.T.Herman, The theory, design, implementation and evaluation of a three-dimensional surface detection algorithm, Comput. Vision Graphics Image Process. 15, 1981, 1–24.
3. L. Chen, Generalized discrete object tracking algorithms and implementations, Melter, Wu, and Latecki ed, Vision Geometry VI, SPIE 3168, pp 184–195, 1997.
4. L. Chen, A note on (alpha, beta)-type digital surfaces and general digital surfaces, in Melter, Wu, and Latecki ed, *Vision Geometry VII*, SPIE Proc. 3454, pp 28–39, 1998.
5. L. Chen, Point spaces and raster spaces in digital geometry and topology, Melter, Wu, and Latecki ed, Vision Geometry VII, 1998.
6. L. Chen, Note on the discrete Jordan curve theorem, Vision Geometry VIII, Proc. SPIE Vol. 3811, 1999. 82–94.
7. Chen, L., Discrete Surfaces and Manifolds. Scientific and Practical Computing, Rockville, 2004
8. Chen, L., Genus computing for 3D digital objects: Algorithm and implementation. In: Kropatsch, W., Abril, H. M., Ion, A.(Eds.) Proceedings of the Workshop on Computational Topology in Image Context (2009)
9. L. Chen, Digital Functions and Data Reconstruction, Springer, NY, 2013.
10. Chen, L., Rong, Y.: Digital topological method for computing genus and the Betti numbers. Topology and its Applications 157(12) 1931–1936 (2010)
11. L. Chen and J. Zhang, Digital manifolds: A Intuitive Definition and Some Properties, Proceedings of the Second ACM/SIGGRAPH Symposium on Solid Modeling and Applications, Montreal, 1993, 459–460.
12. L. Chen and J. Zhang, A digital surface definition and fast algorithms for surface decision and tracking, in R.A. Melter and A.Y. Wu, Vision Geometry II, SPIE Proceedings 2060, 169–178.
13. L. Chen and J. Zhang, Classification of simple digital surface points and a global theorem for simple closed surfaces, in R.A. Melter and A.Y. Wu, Vision Geometry II, SPIE Proceedings 2060.
14. L. Chen, H. Cooley and J. Zhang, The equivalence between two definitions of digital surfaces, *Information Sciences*, Vol 115, pp 201–220, 1999.
15. T. H. Cormen, C. E. Leiserson, and R. L. Rivest, Introduction to Algorithms, MIT Press, 1993.
16. S. Fortune, Voronoi diagrams and Delaunay triangulations, in D.Z. Du and F. K. Hwang ed, Computing in Euclidean Geometry, World Scientific Publishing Co, 1992.
17. R. C. Gonzalez, and R. Wood, *Digital Image Processing*, Addison-Wesley, Reading, MA, 1993.
18. J. E. Goodman and J. O'Rourke eds, Handbook of Discrete and Computational Geometry, CRC Press, New York, 1997.
19. F. Harary, Graph Theory, Addison-Wesley, Reading, Mass., 1969.
20. A. Hatcher, *Algebraic Topology*, Cambridge University Press, 2002.
21. G. T. Herman, "Geometry of Digital Spaces," Birkhauser, Boston, 1998.
22. G. T. Herman and D. Webster, A topological proof of a surface tracking algorithm, Comput. Vision Graphics Image Process. 23, 1983, 162–177.
23. T. Y. Kong and A. W. Roscoe, Continuous analogs of axiomatized digital surfaces, Computer Vision, Graphics, and Image Processing, Vol 29, 60–86, 1985.
24. T. Y. Kong and A. Rosenfeld, Digital topology: introduction and survey, Computer Vision, Graphics and Image Processing, Vol. 48, 1989, 357–393.
25. V. A. Kovalevsky, Finite topology as applied to image analysis, Computer Vision, Graphics and Image Processing, Vol. 46, 1989, pp. 141–161.

26. S. Lefschetz, Introduction to Topology, Princeton University Press New Jersey, 1949.
27. M. Newman, Elements of the Topology of Plane Sets of Points, Cambridge, London, 1954.
28. A. Rosenfeld, "Digital topology," Amer. Math. Monthly, Vol 86, pp 621–630, 1979.
29. A. Rosenfeld, Three-dimensional digital topology, Inform. and Control 50, 1981, 119–127.
30. W. T. Tutte, Graph Theory, Addison-Wesley, Reading, Mass., 1984.

Chapter 8
Discretization, Digitization, and Embedding

Abstract This chapter deals with the relationships among objects in discrete, digital, and continuous spaces. If Chap. 7 can be viewed as covering the theoretical aspect of discrete spaces and its objects, then this chapter can be viewed as covering the practical methods to make actual moves from one space to another. Likewise, Chap. 9 then connects classical mathematics to discrete and digital topology. Two of the most important operations are discretization (obtaining a discrete object from continuous space) and embedding (putting a discrete or digital object into a continuous space). To obtain a discrete object, we need to retain its geometric and tropologic identity, such as distance measurements and genus (holes). In general, a discrete object differs from a continuous object in that it has two basic measures: graph distance and Euclidean distance. The graph distance measures the number of edges between two notes as discrete sampling points. Euclidean distance can be viewed as the weight on the edges. Embedding is putting a discrete object back into a continuous space. For instance, putting a weighted graph into 3D Euclidean space, we must not allow two edges to cross each other. In practice, discretization and embedding are not separated. They are two parts of a unified process called mesh generation in computer graphics. Other embedding methods, such as the piecewise linear reconstruction method and more sophisticated polynomial fitting and B-spline methods are discussed in Chap. 11.

Keywords Discretization · Digitization · Triangulation · Decomposition · Marching cube · Point space · Raster space · Interpretation

8.1 Mesh Generation: Method of Discretization and Triangulation for Continuous Surfaces

Meshing is the standard method for discretizing a continuous surface into piecewise-linear 2-cells. Usually these 2-cells are triangles or convex polygons. When we have a solid object, the key is to represent the boundary surface of the solid using triangles or 2-cells. There are three meshing methods: (1) Cubic-based, (2) Voronoi and Delauney triangulation, and (3) Curvature and normal related advanced methods. Mesh generation becomes a special technology in computer graphics [14].

This chapter begins with meshes and discusses the embedding technologies. According to Whitney and Nash, a manifold can be smoothly embedded into Euclidean

© Springer International Publishing Switzerland 2014   131
L. M. Chen, *Digital and Discrete Geometry,* DOI 10.1007/978-3-319-12099-7_8

space [17]. Therefore, we mainly deal with Euclidean space as we study continuous space in this chapter. We also discuss projections as a special immersion method.

8.1.1 Cubic-Based Boundary Meshes for 3D Continuous Manifolds

For 3D, data collection is usually in digital. For instance, the 3D CT scans form a function or image in 3D digital space. Each voxel or 3D pixel (also called a cube) has a value in the space. Let us assume that we use a clip-level to transfer this image into a binary image or we extract the so called isosurface where the value of each point on the surface is the same. We can then get a solid object in digital space. How do we represent the boundary surface using triangle patches?

8.1.1.1 Marching Cubes

The most commonly used method is called marching cubes [1, 22, 26]. In this technique, the data points (the values are not zero) in a 3D cube will be checked to match some patterns. Of course, we only check the cube on the boundary of the solid object. If it is inside the object, then all of the 8 points on the cube would have the same value.

A cube contains 8 points. Therefore, there are a total of $256 = 2^8$ possible combinations in which some selected points are in the object. The rest of the points are not in the object. We assume the points considered are on the boundary.

It was found that only 15 patterns among the 256 cases are really useful. See Fig. 8.1.

When a pattern is matched, then a set of small triangles would be assigned to this cube. That means we represent the discrete data points with several triangles. Then, we "march" to another cube nearby, also on the boundary of the solid, to find its pattern.

In other words, the algorithm of marching cubes checks the type of local configuration of a boundary point of a solid object. It must be one of 15 cases in 15 cubes in Fig. 8.4. Then, we put the triangles in the cube into the file that stores the surface patches.

This method has been used in many industries and applications. The main idea is to use half point interpolations to convert a digital object into continuous space in terms of triangulation. Different scale factors will be selected as needed to get high or low resolution. It can also be used to get the triangulation of an object.

Marching cubes have two disadvantages: (1) Each cube on the boundary of an object will be replaced by several triangles, resulting in several triangles, and (2) This method is not mathematically proven in terms of preserving the mathematical properties including topological properties of the original object.

For the first problem, we can combine small neighborhoods into one big triangle by calculating the similarities of normals. However, for the second disadvantage, we need special methods in order to fix it. In fact, Wood found that marching cubes

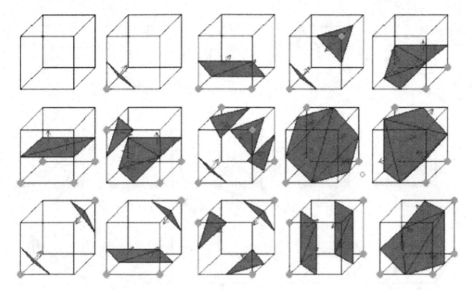

Fig. 8.1 Triangulation using marching cubes made by D. Lingrand

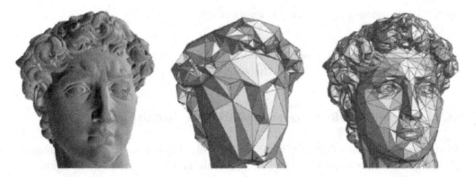

Fig. 8.2 Triangulation generates holes by Z. Wood: **a** A picture without holes using 1 million triangles, **b** A picture with 15,000 triangles and 957 holes, and **c** A picture with 15,000 triangles after deleting all holes

will generate many holes when the original data has some noise [27]. Even if there is no noise, the scaling of discretization and the size of cubes may generate some pathological situations that will create many holes.

This example tells us that a very fine detailed triangulation might not require us to consider topological error. However, we may not be able to hold this in the memory of small computers since there is too much information. For Fig. 8.2b, we know that many triangles are wasted in representing small holes, which might be invisible [27].

Preserving topological properties can not only save space in computer graphics, but can also be the key in geometric processing for image analysis, especially in medical imaging. Dealing with both theoretical and practical uses of geometric methods is the goal of this book. In fact, calculating the number of holes is significant in topology. We present a fast digital algorithm in Chap. 14.

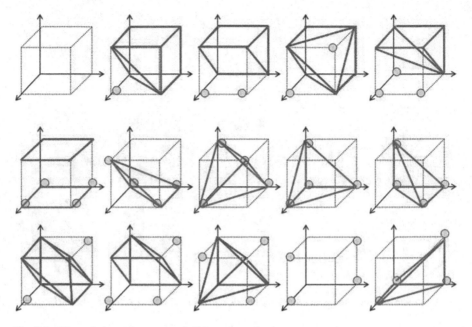

Fig. 8.3 Triangulation using convex-hull boundary of cubes

The small pathological case is related to how we interpret the boundary points and unit cells discussed later in this chapter. Wood refers to it as artifacts in [27].

8.1.1.2   Convex Hull Representation of 3D Cube Boundaries

We developed a method that uses the convex hull in a 3-cell to represent the boundary of an object. Unlike marching cubes, a half scale is used for triangulation, i.e. the triangles can be used as small half sub-cubes. The convex hull method considers the total data points in a 3-cell in order to form a unique 3D shape inside a cube to be attach to the inner cube of the original object.

The convex hull formed by certain vertices of a unit cube in digital space is unique. The idea is the same as that of the partial graph idea in Chap. 7, but this time the inside of the cube is unique [7]. See Fig. 8.3.

This method is valid since the faces of the convex hull would be cancelled in the inner cubes but would remain on the boundary surface of the 3D solid (object).

Unlike the marching cubes method that uses half of the points to form meshes, our method preserves the original cube information. For a higher dimensional object, an m-dimensional object in m-dimensional digital space, the boundary of the object in each m-cube is unique. Therefore, the convex hull of the boundary points in an m-cube is also unique. The upper-face of the convex hull will represent the boundary of the original object.

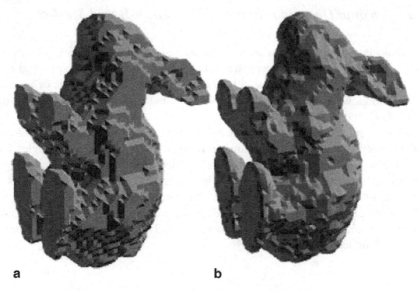

a　　　　　　　　　　　　　　　　　b

Fig. 8.4 Example of triangulation from cubes: **a** Convex-hull boundary method, and **b** Marching cube method

We have tested many examples of using these two triangulations. In Fig. 8.4, we first use a digital filling method to make the rabbit data to be a digital 3D solid data set. The original data is from Princeton's 3D benchmark [25]. Then we use the convex-hull cube boundary method and the marching cube method to do the triangulated surface to see their difference. More examples are shown in [7] for data reconstruction.

In Fig 8.4a, the convex-hull boundary, if there are more than 4 data points, then we will use its complement (blue) configuration.

If we say that marching cubes will get more detailed information on a 3D object, then the convex-hull boundary configuration is a mathematical solution. On the other hand, the marching cube method is a technical solution, and it may generate false results in some local cases.

The question is, can digital geometry generate a method to complete the task? This is the essential question to ask.

Even though we can say that the marching cube method can be viewed as a modification of a digital geometric method, the method is still mainly a numerical process since the half point does not exist in our digital "language."

It is very obvious that we could just embed the digital object in Euclidean space. This does not look good since the boundary surface points are forced to be within only 6 types.

We know that we have 18-adjacency, 26-adjacency, and maybe even 12-adjacency (when we only admit the half diagonal points in a unit square, we will then develop a method to use this knowledge).

In a cube containing 8 points, only the cases containing 4 or more points can form a polyhedron.

8.1.2 Voronoi Decomposition and Curvature-Based Meshes

Voronoi decomposition is often used to decompose a closed surface to obtain the meshes. The idea is the same: represent a flat area using a large polygon and represent a curved area using many small polygons. Based on the principle of Voronoi decomposition, we need to first place the points (sites) in the correct locations of the surface [24].

Curvature-related and Poisson surfaces are also used for surface decomposition. In such a method, we use the curvature to determine the length of a triangle: The bigger the curvature, the smaller the triangle needed. In a flat region, we use bigger triangles. In a curved region, we use smaller triangles. This method also relates to the normal of a flat 2-cell where a Poisson equation is employed [19].

The finite-element method is also used in determining how the triangles are selected in a decomposition for continuous surfaces. This method usually requires a mechanical model [24].

8.2 Digitization and Dual Digital Spaces

In Chap. 4, we already discussed there are two ways to digitize an object into digital representation (Fig. 4.5). We now focus on the relationship of these two digitizations: point digital spaces and raster digital spaces. They are dual digital spaces, just like Delaunay and Voronoi decomposition, where one uses points and another uses an n-cell to discretize the n-manifold. In Chap. 3, we discussed their definitions and gave examples. In Chap. 10, we will give the algorithms to obtain Delaunay and Voronoi decomposition.

This section investigates the relationship and the theoretical framework for point digital spaces and raster digital spaces of digital spaces in digital geometry[4].

In point-spaces, a digital object such as a curve or a surface is represented by a set of elements called points. We have used this method in Chap. 5. A k-D object is defined by $(k - 1)$-cells, inductively. Therefore, they will eventually be defined by points.

On the other hand, in a raster space, a digital object is a subset of a "relation" in the space. For instance, an n-manifold is partitioned into n-cells. Therefore, an $(n - 1)$-cell will be an intersection of two n-cells, where such an intersection is a pair of two n-cells. Then, an $(n - 1)$-manifold, which is a set of $(n - 1)$-cells, will be viewed as a relation of n-cells. Please note that not every relation forms an $(n - 1)$-manifold.

8.2.1 Examples of Two Basic Digital Spaces

Given a set S of points, called sites, we can get the Voronoi diagram of S and its Delaunay triangulation in Euclidean space. The Voronoi diagram can be viewed as raster space, and the Delaunay simplicial decomposition can be viewed as point space. Therefore a point space is a dual space of a raster space.

The following problems can be asked: (1) How can we define the digital curves, surfaces, and manifolds in point spaces or raster spaces? (2) What is the difference and relationship between them? (3) What are the advantages and/or disadvantages of using point spaces or raster spaces in practical computation?

A real image is a continuous function; however, it must be stored in a digital array in computers today.

Mathematically, let E_m be the m-dimensional Euclidean space. Z^m is the set of all points $(a_1, ..., a_m)$ in E_m, where all $a_1, ..., a_m$ are integers. Z^m is called a digital space. Because the computer can only store a finite amount of data, we usually consider a finite subset of Z^m, $\Sigma_m = \{(a_1, ..., a_m) | 0 \leq a_i \leq N; i = 1, ..., m\}$ for a certain N, as a digital space.

Digital geometry and topology deal with adjacency, connectivity, curves, surfaces, manifolds, etc. in Z^m. Unlike continuous geometry and topology, digital objects in digital spaces have concrete meanings. For example, a digital point may have length, but a digital curve may have area. Such properties are dependent on how we define · a point, curve, and so on. Thus, different interpretations in digital geometry and topology can yield different results.

Assume f is a two-dimensional image that is stored in a two-dimensional array in a computer. If (x, y) is an element in the array and $f(x, y)$ is the value on (x, y), then typically, $f(x, y)$ represents the average value of a unit square surrounding the point (x, y), and this unit square is called a pixel $Pixel(x, y)$. (A 3D image can be represented by a function on a 3D array, where the value on a point is the average value of a unit cube called a voxel.)

Let us recall the example in Chap. 4, Fig. 4.5. Suppose that we have a connected region S in Fig. 4.5. If we want to determine the boundary of S, then a problem will appear. How do we find the boundary of S? The most natural way is to consider it as the boundary in Fig. 4.5d, but how do we represent it using pixels? In this case, a line segment is the intersection between two adjacent pixels, and so we can use a set of pixel-pairs to represent the boundary. That is to say, a line-cell (1-cell) is an element of a binary relation of 2-cells. By the way, a point (0-cell) is represented by the intersection of four pixels. However, in continuous spaces, the boundary is a curve. In order to be consistent, we represent a curve by using a set of pixel-pairs. Such a digital space is called a raster-space in this note.

On the other hand, we can define the boundary of S as C shown in Fig. 4.5c. That is to say, a curve can be represented by a set of pixels (or points) in a 2D array. Without loss of generality, we can directly use a point (x, y) to represent a pixel and omit the $Pixel(x, y)$. This digital space is called a point space. A line-cell consists of two adjacent points in this case. *The difference* between a point space and a raster space is how to define i-dimensional cells (or i-cells) for $i = 0, ..., m$ in an m-dimensional space. We define i-cells for both "point spaces" and "raster spaces" in this section. In addition, we also introduce the concept of "connectedness" in this section.

8.2.2   Point Spaces in Direct Adjacency

A point p in Σ_2 has four horizontal and vertical neighbors, namely $(x \pm 1, y)$ and $(x, y \pm 1)$; p also has four diagonal neighbors, namely $(x \pm 1, y \pm 1)$ and $(x \pm 1, y \pm 1)$. We say that horizontal and vertical neighbors are directly adjacent to p (or 4-adjacent to p), and we say that both types of neighbors are (indirectly) adjacent to p (or 8-adjacent to p) in Chap. 4 [3, 5].

In general, the points $p = (x_1, x_2, ..., x_m)$ and $q = (y_1, y_2, ..., y_m)$ are two directly adjacent points in Σ_m if $d_D(p, q) = \sum_{i=1}^{m} |x_i - y_i| = 1$. We say that p and q are general-adjacent (directly or indirectly) points if $d(p, q) = \max_{1 \le i \le m} |x_i - y_i| = 1$. We only consider direct adjacency in this sub-section. Let p and q be a pair of (directly) adjacent points of Σ_m. Intuitively, $\{p, q\}$ is a line-cell. A surface-cell is a set of 4 points that form a unit square parallel to the coordinate planes. A 3-cell (or solid-cell) is a unit cube that has 8 points.

More generally, a k-cell (k-cube) has exactly 2^k points, $p_1, ..., p_{2^k}$ in direct adjacency, where each $p_1, ..., p_{2^k}$ has m components, denoted by $p_i = (x_1^{(p_i)}, ..., x_m^{(p_i)})$. There are $m - k$ components that do not change value for all $p^{(1)}, ..., p^{(2^k)}$. For p_i and $p_j, i \ne j, 1 \le \sum_{t=1}^{m} |x_t^{(p_i)} - x_t^{(p_j)}| \le k$, and $\max_{t=1}^{m} |x_t^{(p_i)} - x_t^{(p_j)}| = 1$. Again, we sometimes refer to a 0-cell as a point-cell, a 1-cell as a line-cell, and a 2-cell as a surface-cell. Two unit-cells u and v are point-adjacent if they share a point, line-adjacent if they share a line-cell, and k-adjacent if they share a k-cell. u and v are called k-connected in S if there exists a path $u_0 = u, u_1, ..., u_n = v$ in S, such that u_i and u_{i+1} are k-adjacent, $i = 0, ..., n - 1$. Figure. 2.10 shows some examples of adjacency and connectedness [8, 9].

A raster space, RS_m, is a partition (or decomposition) of E_m. Such partition splits E_m into m-dimensional unit cells (m-cells), each m-cell contains a point of Z^m in the center of the cell.

For two distinct m-cells r and r', if $r \cap r'$ is an $(m - 1)$-cell, then the $(m - 1)$-cell is called a surface-element[18]. So, a surface-cell can be represented by an element of a binary relation on RS_m. Note that a surface-element is in $(m - 1)$-dimension. In order to be consistent in this note, we call it an $(m - 1)$-cell. If $m = 3$, then a "surface-element" is a surface-cell.

In general, for 2^k m-cells $r_1, ..., r_{2^k}$, if $\cap_{i=1}^{2^k} r_i$ is an $(m - k)$-cell, then an $(m - k)$-cell can be represented by an element of a 2^k-relation on RS_m. A 2^k-relation in direct adjacency can be defined as follows:
$$\mathcal{RS}^{(m-k)} = \{r = (r_1, ..., r_{2^k}) | r_1 \cap .... \cap r_{2^k} \text{ is an } (m - k)\text{-cell}\}.$$
Note, $(r_1, ..., r_{2^k}) = (p_1, ..., p_{2^k})$, where $\{p_1, ..., p_{2^k}\}$ is any permutation of $\{r_1, ..., r_{2^k}\}$. Such a relation is called a "symmetric relation."

Although RS_m can also be represented by Z^m, RS_m has different implications than those of Σ_m. If we say the definition of k-cell in point spaces is "bottom to top," then the definition of k-cells in raster spaces is "top to bottom." They are just in opposite order. In fact, if we let $P = \{p | p$ is 0-cell in $RS_m\}$, then we can generate a similar k-cell structure in point-spaces.

Fig. 8.5 Example of 0-cells, 1-cells, and 2-cells in 2D raster and point spaces: **a** Two spaces, **b** Raster-space 2-cell, **c** Point-space 0-cell, **d** Raster-space 1-cell = the intersection of two 2-cells, **e** Point-space 1-cell = the union of two 0-cells, **f** Raster-space 0-cell = the intersection of four 2-cells, and **g** Point-space 2-cell = the union of four 0-cells

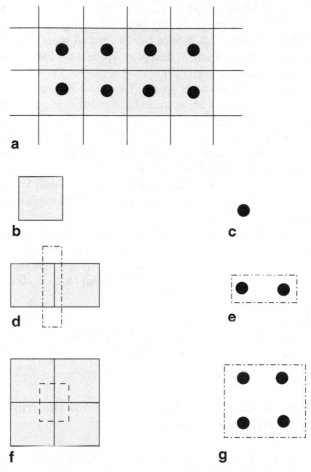

For two t-cells u and v, where $t > k$, they are directly k-adjacent if $u \cap v$ contains a k-cell. u and v are called directly k-connected in S if there exists a path $u_0 = u, u_1, ..., u_n = v$ in S, such that u_i and u_{i+1} are k-adjacent, $i = 0, ..., n - 1$.

It is not difficult to know that RS_m is the Voronoi decomposition of "sites" Z^m. If we link every pair of points, which are $(m - 1)$D-adjacent in Z^m, then we get a Delaunay "triangulation" of the Voronoi diagram. Therefore, the point space Z^m is a dual space of RS_m.

Kovalevsky first gave examples of 0-cells, 1-cells, 2-cells, and 3-cells in raster spaces in [21]. To clarify the relationship of these two spaces, we show the 2D raster, point spaces, and their 0-cells, 1-cells, and 2-cells with direct adjacency in Fig. 8.5.

We can see, in point spaces, a k-dimensional digital object (including digital curves, surfaces, manifolds, etc.) is represented by a set of elements in Z^m. In raster spaces, a k-dimensional digital object is represented by a set of elements in $2^{(m-k)}$-relation on RS_m.

for the relationship between point spaces and raster spaces

we can see three interesting facts concerning point space and raster space. First, a raster space RS_m with defined k-cells is a dual space of a point space. A raster-space is a Voronoi diagram, and its dual point-space is the Delaunay simplicial decomposition of the Voronoi diagram [19].

Second, for 3D point spaces, Chen and Zhang obtained the following: Simple surface points in $(6, 26)$-surfaces have exactly six types [10]. A regular inner surface point based on Definition 5.5 was introduced to get the results in Chap. 5. A $(6, 26)$-surface point defined in Definition 5.2 is equivalent to a regular inner surface point. Here, "regular" means all surface-cells containing p in S are line-connected in $S \cap N_p$. More details can be found in Chap. 5.

For 3D raster spaces, if V is a subset of RS_3, then V is a set of voxels. Latecki examined the same results as the six types of simple surface points for "Well-composed" sets [6].

Therefore in most of cases, the raster space representation are the same as The point space representation such as the example shown in Fig. 8.5.

However, we do have the following example that shows that the two representations are not equivalent (Fig. 8.6). We want to think about: Can a set in raster space be represented by its 0-cells?

Example 8.1 This example shows that we may need a refinement in order to make an appropriate match. That is also true, in Chap. 7, we want to use the point subset to represent a cell. See more discussion in [6, 4].

8.3   Discrete Manifolds in Voronoi Diagrams and Delaunay Decompositions

In theory, for a smooth m-manifold M, we can get its piece-wise linear decomposition. If we isometrically embed M into an Euclidean space, then we can get its digital decomposition. Or we can randomly arrange n points with a probabilistic distribution in M, so we can get the Voronoi Diagram and Delaunay triangulation.

The following method will can be used to get Voronoi or Delaunay decomposition of M.

We know that if given n points, a set P, in the m-dimensional Euclidean space, we can get a Voronori diagram and Delaunay triangulation. In [16], Bern explained an algorithm that can get the Delaunay triangulation of a point set in d dimensions. This is done by obtaining the convex hull of these n points (lifting them into $(d + 1)$ dimensions). Then we can get its dual diagram, the Voronoi decomposition. Here, the convex hull is the smallest convex that contains all points in P. Two algorithms for computing the convex hull will be introduce in Chap. 9.

Such a decomposition is a partition of E_d. We can use the partition on M to get the decomposition of M by interesting M into the Euclidean space. Each Voronoi

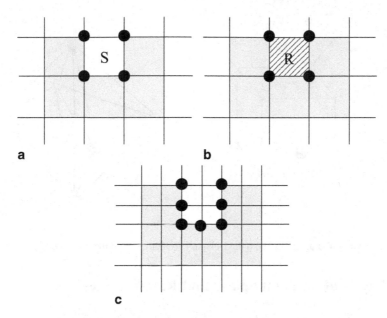

Fig. 8.6 A pixel set in raster space that is hard to be represented by its 0-cells: **a** The pixel set V, where s is not in V; **b** s must be in V if 0-cells are used to define a 2-cell in point spaces; and **c** A refined V

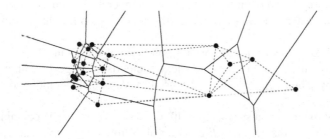

Fig. 8.7 Real data example of the Voronoi diagram and Delaunay triangulation by D. Mount

region will be called a m-cell. The boundary of the partition on M would be a set of $m - 1$ piece wise linear cells.

In Chap. 9, we will design algorithms for the Delaunay triangulation and Voronoi diagram. We showed a figure in Chap. 3 for those decompositions (Fig. 3.4). In Fig. 8.7, we show the 2D the Voronoi diagram and the Delaunay triangulation for a real data set [7].

For 3D, Fortune showed an example of the Voronoi diagram and Delaunay decomposition [15, 16] (Fig. 8.8).

When we treat the decompositions as we extract k-cells from them, we can define the cell complexes and m-discrete manifolds. We have had extensive discussion about this in Chap. 7. In next section, we are specifically interested in the relationships of the representation of these two decompositions.

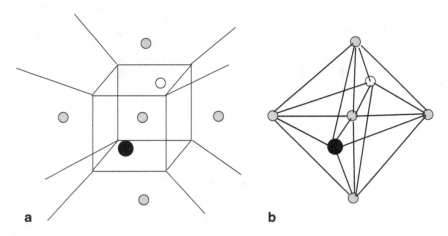

Fig. 8.8 a A Voronoi diagram in 3D; **b** the Delaunay triangulation of the same sites

8.4   Manifolds in Point Spaces and Raster Spaces*

The point space and the raster space are dual digital spaces, just like the Voronoi diagram and the Delaunay triangulation.

Now we can define manifolds in these two digital spaces. In point spaces, we build the cell structure by first defining points, then to upper level such as m-cells. On the other hand, in raster spaces, we first identify m-cells, then find lower dimensional cells.

We have had extensive discussions on point digital spaces in Chap. 5. Now, we just summarize the definition here for next comparison. A general approach was considered by Chen [2], and it can be used in Voronoi diagrams and Delaunay decompositions.

A Voronoi diagram is a special "graph" $G = (V, E)$, where V contains all vertices of the Voronoi diagram and E contains all Voronoi edges (Voronoi 1-cells). A Delaunay decomposition is also a special "graph" $G = (V, E)$, where V is the set of sits, and E is the set of Delaunay edges (Delaunay 1-cells). We also have Voronoi 2-cells, Voronoi 3-cells, Delaunay 2-cells, and Delaunay 3-cells.

For a graph G, let $\tau(G)$ be a group of defined sets of 0-cells (V), 1-cells (E), 2-cells,..., k-cells,...., where $\tau(G)$ is called a topological structure of G.

Definition G iven a G and $\tau(G)$, the partial graph[1] D of G is a k-dimensional digital manifold (or k-manifold) if and only if:

(1) Any two k-cells in D are $(k - 1)$-cell-connected in D,
(2) Each $(k - 1)$-cell in D is included in one or two k-cells in D, and
(3) D does not contain any $(k + 1)$-cell.

---

[1] It was called subgraph in [6, 4], but meaning is the same as in this book.

In addition to the above definition, the boundary of a k-manifold D, denoted by ∂D, is a subset of $(k-1)$-cells contained in D so that each element of ∂D is exactly contained by one k-cell in D.

For raster spaces, let V be a set, R be a symmetric binary relation, and τ be a partition of R. $H = (V, \tau)$ is a raster digital space. When $\tau = R$, i.e. partitioning to a single element, then this definition is equivalent to the original definition by Herman [18].

Definition 8.2 An n-D simple digital manifold is a subset S of V, and for any two elements $p, q \in S$, there is a path $(p, p_1), ..., (p_{n-1}, q) \in \tau_S$.

S is called simply $(n-1)$-D connected if for any two elements $p, q \in S$, there is a path $(p, p_1), ..., (p_{n-1}, q) \in \tau_S$ [18]. In fact, simply $(n-1)$-D connectedness is equivalent to the connectedness in Chap. 7.

Let us consider the collection of $V, R, ..., R^{(n)}, \tau, ... \tau^{(n)}$, where V is a set and R is a symmetric binary relation on V, $R^{(i)}$ is a binary relation on $R^{(i-1)}$, and $\tau^{(j)}$ is a partition of $R^{(i)}$. We say that $G = (V, R, ..., R^{(n)}, \tau, ..., \tau^{(n)})$ is an n-dimensional topological structure on V. Given a topological structure $G = (V, R, ..., R^{(n)}, \tau, ..., \tau^{(n)})$. Let S be a subset of V and
$$G_S = (S, R_S, ..., R_S^{(n)}, \tau_S, ..., \tau_S^{(n)}),$$
where $R_S^{(i)} = \{r = (a, b) | (r \in R^{(i)}) \& (a, b \in R_S^{(i-1)})\}$, and $\tau_S^{(i)} = \{r = (a, b) | (r \in \tau^{(i)}) \& (a, b \in \tau_S^{(i-1)})\}$.

If S is a subset of $\tau^{(i)}$, then for $j > i$, let $S = \tau_S^{(i)}$, $R_S^{(j)} = \{r = (a, b) | (r \in R^{(j)}) \& (a, b \in R_S^{(j-1)})\}$, and $\tau_S^{(j)} = \{r = (a, b) | (r \in \tau^{(j)}) \& (a, b \in \tau_S^{(j-1)})\}$.

Then, the boundary of S is defined as
$$\partial S = \{r = (a, b) | (r \in \tau^{(i+1)}) \&, \text{ where only one of } a, b \in S\}.$$

Definition 8.3 An i-dimensional manifold is a subset S of $\tau^{(n-i)}$, and the following is true:

(1) S is $(i-1)$-cell connected, i.e. for any two elements $p, q \in R_S^{(n-i)}$ there is a path $(p, p_1), ..., (p_{n-1}, q) \in R_S^{(n-(i-1))}$.
(2) $\tau_S^{(n-(i-1))} = R_S^{(n-(i-1))}$, i.e. $\tau_S^{(n-(i-1))}$ is the single element partition. (Intuitively, any $(i-1)$-cell (p, p') is contained by at most two i-cells p, p' in S.)

For Voronoi diagrams and Delaunay decompositions, we can naturally get their topological structures $G = (V, R, ..., R^{(n)}, \tau, ..., \tau^{(n)})$.

Raster spaces, as defined in this chapter, are more meaningful in nature but they are difficult to represent in computers. On the other hand, point spaces are easy to represent, but sometimes it is difficult to find meaning in continuous mathematics. These are dual spaces to each other, and for that reason, both of them possess the "dual properties."

8.5 Object Interpretation and Error Analysis: An Exploration

Digitization and discretization contain errors. How big the error is will provide an upper limit on the accuracy of the process. In fact, we try to restrict the errors to a range. For example, when a line is digitized. The length of the digital curve is sometimes difficult to deal with. There are two ways: (1) Count all vertical or horizontal directions and diagonal directions on the digital curve. (2) Use the least squares method to perform an interpolation on the digital points. We then consider the length of the continuous curve.

In 2D, a digitized connected region can converge to the original area, but its digitized boundary may not converge to the original boundary of the region. This is very interesting in high dimension. The boundary can be arbitrarily bigger than the original boundary, but the volume is converging to the original region.

8.5.1 Line Digitization and Error Estimation

In Chap. 4, we presented the Bresenham algorithm for line digitization. Such a digitization will generate some errors in practice. Given a line segment in 2D Euclidean plane with the slope $k <= 1$ (otherwise, we swap the x and y axis), the best digitization is using Bresenham's algorithm to get the line. The characteristics of Bresenham's line is that it is an 8-adjacent curve. If we are given a Bresenham line segment, then the question we can ask is: What is the length of the segment? If we do not know this digital curve is a line, how do we calculate the length of the curve.

Consequently, there are three ways for answering these two questions: (1) Count the points on the digital curve. This way, we can get a graph length that is not embedded Euclidean length. The accuracy in Euclidean space is very low. (2) Count the length locally, i.e. count all diagonal edges as $\sqrt{(2)}$ points. (3) Do a least squares fitting and then calculate the fitted line length.

For the first case (1), we know that the worst case is when the slope is 1 and the error ratio is $1/\sqrt{2}$. For (2), we can get the exact error analysis. Let $p1 = (x1, y1)$ and $p2 = (x2, y2)$. Consider p_i and p_{i+1}. Assume that $a = x2 - x1 \geq b = y2 - y1$, so let $b = ka$, where $k \leq 1$ and $c = \sqrt{(a^2 + (ka)^2)}$. We can assume also that $p1$ and $p2$ are integer points (otherwise use center truncation, i.e. $x = [v + 0.5]$). The counterpart of line "c" (a hypotenuse line) will be a Bresenham line that contains a points in which $b = ka\ sqrt(2)$ points and $a - ka$ straight points are parallel to the x-axis.

Let $n = a$. We count the local length, meaning that at a local 1' point, we count the length as 1 and at a local $\sqrt{2}$ point, we count the length as $\sqrt{2}$. Therefore, the $digital/Bresenham$ length will be $B_l = (n - kn) + kn\sqrt{(2)} = n((1 - k) + k\sqrt{(2)})$. If we want to estimate the ratio of B_l/c, then:

$$n((1 - k) + k\sqrt{(2)})/\sqrt{(n^2 + (kn)^2)} = ((1 - k) + k\sqrt{(2)})/\sqrt{(1 + (k)^2)}$$

We hope that the above expression is equal to 1. In this case, we can solve the equation

$$\frac{((1-k)+k\sqrt{(2)})}{\sqrt{(1+(k)^2)}} = 1$$

What we get is:

$$((1-k)+k\sqrt{(2)})^2 = (1+(k)^2)$$

$$((1-k)+k\sqrt{(2)})^2 = (1-k)^2 + 2(1 \quad k)k\sqrt{(2)} + 2k^2$$

$$1 - 2k + k^2 + 2(1-k)k\sqrt{(2)} + 2k^2 = (1+(k)^2)$$

$$-2k + 2(1-k)k\sqrt{(2)} + 2k^2 = 0$$

$$-2k + 2k\sqrt{(2)} + 2(-k)k\sqrt{(2)} + 2k^2 = 0$$

$$-2k + 2k\sqrt{(2)} + 2(-k)k\sqrt{(2)} + 2k^2 = 0$$

$$-2k + 2k\sqrt{(2)} + 2(-k)k\sqrt{(2)} + 2k^2 = 0$$

So $k = 0$ is a solution. or if $k \neq 0$ we have

$$-1 + \sqrt{(2)} - k\sqrt{(2)} + k = 0$$

$$\sqrt{(2)} - 1 = k\sqrt{(2)} - k$$

i.e. $k = 1$

That is to say, there are two occasions such that the Euclidean length of a line segment equals the interpreted length of a discrete line (Bresenham line): $k = 0$ and $k = 1$.

The question is, what makes the largest difference between these two measurements? A good guess would be $arct\, g(k) = 22.5°$ degrees of the angle.

One way is still to see the

$$maximum = f(k) = \frac{((1-k)+k\sqrt{(2)})^2}{(1+(k)^2)}$$

$$\frac{df}{dk} = 0 = \frac{d}{dk} \frac{((1+(\sqrt{(2)}-1)k)^2)'((1+(k)^2)) - (1+(k)^2)'(1+(\sqrt{(2)}-1)k)^2}{((1+(k)^2))^2}$$

$$= 2(1+(\sqrt{(2)}-1)k)(\sqrt{(2)}-1)(1+(k)^2) - 2k(1+(\sqrt{(2)}-1)k)^2$$

So $(1+(\sqrt{(2)}-1)k) = 0$, i.e. $k = 1/(1-sqrt(2)) = -(\sqrt{(2)}+1)$, which is not a valid solution.

On the other hand,

$$2(\sqrt{(2)}-1)(1+(k)^2) - 2k(1+(\sqrt{(2)}-1)k) = 0$$

$$(\sqrt{(2)}-1) + (\sqrt{(2)}-1)(k)^2 - k + -k(\sqrt{(2)}-1)k) = 0$$

$$k = (\sqrt{(2)}-1)$$

So $tg(\alpha) = k = (\sqrt{(2)} - 1)$, where the angle is 22.5°.
The ratio is

$$\sqrt{(f(k))} = \sqrt{(f(\sqrt{(2)} - 1)}$$

$$= \sqrt{(\frac{(1 + (\sqrt{(2)} - 1)(\sqrt{(2)} - 1))^2}{1 + (\sqrt{(2)} - 1)^2})}$$

$$= \sqrt{((1 + (\sqrt{(2)} - 1)^2)} = \sqrt{(4 - 2\sqrt{(2)})} = 1.08239$$

That means the interpreted length of the discrete segment is longer than Euclidean length. However, the maximum ratio is 1.08 when we did the local counting. Therefore,

Lemma 8.1 *The maximum error ratio of using local counting on horizontalvertical and diagonal types of 1-cells is 1.08.*

Next, we can consider the average ratio of differences on digitization and embedding. We have

$$\int_0^1 \sqrt{(f(k))}dk = \int_0^1 \frac{((1 - k) + k\sqrt{(2)})}{\sqrt{((1 + (k)^2))}}dk$$

$$= \int_0^1 \frac{1}{\sqrt{((1 + (k)^2))}}dk + \cdots + (\sqrt{(2)} - 1)\int_0^1 \frac{k}{\sqrt{((1 + (k)^2))}}dk$$

$$= ln((\sqrt{(2)} + 1) - 2\sqrt{(2)} + 3$$

$$= 1.05294646227$$

For the general discrete model,

$$\Sigma_{m=0}^{m=n} \frac{1}{n + 1} \cdot \frac{((n - m) + m\sqrt{(2)})}{\sqrt{(n^2 + (m)^2)}}$$

when $n = 5$, *value* $= 1.04239$; $n = 10$, *value* $= 1.04766$; $n = 100$, *value* $= 1.05242$, $n = 1000$. the value is 1.05294. $n = 5000$, the value is also 1.05294

Sometimes, we do not want to count the last sample point. Then, we have

$$\Sigma_{m=0}^{m=n-1} \frac{1}{n} \cdot \frac{((n - m) + m\sqrt{(2)})}{\sqrt{(n^2 + (m)^2)}}$$

When $n \geq 5000$, the value is also 1.05295.

Therefore, the later model is closer to the continuous model. The average difference ratio is only 5 % of the actual Euclidean line.

For case (3), we can leave this as an open problem: A fitted curve in what scenario will be the best fit to the original boundary if the guiding point are all digital?

8.5.2 2D Region: Area and the Boundary Length

Based on the discussion in the previous subsection, digitization will approximate the area, but not the edge. Edge approximation requires localized digitization [20, 23]. We can predict that the fitting on the boundary of a region will be the best for curve length estimation. However, in such a case, we might not be able to get the area of the region accurately.

Another question we can study is if we can use the uncertainty principle: In digital space, if we can get the right area, then the length of the boundary will never converge. If we get the boundary right, then we cannot preserve the area of the region that will converge to the original.

Let us assume that E_a denotes the difference (error) in area between the original object and the digitized object, and E_b denotes the difference (error) in length between the original object and the digitized object. We know that E_a can be zero. However, E_b can never be zero in digital space.

If it is not in digital space, then there is an instance where E_b can be arbitrarily bigger when E_a is approximating zero. Is there an inequality such as $E_b \times E_a < \varepsilon(\delta) \cdot area \cdot perimeter$ for any shape including the refinement scale δ?

The perimeter of a region in discrete space is only 1.08 times the original. What is the average perimeter for a piecewise linear polygon? What would the probability model be? It is difficult to do the same thing in 3D. We can only do some actual experimental comparisons. This question is related to geometric measure theory. See Federer's book on rectifiable curves [13].

8.6 Remark: Embedding to Euclidean Space

Embedding a digital object into Euclidean space is usually related to modeling. The best way is to find a function (or collection of functions and equations) to approximate the original object.

A manifold can isomorphically be embedded into Euclidean space. Nash's theorem also stated that a smooth manifold can be embedded into Euclidean space.

In Chap. 12, we will consider so called manifold learning that can be viewed as a technique for embedding. We want to find the smallest dimension that can hold a manifold. In other words, we want to delete all unnecessary components in a very high dimensional space.

For algorithmic considerations [11], including approximation algorithms, can be used in this type of research. We will give examples in Chap. 15 for approximation of digitization.

References

1. Pierre Alliez, Laurent Saboret, and Gael Guennebaud. Surface reconstruction from point sets. In CGAL User and Reference Manual. CGAL Editorial Board, 4.4 edition, 2000.
2. L. Chen, Generalized discrete object tracking algorithms and implementations, Melter, Wu, and Latecki ed, Vision Geometry VI, SPIE 3168, pp 184–195, 1997.
3. L. Chen, A note on (alpha, beta)-type digital surfaces and general digital surfaces, in Melter, Wu, and Latecki ed, *Vision Geometry VII*, SPIE Proc. 3454, pp 28–39,1998.
4. L. Chen, Point spaces and raster spaces in digital geometry and topology, Melter, Wu, and Latecki ed, Vision Geometry VII, 1998.
5. L. Chen, Note on the discrete Jordan curve theorem, Vision Geometry VIII, Proc. SPIE Vol. 3811, 1999. 82–94.
6. L. Chen, *Discrete Surfaces and Manifolds: A theory of digital-discrete geometry and topology*, 2004. SP Computing.
7. L. Chen, Digital Functions and Data Reconstruction, Springer, NY, 2013.
8. L. Chen, and Y. Rong, Digital topological method for computing genus and the Betti numbers, Topology and its Applications, Volume 157, Issue 12, 2010, Pages 1931–1936.
9. L. Chen and J. Zhang, Digital manifolds: A Intuitive Definition and Some Properties, Proceedings of the Second ACM/SIGGRAPH Symposium on Solid Modeling and Applications, Montreal, 1993, 459–460.
10. L. Chen and J. Zhang, Classification of simple digital surface points and a global theorem for simple closed surfaces, in R.A. Melter and A.Y. Wu, Vision Geometry II, SPIE Proceedings 2060.
11. T. H. Cormen, C.E. Leiserson, and R. L. Rivest, Introduction to Algorithms, MIT Press, 1993.
12. H.S.M. Coxeter, Introduction to geometry, John Wiley, 1961.
13. H. Federer, Geometric Measure Theory, Springer, Berlin 1969. Page 202.
14. James D. Foley, Andries Van Dam, Steven K. Feiner and John F. Hughes, Computer Graphics: Principles and Practice. Addison-Wesley. 1995.
15. S. Fortune, Voronoi diagrams and Delaunay triangulations, in D.Z. Du and F. K. Hwang ed, Computing in Euclidean Geometry, World Scientific Publishing Co, 1992.
16. J.E. Goodman, J. O'Rourke, Handbook of Discrete Geometry, CRC, 1997.
17. Han, Qing; Hong, Jia-Xing (2006), Isometric Embedding of Riemannian Manifolds in Euclidean Spaces, American Mathematical Society, ISBN 0-8218-4071-1
18. G.T. Herman, "Geometry of Digital Spaces," Birkhauser, Boston, 1998.
19. M. Kazhdan, M. Bolitho, and H. Hoppe. Poisson Surface Reconstruction. In Symp. on Geometry Processing, pages 61–70, 2006.
20. R. Klette and A. Rosenfeld, Digital Geometry, Geometric Methods for Digital Picture Analysis, series in computer graphics and geometric modeling. Morgan Kaufmann, 2004.
21. V. A. Kovalevsky, Finite topology as applied to image analysis, Computer Vision, Graphics and Image Processing, Vol. 46, 1989, pp. 141–161.
22. William E. Lorensen, Harvey E. Cline: Marching Cubes: A high resolution 3D surface construction algorithm. In: Computer Graphics, Vol. 21, Nr. 4, July 1987
23. A. McAndrew and C. Osborne, Algebric methods for multidimensional digital topology, vision Geometry II, R. Melter, and A. Y. Wu, ed. SPIE processing 2060, 1993.
24. Kenji Shimada and D. Gossard. Automatic triangular mesh generation of trimmed parametric surfaces for finite element analysis. Computer Aided Geo-metric Design, 15(3):199–222, 1998.
25. P. Shilane, P. Min, M. Kazhdan, and T. Funkhouser, The Princeton Shape Benchmark, Shape Modeling International, Genova, Italy, June 2004
26. Hang Si, Tetgen: A quality tetrahedral mesh generator and a 3D Delaunay triangulator. http://wias-berlin.de/software/tetgen/
27. Z. J. Wood. Computational Topology Algorithms for Discrete 2-Manifolds. PhD thesis, California Institute of Technology, 2003.

Chapter 9
Combinatorial Topology and Digital Topology

Abstract Topology is the study of the equivalence between two general spaces under continuous mappings. Two spaces are called homemorphic if there is an invertible continuous function between them. Triangles or simplexes are used in topological analysis of a space since we want to decompose a complex space into some simpler shapes to better understanding the whole structure of a space. This type of topology is called combinatorial topology. Combinatorial topology is the old name for algebraic topology before the theory of homology was developed.

In this chapter, we briefly introduce combinatorial topology and (modern) algebraic topology for later chapters. Then, we move to use these tools in digital topology problems. We are specifically interested in using Euler characteristic to analyze digital curves and surfaces. For the other two important problems related to discrete and digital topology: Jordan curve theorem and digital genus computation, we will discuss these in Chaps. 14 and 15.

Keywords Topology · Simplex · Simplicial complex · Cell complex · Finite topology · Euler characteristic · Digital topology

9.1 Basic Concepts of Topology

Comparing to geometry, measurement of distance and volumes, topology studies the relationship between two spaces, especially the equivalence between two general spaces under continuous mappings. Two continuous spaces are called homeomorphic if there is an invertible continuous function between them. The simplicial approximation theorem guarantees that continuous maps exist from a finite simplicial complex to a "refined" simplicial complex by barycentric subdivision that can be approximated. This theorem basically says that we can use the simplexes to decompose a continuous topological space without change the topological properties of the space.

We have already introduced the concept of point set topology in Chap. 3. Let X be a set and τ be a collection of subsets of X. τ is called a topology on X if: (1) The empty set and X are in τ, (2) The intersection of a finite number of elements of τ is in τ, and (3) The union of an arbitrary number of elements of τ is in τ. The elements of X are usually called *points* and the elements in τ (subsets of X) are called open sets. The complement of an open set A, $X - A$, is called a closed set. Note that a

© Springer International Publishing Switzerland 2014   149
L. M. Chen, *Digital and Discrete Geometry*, DOI 10.1007/978-3-319-12099-7_9

subset $A \subset X$ may be open, closed, both, or neither. If $|\tau|$ is finite, then the open set is the same as the closed set.

The easiest example of topology is the $<R = \{real\, numbers\}, \tau = \{open\, intervals\}>$.

A topological space can also refer to a function space in which each element of X is a function.

A function on topological space usually means that the function is on the base set X. We can also define a function between two topological spaces (X, τ) and (Y, τ'). For instance, $f : X \rightarrow Y$.

Intuitively, we say that two objects are topologically equivalent if there is a process that can continuously change one object into another. It can be defined as a continuous one-to-one onto function. We also say that these two objects have homeomorphism [2].

Definition 9.1 (X, τ) and (Y, τ') are said to be homeomorphic or topologically equivalent if there exists a continuous and invertible function f between these two spaces. (f^{-1} is also continuous.)

Surfaces and manifolds in Euclidean space are special types of topological spaces. A surface usually refers to a 2D manifold. A mathematical definition of a surface is a structure where at each point on the surface, there is a neighborhood that is similar to a disk. More precisely, the neighborhood is homeomorphic to 2D Euclidean space. In discrete space, we can view a surface as the discretization of a continuous surface. This means that there is a partition on the surface that is piecewise linear [30]. In other words, a complex or general surface can be viewed as gluing together of simple 2D shapes such as triangle, rectangles, and even polygons.

We can extend the definition of surfaces to an n-dimensional manifold: a (topological) n-manifold is a topological space $M = (X, \tau)$. Each element (point) of X has an open nD neighborhood U_x that is continuously equivalent or homeomorphic to an nD Euclidean space. This means there exists an invertible continuous function between them. For instance, there is a $f_x : U_x \rightarrow E_n$ where f_x^{-1} is also continuous.

In this book, we will not concern extensively smooth manifolds. We just refer a smooth n-manifold is a manifold where there is no bent angle in each local location in the space.

9.2   Triangulation, Simplicial Complexes, and Cell Complexes

Triangles or simplexes are used decompose a topological space for two reasons: (1) Understand the local configuration of the space, (2) Assist to find the global topological invariants. This approach is valid because of the simplicial approximation theorem, proved by L.E.J. Brouwer. This theorem states: the continuous maps from a finite simplicial complex to a simplicial complex can be approximated by barycentric subdivision. The meaning of this theorem is that any continuous shape, e.g. a curve or surface, in a simplicial decomposed space can be approximated by a sequence of

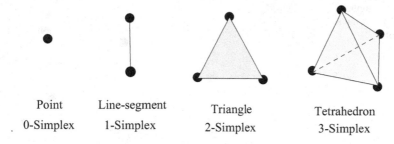

Point | Line-segment | Triangle | Tetrahedron
0-Simplex | 1-Simplex | 2-Simplex | 3-Simplex

Fig. 9.1 Examples of simplexes

piecewise linear (sub-)complexes. (This approximation is a type of deformation or homotopy that is defined later in this chapter.)

Partitioning a 2D region into triangles is called triangulation. However, only triangles cannot build a topological structure and we need to use edges and vertices in a combined manner. Due to the fact that points are 0-simplexes and edges are 1-simplexes, the collection of those simplexes is called simplicial complex.

For any such a decomposition or a partition, we will generate vertices, edges, and triangles, the combination of which is called a complex. However, the joint parts shared by two triangles must be an edge or point in the complex.

9.2.1 Definition of Simplicial Complexes

A point is a 0-simplex, a line segment is a 1-simplex, a triangle is a 2-simplex, and a tetrahedron is a 3-simplex. See Fig. 9.1.

Mathematically, An m-simplex Δ is defined as a shape that has $m + 1$ vertex points $u_0, u_1, \cdots, u_m$ in R^n satisfying:

(1) $(u_1 - u_0), \cdots, (u_m - u_0)$ are linearly independent, and
(2) $\Delta = \{\lambda_0 u_0 + \cdots + \lambda_m u_m | \Sigma_{i=0}^m \lambda_i = 1\}$.

where $\lambda_i \geq 0$, $0 \leq i \leq m$. For any set S that has above property, we call $(u_0, \cdots, u_m) \in R^n$ affine-independent. The S is called an affine-shape. In fact, the simplex is the convex hull of $u_0, \cdots, u_m$.

A subset of $u_0, \cdots, u_m$ that is also a simplex is called a face of S. Therefore, the k-face of S contains $k+1$ vertices of $u_0, \cdots, u_m$. For example, S with vertices $u_0, \cdots, u_5$ is a 4-simplex. F_2 containing u_0, u_1, u_2 is a 2-simplex, which is the convex-hull of u_0, u_1, u_2. where F_2 is a 2-face of S.

A simplicial complex K in R^n is a set of simplexes in R^n such that

(1) Every face of a simplex $S \in K$ is an element of K, and
(2) The intersection of any two simplexes S_1 and S_2 of K is a face in both S_1 and S_2.

The intersection is just a face and not a union of a set of faces. But a face could contain many low dimensional faces. Therefore, the standard definition should be changed from "the intersection of any two simplexes S_1 and S_2 of K is a face in each

Fig. 9.2 A simplicial
complex that is not a
3-manifold

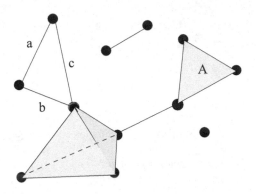

S_1 and S_2" to "the intersection of any two simplexes S_1 and S_2 of K is a k-face in each of S_1 and S_2."

This intersection can be empty. We use (-1)-face to indicate the empty intersection because k could never be -1. Based on convention, we do not use $\emptyset$-face, which refers to a unique cell in the space. In summary, a k-simplex is the convex hull of $k + 1$ affinely independent points [20]. A simplicial complex is a set of simplexes that represents topological space.

Basically, a simplicial complex can be viewed as a topological space when thinking about the topology τ as a special subset of points who makes points, line segments, triangles, and k-simplexes. Researchers in combinatorics also invented the so called abstract simplicial complex [26].

A simplicial k-complex $\mathcal{K}$ is a simplicial complex where the largest dimension of any simplex in $\mathcal{K}$ equals k. A k discrete manifold is a k-complex, but a k-complex might not be a discrete manifold, as shown by the following example. See Fig. 9.2.

In Fig. 9.2, we can also see that a, b, and c are three edges (three 1-cells). They are not the edge of a 2-cell and there is no 2-cell bounded by a, b, and c. However, A is a 2-cell. This complex has three components. In addition, a simplicial 2-complex must contain at least one triangle, and must not contain any tetrahedra or higher-dimensional simplices.

9.2.2 More Examples of Simplicial Complexes and Other Natural Complexes

An arbitrary continuous space can be approximated by a set of simple discrete shapes. These shapes are called cells, and they are usually defined as convexes. The simplest convex in 2D is triangle, that is called a simplex in high dimension. Such an approximation is made by a partition meaning that two k dimensional-cells does not overlap, i.e. their intersection is not an k dimensional object.

Beyond triangulations and rectangle decomposition, the Voronoi Diagrams and Delaunay Decompositions are most popular. We like to discuss the meanings of them below.

Given n points in m-dimensional Euclidean space, we can obtain a Voronoi diagram and Delaunay triangulation. Let P be a set of points in Euclidean space E_m. P is called sites. the Voronoi diagram of P partitions E_m into regions. Each region contains just a site. so that any point in the region containing site $p \in P$ are closer to p than to any other site in P.

The Delaunay triangulation of P is the unique triangulation of P so that there are no elements of P inside the circumsphere of any triangle. Here, "triangulation" has an extended meaning: it decomposes the convex hull of P into simplexes using S as vertices [14]. In addition, Delaunay triangulation does not need to be a decomposition where each face is a triangle.

Voronoi diagrams are a central subject in computational geometry. Some efficient algorithms have been developed to compute the Voronoi diagram and Delaunay decomposition [14, 15]. In [5], we show how to define a discrete surface (or manifold) in Voronoi diagrams or Delaunay decompositions. For how algorithms calculate the Voronoi diagrams and Delaunay decompositions, see Chap. 10.

In 3D, a Voronoi diagram is a special "graph," $G = (V, E)$, where V contains all vertices of the Voronoi diagram and E contains all Voronoi edges (Voronoi 1-cells). A Delaunay decomposition is also a special "graph," $G = (V, E)$, where V is the set of sits and E is the set of Delaunay edges (Delaunay 1-cells). We also have Voronoi 2-cells, Voronoi 3-cells, Delaunay 2-cells, and Delaunay 3-cells.

For a graph G, let $\tau(G)$ be a set of defined sets of 0-cells (V), 1-cells (E), 2-cells (U_2), ... , k-cells (U_k), ... by Voronoi cells or Delaunay cells. Let $\tau(G) =<$ $G, U_2, ..., U_n>$ be a topological structure of G. Based on the Voronoi cells or Delaunay cells, we can get all the matches for the definition of discrete surfaces in Sect. 7.5 and discrete manifolds in Sect. 7.4.

A Voronoi diagram gives a good example of the cell complex. We will define the cell complex next.

9.2.3 Cell Complexes*

We have discussed simplicial complexes and digital complexes (cube complexes in Chap. 5). The more general complexes can be defined by replacing triangles or simplexes with any other shape. These are called cell complexes.

A more general form of cell complexes is called the CW complex. However, it is hard to use a computer to represent CW Complexes. In fact, there is no need to require that a cell have a polygonal boundary. It can be any curve as long as when two cells meet, they must have a simply connected component as the intersection (in other words, part of the boundary of the two cells is also a cell.).

We know that a kD unit ball is represented as,

$$B^k = \{x \mid x \in R^k \& \|x\| \leq 1\}.$$

The boundary of B^k is a unit sphere,

$$S^{k-1} = \{x \mid x \in R^k \& \|x\| = 1\}.$$

Basically, a k-cell is a deformation of a unit k-ball; an open k-cell is a topological space that is homeomorphic to an open ball. Mapping (called "attaching") the boundary of B^k to a point will result in a S^k.

A space M is called a CW-complex, which stands for complexes with weak topology, if there is a nested sequence of topological spaces

$$M^{(0)} \subset M^{(1)} \subset \cdots \subset M^{(m)} = M$$

where $M^{(0)}$ is a set of discrete points. Just like V in $G = (V, E)$ in Chap. 7, we attach 1-balls B^1 to $M^{(0)}$ along their boundaries S^0. If B^1 is a line-segment and it is attached to two points in $M^{(0)}$, then the interiors of the attached 1-ball is called a 1-cell. When we get an $M^{(1)}$ that is exactly like graph $G = (V, E)$, then we can attach 2-balls B^2 to $M^{(1)}$ along their boundaries S^1. If B^1 is a circle, then ideally we can attach a 2-ball to a simple cycle in $M^{(1)}$. Then, we can get $M^{(2)}$.

To summarize, for $k > 0$, $M^{(k)}$ is the result of attaching a set of k-balls to $M^{(k-1)}$ by mapping (gluing) their boundaries S^{k-1} to $M^{(k-1)}$. $M^{(k)}$ is called a k-skeleton of M. The k-cell is the mapping (deformation) of the interior part of the k-ball.

A CW complex is called regular if the gluing maps are 1-to-1 continuous onto, which is called a homeomorphism. The complex restricts each cell where a regular CW complex meets each vertex of M at most once.

A fundamental theorem called the CW-approximation theorem says that every topological space X has a CW complex to weakly approximate X [19, 20].

It is clear that the definition of discrete manifolds (Chap. 7) is a special case of the CW complex. However, even though the CW complex has great mathematical properties, we cannot represent a CW complex in computers since we cannot represent a general continuous map (gluing map) in computers.

9.2.4 Euler Characteristic of Simplical Complexes

A property of a space is called topological invariant if it does not change under the invertible continuous mapping. For a connected (and orientable) surface, the genus g also indicates the number of handles or tunnels on the surface. Therefore, genus is a topological invariant. A sphere does not have any handle or tunnel, so its genus is zero. A donut has one handle, its genus is one. See Fig. 7.9.

Euler characteristic is another topological invariant that is defined for a simplicial complex even for any finite CW complex:

$$\chi = k_0 - k_1 + k_2 - k_3 + \cdots, \tag{9.1}$$

where k_i indicates the number of cells in dimension i in the complex. The Euler characteristic has fundamental importance in combinatorial topology. For a surface, we know that

$$\chi = 2 - 2g - b \tag{9.2}$$

where b is the number of the boundary components. So we have $\chi = 2 - 2g$ for closed surfaces. We have presented and used the most popular case of this theorem for the planar graph $G = (V, E)$ in Chaps. 4 and 6. In such a case, we have the following facts: (1) $g = 0$, (2) $b = 0$, (3) $k_0 = |V|$, $k_1 = |E|$, and $k_2 = |F|$, and (4) the rest of k_i, $i \geq 3$ is 0. Therefore we got, $k_0 - k_1 + k_2 = 2$, thus we have Euler's formula for planar graphs (4.2).

$$|V| + |F| - |E| = 2 \tag{9.3}$$

9.3 Basic Algebraic Topology: Homotopy Groups and Homology Groups*

Two curves on a surface are called homotopy if one curve can be changed to the other curve in a continuous way. In engineering, this concept is called deformation. Fundamental groups is about an algebraic group relating to "homotopic" curves, an essential concept connecting algebra and geometry. Homotopy groups are an extension of fundamental groups that deal with continuous changes between surfaces and higher dimensional manifolds (in a topological space). Although homotopy groups are strong in the topological sense, they are difficult to calculate and even not computable in the general case. Researchers developed simpler groups called homology groups that are easier to calculate.

Both the homology group and the homotopy group are topological invariants to topological spaces. Thus, if space X and Y have different homology groups or homotopy groups, then X and Y are not homeomorphic. However, we can not say that X and Y are homeomorphic even if their homotopy groups are the same.

We have another example of how to understand topology and its invariants: there is a space that has one hole and another space that has two holes. These two spaces are not topologically equivalent and they also have different homology groups. In practice, calculating the number of holes for two 2D pictures will tell us the topological equivalence between them. However, in higher dimensional spaces, even two 2D surfaces have the same number of holes, but they may not be topologically equivalent.

In this section, we give an overview of the algebraic aspects of topology. This section contains some profound knowledge in topology [1, 19].

9.3.1 Review of Groups in Algebra

We here review the basic concept of groups in algebra. $G = (S, \cdot)$ is called a group if it contains a base set S and a binary operator $\cdot$ with the following properties:

(a) If $a, b \in G$, then $a \cdot b \in G$ (closure property),
(b) If $a, b, c \in G$, then $(a \cdot b) \cdot c = a \cdot (b \cdot c)$ (associative property),
(c) There exists an $e \in G$ such that for every $a \in G$, $e \cdot a = a \cdot e = a$ (identity property), and
(d) If $a \in G$, there exists $b \in G$ such that $a \cdot b = b \cdot a = e$ (inverse element property).

If $a \cdot b = b \cdot a$, then the group is called an Abelian group (commutative property). A subgroup H of G is called a normal subgroup if for any $a \in G$ we have,

$$aH = Ha$$

meaning that $a \cdot h$, where $h \in H$ would be an element of Ha, and vice versa. aH is a (left-) coset. If we define the quotient set as $G/H = \{aH | a \in G\}$, then we can prove that G/H is a group when we treat H as the *identity* of G/H. G/H is called the quotient group with respect to H. In algebra, cosets create the partition of a set.

Let G and H be two groups. a function $f : G \rightarrow H$ is called a (group-) homomorphism from G to H for all a and b in G. We have

$$f(a \cdot b) = f(a) \cdot f(b).$$

We can further define the kernel of f denoted by **Ker**(f). This kernel is a subset of G such that for any $a \in$ **Ker**(f), we have $f(a) = e_h$, where e_h is the identity of H.

$$\mathbf{Ker}(f) = \{a \in G | f(a) = e_h\}.$$

We also denote the image of f, $f(G)$, as **Im**(f). We have the following basic result, which is not difficult to prove.

Proposition 9.1 *(1) The kernel of a homomorphic mapping $f : G \rightarrow H$ is a normal subgroup of G. (2) The image of f is a subgroup of H.*

9.3.2 Fundamental Groups and Homotopy Groups

Let X be a topological space and $x \in X$ be a point. A closed curve starting and ending at x is called a loop $loop(x)$. We call x a base point. Given the direction of the loop, clockwise or counterclockwise, the loop now is a path.

Now, we can define a group that is formed by the closed curves passing x. Let us define the group product $a \cdot b$ of loop a and loop b as the path of a followed by the path of b.

a and b are equivalent if a can be deformed to b (a and b are homotopic). The identity element is the point x which is treated as a special loop. a's inverse element is defined as the inversed path of a.

All loops with base point x is a group denoted by $\pi_x(X, \cdot)$. It is called the fundamental group. The fundamental group was defined by Poincare in 1895 [1]. For a (path-) connected space X, we can prove that if $x, y \in X$, $\pi_x(X, \cdot)$ and $\pi_y(X, \cdot)$ are the same (isomorphically equivalent) [1].

Formally, a loop is a function from 1-sphere (circle) S^1 to X. Functions f and g are said to be homotopic if there is a mapping $\mathbf{H}$ such that

$$\mathbf{H} : X \times [0, 1] \to Y,$$

where $\mathbf{H}(x, 0) = f(x)$ and $\mathbf{H}(x, 1) = g(x)$. In addition, $\mathbf{H}(x, t)$ is continuous. So the fundamental group is formed by loops under the homotopic. If any two loops are homotopic on X, then this space is called simply connected. Therefore, the fundamental group of a simply connected Space has just one element.

In other words, one can contract a loop to be a simple point in simply connected spaces.

The generalization of the fundamental group to the homotopy groups is to use higher dimensional spheres instead of circles.

Definition 9.2 All maps (functions) from the n-sphere S^n to X can be classified into the set of homotopy classes where any two elements of a homotopy class are homotopic. This set forms a group called the nth homotopy group of a topological space X, denoted by $\pi_n(X)$.

In fact, the fundamental group $\pi_1(X)$ can be very complicated. For $n > 1$, the homotopy group $\pi_n(X)$ is an Abelian group [19]. Even though the homotopy group is an Abelian group ($n > 1$), this definition did not suggest any thinking path to obtain a homology group for the specific manifold. However, for some special manifolds, we can get their homotopy groups.

Example 9.1 (1) $\pi_1(S^1) = Z$. This is because a loop cannot be contracted on S^1, and we can only loop it multiple times. Therefore, it is equal to Z. (2) $\pi_1(S^2) = \{e\}$ because any loop on S^2 can be contracted to a single point, i.e. every loop is homotopic to an infinitively small Circle, that is a point. (3) $\pi_2(S^2) = Z$. The reason is the same as that of (1). □

9.3.3 Homology Groups

Let X be a topological space that can be viewed as a general cell complex. A cell complex does not have to be a discrete manifold. Homology groups will be appeared as a group sequence $H_0, H_1, \cdots, H_n$ where n is the highest dimension of n-cell in X. Homology groups are hard to define. But they have simple meanings: H_0 is the number of connected components and H_1 is the number of holes.

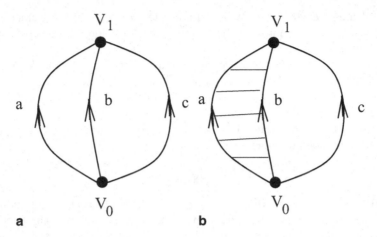

Fig. 9.3 Examples of cell complexes and their homology groups: **a** Two 0-cells and three 1-cells, and **b** two 0-cells, three 1-cells, and one 2-cell

We can use simplicial complex X as a base to introduce homology groups. A k-simplex Δ_i has $k + 1$ vertices $v_0, \cdots, v_k$. The orientation of Δ_k is determined if the orientation of each edge (1-simplex) is determined. We use $[v_0, \cdots, v_k]$ to represent the cell and its orientation by assigning the arrow of the edge from v_i to v_j if $j > i$ (in the vector $(v_0, \cdots, v_k)$). So the boundary of Δ_k would be $k + 1$ Δ_{k-1}s: $[v_0, \cdots, v_i, vi + 1, \cdots, v_k]$ for all $i = 0, \cdots, k$.

Define $\partial[v_0, \cdots, v_k] = \Sigma_i^k(-1)^i [v_0, \cdots, v_i, vi + 1, \cdots, v_k]$ as a convenient way of representation. For instance, $\partial[v_0, v_1] = [v_1] - [v_0]$ means we have two boundary points. However, $\partial[v_0, v_1, v2] = [v_1, v2] - [v_0, v_2] + -[v_1, v_2]$ indicates the three edge directions and the orientation of the boundary clockwise or counter-clockwise.

Let $C_n(X)$ be the Abelian group with all n-cells (or n-simplexes) $a_t^{(n)}$, $t = 1, \cdots, N$, in X that are considered: $\Sigma n_i \cdot a_i^{(n)}$ where n_i is an integer. $C_n(X)$ is called an n-chain and an element of $C_n(X)$ is a simple path of n-cells, just like we defined in Chap. 7. The following example explains how we get $C_n(X)$ [19].

Example 9.2 In Fig. 9.3a, $a - b$ is a path or chain. So $2a - 2b$ is a path with two cycles from v_0 to v_1 and back to v_0. In Fig. 9.3b, we add a 2-cell A in the clockwise direction. We can see that $C_2(X) = 0$ in Fig. 9.3a since there is no 2-cell, and $C_1(X)$ is the set of elements like $n(a) + m(b)$ where n, m are integers. $C_0(X)$ can also be represented as a set of $n(v_0) + m(v_1)$. In Fig. 9.3b, $C_2(X)$ has the elements listed as $n(A)$.   □

Now we introduce boundary operators ∂_k that are boundary homomorphic mappings from $C_k(X)$ to $C_{k-1}(X)$, i.e. $\partial_k : C_k(X) \rightarrow C_{k-1}(X)$. We want to map an n-Chain (a chain of n-cells), P_n, to its own boundary. Specifically, if P_n is a set of $[v_0, \cdots, v_n]$, then

$$\partial_n P_n = \Sigma_i(-1)^i P_n|[v_0, \cdots, v_i, vi + 1, \cdots, v_n] \tag{9.4}$$

where $P_n|[v_0, \cdots, v_i, vi+1, \cdots, v_n]$ is the restriction.

Since the boundary of an n-cell is closed, therefore, the boundary of the boundary is empty. It is easy to see that

Theorem 9.1

$$\partial_{(n)}(\partial_{(n+1)}) = 0 \tag{9.5}$$

This theorem means that the boundary of $C_{n+1}(X)$ is always closed, and every closed shape will map to 0. Therefore, $Im(\partial_{n+1}) \subset Ker(\partial_n)$. This means that every boundary is closed but not every closed shape is the boundary of a higher dimensional cell. However, if X is simply connected, we can say that they are equal.

The chain complex is a sequence of Abelian groups $C_0, C_1, C_2, \ldots C_n, \ldots$ It is connected by homomorphisms $\partial_n : C_n \to C_{n-1}$ [19].

$$\cdots \xrightarrow{\partial_{n+1}} C_n \xrightarrow{\partial_n} C_{n-1} \xrightarrow{\partial_{n-1}} \cdots \xrightarrow{\partial_2} C_1 \xrightarrow{\partial_1} C_0 \xrightarrow{\partial_0} 0$$

The homology groups are defined as follows:

$$H_n(X) := \frac{\mathbf{Ker}(\partial_n)}{\mathbf{Im}(\partial_{n+1})}$$

Example 9.3 We still use Fig. 9.3 to be example for analyzing the structure of homology groups.

In Fig. 9.3a, $H_1(X) = \mathbf{Ker}(\partial_1)$ since $\mathbf{Im}(\partial_2) =$, i.e. no 2-cells in $\mathbf{Im}(\partial_2)$. $\mathbf{Ker}(\partial_1)$ contains all the closed cycles since only closed cycles can map to the identity element in C_0. The cycle is in the form of $n(a - b) + m(b - c)$. In addition, the group $\{n(a - b) + m(b - c)\}$ is the normal subgroup of the group $\{n(a) + m(b) + l(c)\}$, which is the general form of C_1. Therefore, $H_1(X) = \mathbf{Ker}(\partial_1) = Z \times Z$. $H_0 = \mathbf{Ker}(\partial_0)/\mathbf{Im}(\partial_1)$; $\mathbf{Ker}(\partial_0)$ has the Abelian group generated by two points since ∂_0 does not map 0-cells anywhere. $\mathbf{Im}(\partial_1)$ is the boundary of all edges, which is still the same Abelian group generated by these two points. Therefore, $H_0 = 1$ is the single element that is also the identity.

In Fig. 9.3b, $H_2(X) = Z$ since there is only one 2-cell. However, $\mathbf{Im}(\partial_2)$ is a group generated by $a - b$, where $a - b$ is the boundary of A. Therefore, $\mathbf{Im}(\partial_2) = \{n(a - b)|n \in Z\}$. $\mathbf{Ker}$ is the same as in Fig. 9.3a and $H_1(X) = \mathbf{Ker}(\partial_1)/\mathbf{Im}(\partial_2) = (Z \times Z)/Z = Z$. It means that Fig. 9.3b contains only one hole by the edge cycle.
□

The geometric meaning of $\mathbf{Ker}(\partial_n)$ is the collection of all "cycles" made by n-cells or "closed n-manifolds." This is because only one object without a boundary can map to zero, the identity element in C_{n-1} which is an Abelian group.

Again, an $n + 1$ object (manifold) has a closed boundary in C_n. Therefore, $\mathbf{Im}(\partial_{n+1}) \subset \mathbf{Ker}(\partial_n)$. When every n-"cycle" is the boundary of an $(n + 1)$-manifold, the space with the property of $\mathbf{Im}(\partial_{n+1}) = \mathbf{Ker}(\partial_n)$, has no n-hole. Therefore, $H_n(X)$ indicates the number of n-holes in the space X. Homology groups are powerful tools for topological structures. We use homology groups in Chap. 14 to find the genus of 3D objects.

9.4   Digital Topology: An Introduction

In this section, we will first introduce the concept of finite topology that makes use of classical topological method such as cell complexes in a discrete space especially for digital images in digital space. Then, we will present a unified method for topological analysis in 2D and 3D digital space by using the Euler theorem for planar graphs. The topics on more advanced digital topology will be discussed in Chaps. 14 and 15. All results discussed in this section will not only satisfy all requirements of classical topology, but also satisfy the methods of digital manifolds and discrete manifolds given in Chaps. 5 and 7.

9.4.1   Finite Topology and Grid-Cell Topology

Finite topology was proposed by Kovalevsky for creating a type of cellular topology for a finite set, especially for grid-cell spaces. This topology is based on cell complexes. It was a nice bridge to topologies between digital space to continuous space. It was specifically for describing the structure of images [25]. This method first appeared in P. S. Alexandrov famous book in 1930's [1]. The method was used as an example of the grid-cell complexes.

Kovalevsky's method is to encode images into cellular complexes. However, cellular complexes defined in [25] through using open sets is not a discrete mathematical method defined in Chap. 7. Finite topology maps a digital image to a continuous space in order to make an encoding, which could cause additional problems in computations or algorithm design. The cell in Kovalevsky's finite topology may be a polyhedron or faces of a polyhedron.

Definition 9.3 A (abstract) cellular complex $C = (E, B, dim)$ is a set E of (abstract) elements provided with an antisymmetric, irreflexive, and transitive binary relation $B \subset E \times E$ called the bounding relation (or the face relation) and with a dimension function $dim : E \rightarrow \{0, 1, 2, ...\}$ such that $dim(e_1) < dim(e_2)$ for all pairs $(e_1, e_2) \in B$. $e \in E$ with $dim(e) = i$ is called an i-dimensional element or an i-cell. A complex is called k-complex if the dimensions of all its elements is less than or equal to k.

The advantage of this definition is to ignore the continuous space, just use discrete objects to define the topology. The disadvantage is that we still need human interpretation to determine whether an actual image component is an i-complex. In other words, given a set of pixels in 2D or voxels in 3D, deciding if the set is a 1-complex, 2-complex, or 3-complex will have results that depend on which cells are pre-included in E. Kovalevsky called this problem the image encoding problem. Therefore, the encoding is key to generate a topology. In Chaps. 5 and 7, we have given the unified method for defining the topology of discrete cells. We can also say that the method we used in Chaps. 5 and 7 are realizations of the topologic spaces defined in Definition 9. 3. The realization of a cell complex was also discussed in continuous space. It is different from the discrete realization here.

9.4.2 Topology of Curves in 2D Digital Space

In this section, we will present some basic properties of curves in a 2D discrete plane or a closed 2D discrete manifold. All results discussed in this section will not only satisfy all requirements of classical topology, but also satisfy the methods of digital manifolds and discrete manifolds given in Chaps. 5 and 7.

9.4.2.1 Case Study: A Simple Topological Theorem of Digital Curves

There are many ways of dividing a plane into smaller pieces or 2-cells. Some examples were presented in Chap. 2. Following the edges of each 2-cells, we can get a discrete curve. Since the most popular 2-cells are triangles and squares. We here use squares as an example to give a topological theorem in discrete space.

A polygon is called regular if every edge is equal in terms of lengths. A square is also called a regular 4-polygon. We can extend this concept to any type of regular polygons. Squared decompositions are not only convenient in digital computers, but we can also derive some interesting topologic properties.

Here, we present a simple theorem of digital curves using the classic Euler formula of planar graphs.

We first give an intuitive explanation of this theorem. Let us assume that a simple closed digital curve C contains at least one point inside the curve in direct adjacency defined in Chap. 2. This curve only contains three types of points: (1) The outward point (CP_2), adjacent to two points on C and not adjacent to any point in the inner area of C, (2) The straight point, CP_3, adjacent to two points on C with a neighbor in the inner area of C, and (3) The inward point CP_4 adjacent to two points on C with two neighboring points in the inner area of C. The theorem says that,

Lemma 9.1 $CP_2 = CP_4 + 4$.

This lemma has multiple proofs. The meaning of this lemma is that for a simple digital curve, the outward points are always 4 more than the inward points. This result can be used directly in calculating how many holes there are in a digital image. As we know, Euler's planar graph theorem states: In a planar graph, if V, E, and F represent the set of vertices, edges, and faces, respectively, then $|E| = |F| + |V| - 2$. We can use this theorem to prove above lemma.

Proof Let IN_C be the set of points inside of the curve not including any point on the boundary C. It is easy to know that the vertices contain all points in IN_C and on the boundary C. So,

$$|V| = |CP_4| + |CP_3| + |CP_2| + |IN_C|.$$

Each point in IN_C will have 4 edges to link with. However every edge was used twice. Also consider three types of points on boundary curve; therefore,

$$|E| = (4 \cdot IN_C + 4 \cdot |CP_4| + 3 \cdot |CP_3| + 2 \cdot |CP_2|)/2.$$

For the same reason, we consider each point in IN_C will be included in 4 2-cells (faces). However a point CP_4 only is only included in three faces, and so on. Considering that a face contains four points and outside of the curve is also a face in planar graphs, we have

$$|F| = (4 \cdot IN_C + 3 \cdot |CP_4| + 2 \cdot |CP_3| + |CP_2|)/4 + 1.$$

Therefore, using formula $|E| = |F| + |V| - 2$, we have:

$|CP_4| + |CP_3| + |CP_2| + |IN_C| + (4 \cdot IN_C + 3 \cdot |CP_4| + 2 \cdot |CP_3| + |CP_2|)/4 + 1 - 2$
$= (4 \cdot IN_C + 4 \cdot |CP_4| + 3 \cdot |CP_3| + 2 \cdot |CP_2|)/2.$

thus, $|CP_2| = |CP_4| + 4.$ $\qquad\qquad\qquad\qquad\qquad\qquad\qquad\qquad\qquad\qquad\square$

Rosenfeld presented a result that states: Any simple closed curve must have five points in a 2D digital plane with direct adjacency (or 4-connected) [24]. In differential geometry there is a famous four vertex theorem: Every simple closed convex curve in a plane has at least four vertices, where the vertex is a point whose curvature has a local maximum or minimum. This theorem in digital plane will be there must be 4 points in $CP_2 \cup CP_4$. The curvature of a point in CP_3 (the straight line point) is zero that is not maximum neither minimum.

Since boundary curve C does not contain any 4-regular polygons, so any two outward corner points cannot be adjacent in our digital curve definition. There are at least four outward corner points. The positions between the two outward corner points must be filled by the points in CP_3 or CP_4. Thus, $|C|$ must be greater than or equal to 8.

Using the theorem proved above, we can easily prove Eq. (4.4) for the hole counting problem [8].

9.4.2.2   Euler's Formula and Other Discrete Curves

We can use Euler's Formula for planar graphs to study other types of discrete curve. If only one type of polygons can be used, there are only three ways to divide a plane into regular polygons. These polygons are regular triangles (3-regular-polygon), squares (4-regular-polygon), and 6-regular-polygons.

In Lemma 9.1, we can see that a closed digital curve has at least eight points in a 4-regular-polygon decomposition plane. Using the same technique, we can prove that there are least six points in a closed discrete curve in the 3-regular-polygon decomposition plane, and at least eight points in a 4-regular-polygon decomposition plane, and at least 12 points in a closed discrete curve in the 6-regular-polygon decomposition plane [6]. Figs. 9.4 and 9.5.

Let us consider the 3-regular-polygon case next. If C is a closed curve, then there are five kinds of simple curve points in $IN_C \cup C$. We call a point p on C a CP_i point if p has i adjacent points in $IN_C \cup C$.

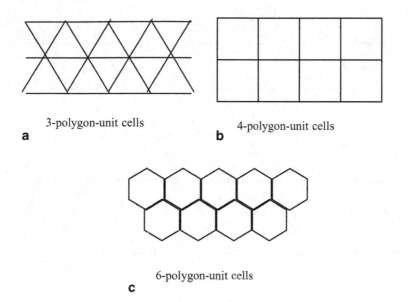

3-polygon-unit cells

a

4-polygon-unit cells

b

6-polygon-unit cells

c

Fig. 9.4 Digital planes made by regular polygon cells

Fig. 9.5 An angle in a regular
polygon

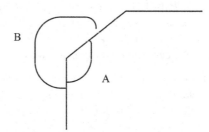

Lemma 9.2 *For regular 3-polygon decomposition, any closed curve must hold the
following property:* $|CP_3| = |CP_5| + 6$.

Proof According to Fig. 9.6, we know $|CP_1| = 0$ and $|CP_i| = 0$ if $i > 6$. In fact,
$CP_2 = 0$ (point A in Fig. 3.7) and $CP_6 = 0$ (point E in Fig. 9.6) because C cannot
contain any 3-regular-polygons. The points in CP_4 are said to be straight line points
such as point C in Fig. 9.6. The points of CP_3 are said to be outward-corner points
such as point B in Fig. 9.6. The points of CP_5 are said to be inward-corner points
such as point D in Fig. 9.6.

   Since the vertices and edges generated by these 3-regular polygons form a planar
graph, we can use Euler's planar graph theorem to find the relationship between CP_3
and CP_5. This theorem says: In a planar graph, if V, E, and F represent the set of
vertices, edges, and faces, respectively, then $|E| = |F| + |V| - 2$. In our case
$|V| = |CP_5| + |CP_3| + |CP_4| + |IN_C|$,
$|E| = (6 \cdot IN_C + 5 \cdot |CP_5| + 4 \cdot |CP_4| + 3 \cdot |CP_3|)/2$,

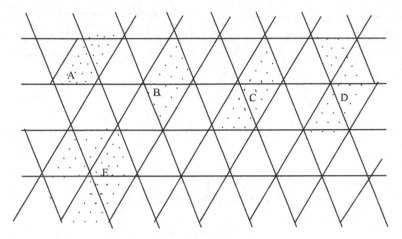

Fig. 9.6 Vertices classification of regular polygon cells in 3-polygon

and
$$|F| = (6 \cdot IN_C + 4 \cdot |CP_5| + 3 \cdot |CP_4| + 2 \cdot |CP_3|)/3 + 1.$$
Therefore,
$$|CP_5| + |CP_3| + |CP_4| + |IN_C| + (6 \cdot IN_C + 4 \cdot |CP_5| + 3 \cdot |CP_4| + 2 \cdot |CP_3|)/3 + 1 - 2$$
$$= (6 \cdot IN_C + 5 \cdot |CP_5| + 4 \cdot |CP_4| + 3 \cdot |CP_3|)/2.$$
So $|CP_3| = |CP_5| + 6.$ □

We can also prove the case for regular 6-polygons. That is

Lemma 9.3 *For regular 6-polygon decomposition, any closed curve must hold the following property: $CP_2 = CP_3 + 6$, $|CP3| \geq 3$.*

See detailed proof in [6].

9.4.3 Topology of Digital Surfaces: An Application of Euler's Theorem

Topology of surfaces is already a well established theory in topology. Any 2D surface can be categorized to be a surface with n handles and m Mobius strips. It means that every surface must be homemorphic to a sphere where we attach n handles (just like the handle of a cup) and m Mobius strips.

Nevertheless, how do we decide that a given set is a surface with n handles and m Mobius strips is a difficult problem. In Chap. 6, we have design an algorithm that can decide if a surface is orientable. In this section, we will present a simple theorem that will be able to help us to decide if a digital surface is simply connected, i.e. there is no handles and Mobius strips. This theorem states that: $|M_3(S)| = 8 + |M_5(S)| + 2|M_6(S)|$ where M_i is the set of surface points that has i neighbors on the surface.

Therefore, in digital space, we could explore more topological invariants in some special cases.

Let p be a point on the digital surface S. p is called the i-point if p is 6-adjacent to i points in $S \cap N_p$. Let $M_3(S)$, $M_4(S)$, $M_5(S)$, and $M_6(S)$ be the sets of 3-points (also called corner points), 4-points, 5-points, and 6-points, respectively. There is no 7-point in a simple surface S [10]. We assume that This surface is a closed surface in the rest of the section.

Lemma 9.4 *(1)* $|V| = |M_3(S)| + |M_4(S)| + |M_5(S)| + |M_6(S)|$;
(2) $|E| = (3|M_3(S)| + 4|M_4(S)| + 5|M_5(S)| + 6|M_6(S)|)/2$;
(3) $|F| = (3|M_3(S)| + 4|M_4(S)| + 5|M_5(S)| + 6|M_6(S)|)/4$.

Proof It is easy to know that $M_0(S)$, $M_1(S)$, $M_2(S)$, and $M_k(S)$, $k \geq 7$ cannot be a surface point. Therefore (1) is true.

Next, we consider each vertex's contribution to the edges E in planar graph G. In $G = (V, E)$, every point in $M_3(S)$ is connected with three edges, every point in $M_4(S)$ is connected with four edges, every point in $M_5(S)$ is connected with five edges, and every point in $M_6(S)$ is connected with six edges. On the other hand, every edge has two end points. So,

$$|E| = (3|M_3(S)| + 4|M_4(S)| + 5|M_5(S)| + 6|M_6(S)|)/2.$$

we have proved case (2).

For case (3), we count the contribution of each vertex to the faces F in G. In $G = (V, E)$, every point in $M_3(S)$ is connected with three faces, every point in $M_4(S)$ is connected with four faces, every point in $M_5(S)$ is connected with five faces, and every point in $M_6(S)$ is connected with six faces. On the other hand, every face has four end points. So,

$$|F| = (3|M_3(S)| + 4|M_4(S)| + 5|M_5(S)| + 6|M_6(S)|)/4.$$

We have completed the proof. □

Theorem 9.2 *The digital form of Euler's formula on surfaces is the following:*

$$|M_3(S)| = 8 + |M_5(S)| + 2|M_6(S)|. \tag{9.6}$$

Proof According to Euler's formula, we have: $|M_3(S)| + |M_4(S)| + |M_5(S)| + |M_6(S)| + (3|M_3(S)| + 4|M_4(S)| + 5|M_5(S)| + 6|M_6(S)|)/4 = (3|M_3(S)| + 4|M_4(S)| + 5|M_5(S)| + 6|M_6(S)|)/2 + 2$. Thus,

$$|M_3(S)| + |M_4(S)| + |M_5(S)| + |M_6(S)| =$$
$$(3|M_3(S)| + 4|M_4(S)| + 5|M_5(S)| + 6|M_6(S)|)/4 + 2.$$

Then, $|M_3(S)| = 8 + |M_5(S)| + 2|M_6(S)|$. □

According to Theorem 9.2, we have:

Theorem 9.3 *Any simple closed surface has at least eight corner surface points.*

Furthermore,

Theorem 9.4 *Any simple closed surface has at least 14 points.*

Proof According to the definition of digital surfaces, a digital surface does not contain any 3-cell (cube). Therefore, any two corner surface points (points in M_3) cannot be adjacent. According to Theorem 9.2, each simple closed surface has at least eight corner points, each of which is connected to three edges that cannot connect with corner surface points. So, there are $24 = (3 \times 8)$ points, some of them are doubly counted, each of which is not a corner surface points. In conclusion, the number of points in a closed surface is:

$$|M_3(S)| + |M_4(S)| + |M_5(S)| + |M_6(S)| =$$
$$8 + |M_5(S)| + 2|M_6(S)| + |M_4(S)| + |M_5(S)| + |M_6(S)| =$$
$$8 + |M_4(S)| + 2|M_5(S)| + 3|M_6(S)|$$

and

$$4|M_4(S)| + 5|M_5(S)| + 6|M_6(S)| \geq 24.$$

According to Theorem 9.2, we consider the following integer programming problem:

$$8 + |M_4(S)| + 2|M_5(S)| + 3|M_6(S)| = min$$
$$4|M_4(S)| + 5|M_5(S)| + 6|M_6(S)| \geq 24;$$

We have $|M_4(S)| = 6$ and $|M_5(S)| = |M_6(S)| = 0$. That is, any simple closed surface has at least 14 points.   □

9.5   Remark

In this chapter, we discuss both the topology of simplicial complexes and finite topology for digital space. In computer graphics, a discrete manifold usually means a meshed object. The simplicial approximation theorem proven by Brouwer is a foundational result. This theorem means that a continuous manifold is somewhat equivalent to a subclass of simplicial complexes.

Computationally, use polygons or triangles for decomposition will not add much time since a polygon can be partitioned into triangles in linear time [4, 13, 16, 27, 29]. However, for data reconstruction purposes, the different triangulation may yield much different result [7].

The homology group part in this chapter will be used in Chap. 14. More detailed discuss can be found in Hatcher's popular book [19].

Fundamental groups for digital spaces were first studied by Khalimsky [22], and then Kong [23]. The path or curves are formed by digital points. The definition is used to simulate the classic fundamental groups in digital space by defining the digital curves and their motions.

Euler's formula for planar graphs [18] plays an important role in this chapter in digital topology. The earlier results using Euler's formula in digital topology can be found in [11]. More detailed results can be found in [6].

Modern geometry and algebra are combined in many ways including information security. For example, the most effective encryption method today is called elliptical curve cryptography that uses groups on elliptical curves to encode and decode information[3]. This group is a finite Abelian group. The Abelian group is particularly important to geometric structures. This geometric curve generates an Abelian group that is a generalization of currently used RSA that uses prime numbers. A problem called the hidden subgroup problem that is related to quantum computing is significant [17]. In particular, decompsition of finite Abelian groups in both regular computers and quantum computers are interesting too [9, 12, 21]. Decomposition of finite Abelian groups is the generalization of Shor's work for prime factorization in quantum computers [12, 28]. This work might relate to fundamental groups and other homotopy-homology groups on the topological structure of a manifold. There might be some future problems in discrete from of elliptical curve cryptography for the approximation of some type of encoding and decoding.

References

1. P. S. Alexandrov, Combinatorial Topology, New York: Dover, 1998.
2. M. A. Armstrong, Basic Topology, rev. ed. New York: Springer-Verlag, 1997.
3. I. Blake, G. Seroussi, and N. Smart, editors, Advances in Elliptic Curve Cryptography, London Mathematical Society 317, Cambridge University Press, 2005.
4. B. Chazelle, Triangulating a Simple Polygon in Linear Time. Discrete and Computational Geometry, Vol 6, 485–524, 1991.
5. L. Chen, Point spaces and raster spaces in digital geometry and topology, Melter, Wu, and Latecki ed, Vision Geometry VII, 1998.
6. L. Chen, Discrete Surfaces and Manifolds. Scientific and Practical Computing, Rockville, 2004
7. L. Chen, *Digital Functions and Data Reconstruction*, 2013. Springer, New York.
8. L. Chen, Determining the number of holes of a 2D digital component is easy, http://arxiv.org/abs/1211.3812, Nov. 2012.
9. L. Chen and B. Fu, Linear and Sublinear Time Algorithms for the Basis of Abelian groups, Theoretical Computer Science archive Volume 412 Issue 32, July, 2011, pp 4110–4122
10. L. Chen and J. Zhang, Classification of simple digital surface points and a global theorem for simple closed surfaces, in R.A. Melter and A.Y. Wu, Vision Geometry II, SPIE Proceedings 2060.
11. L. Chen, H. Cooley and J. Zhang, The equivalence between two definitions of digital surfaces, *Information Sciences*, Vol 115, pp 201–220, 1999.
12. K. H. Cheung and Michele Mosca, Decomposing finite abelian groups, Quantum Inf. Comput., 1 (2001), no. 3, 26–32
13. T. H. Cormen, C.E. Leiserson, and R. L. Rivest (1993), Introduction to Algorithms, MIT Press, Cambridge, MA, 1993.
14. S. Fortune, Voronoi diagrams and Delaunay triangulations, in D.Z. Du and F. K. Hwang ed, Computing in Euclidean Geometry, World Scientific Publishing Co, 1992.

15. Garey, M. R.; Johnson, D. S.; Preparata, F. P.; and Tarjan, R. E. "Triangulating a Simple Polygon." Inform. Process. Lett. 7, 175–179, 1978.
16. J. E. Goodman, J. O'Rourke, Handbook of discrete and computational geometry, CRC Press, Inc., Boca Raton, FL, 1997
17. S. Hallgren, A. Russell, and A. Ta-Shma, The hidden subgroup problem and quantum computation using group representations. SIAM Journal on Computing 32, No.4, 916–934. 2003.
18. F. Harary, *Graph Theory*, Addison-Wesley, Reading, 1972.
19. A. Hatcher, *Algebraic Topology*, Cambridge University Press, 2002.
20. E. Jeff, computational topology, *jeffe@cs.illinois.edu*.
21. G. Karagiorgos and D. Poulakis, Efficient Algorithms for the Basis of Finite Abelian Groups, Discrete Mathematics, Algorithms and Applications, 3(4),(2011) 537–552.
22. E. Khalimsky, Motion, deformation, and homotopy in finite spaces, Proceedings IEEE International Conference on Systems, Man, and Cybernetics (1987), 227–234.
23. T.Y. Kong, A digital fundamental group, Computers and Graphics 13 (1989), 159–166.
24. T.Y. Kong and A. Rosenfeld, Digital topology: introduction and survey, Computer Vision, Graphics and Image Processing, Vol. 48, 1989, 357–393.
25. V. A. Kovalevsky, Finite topology as applied to image analysis, Computer Vision, Graphics and Image Processing, Vol. 46, 1989, pp. 141–161.
26. J.R. Munkres, Elements of Algebraic Topology. New York: Perseus Books Pub., 1993.
27. F. P. Preparata, M. I. Shamos, Computational geometry: an introduction, Springer-Verlag New York, Inc., New York, NY, 1985
28. P. W. Shor, Polynomial-time algorithms for prime factorization and discrete logarithms on a quantum computer, SIAM J. Comput. 26 (1997), no. 5, 1484–1509.
29. R. Tarjan, and van Wyk C. "An Algorithm for Triangulating a Simple Polygon." SIAM J. Computing 17, 143–178, 1988.
30. W. Thurston, Three-Dimensional Geometry and Topology, Volume 1, Princeton University Press, 1997.

Part IV
Geometric Computation and Processing

Chapter 10
Geometric Measurements and Geometric Computing

Abstract In ancient times, the need for measuring land resulted in the development of geometry, much like the need for counting yielded arithmetic. The easiest example is to measure the distance between two points as we discussed in Chap. 3. In this chapter, we cover basic geometric measurements including curve length, surface area, and solid volumes in classical topics of geometry.

The second main topic of this chapter is geometric computing, using algorithms to solve geometric problems in 2D or 3D Euclidean spaces. This area is called computational geometry or algorithmic geometry. We present some basic methods and techniques for the following problems: convex hull, closest pair, and Delaunay and Vonoroi diagrams we discussed in Chap. 3. This chapter covers the basic knowledge for the later chapters of the book.

Keywords Measurement · Metric · Length and area · Lp space · Polygon · Convex hull · Delaunay triangulation · Voronoi decomposition · Algorithms

10.1 Overview of Measurements in Different Spaces

In Euclidean space, the most fundamental theorem is the Pythagorean theorem: in a right triangle, the square of the hypotenuse equals the summation of the squares of both legs.

As we know from Chap. 3, the concept of metric was created because people wanted to measure the distance between two points. Such a simple concept may yield different answers in different spaces, for instance in Euclidean space and spheres.

The main difference between geometry and topology is the measurements. Strictly speaking, geometry must have a metric but topology does not have a metric and does not focus on the distance measure. However, the common ground is that they both deal with continuous neighborhoods and differentiability of spaces.

Measurements in discrete geometry can be similar to those in a continuous space if we embed discrete objects into a Euclidean space. Algorithmic geometry is dealing with this case. We discuss this further in the second half of this chapter.

However, in digital geometry, as we have seen in Chaps. 4 and 6, we are mainly interested two metrics: direct and indirect adjacency. The distance measure we mainly use is graph-based distance (length of the path), not Euclidean distance in many

© Springer International Publishing Switzerland 2014   171
L. M. Chen, *Digital and Discrete Geometry*, DOI 10.1007/978-3-319-12099-7_10

instances. This is because digital geometry was invented for image processing and computer vision in large "portions."

It is obvious that we need to also consider the Euclidean length and area of a geometric object in digital space (as embedded into Euclidean space). This way deals with approximation, especially of the boundary of the object in Euclidean space. For example, we can consider boundary length as a factor when we do image segmentation (See Chap. 4 for the definition), which uses the variational method in continuous spaces. See Chap. 12.

This chapter provides the necessary background of the measurements in Euclidean space and its computation [21, 22]. We also provide some advanced knowledge of geometric measurements for later chapters, especially the measurement of curvatures [8, 15, 16].

10.2   Basic Measurements in Euclidean Space

The basic measurements for geometric shapes are length, area and volume. Length refers to the distance between two points or could also refer to the longest dimension of an object. Area is the quantity of the number of unit squares, a value that can be fractional. Volume is the number of unit-cubes.

10.2.1   Length, Area, and Volume

In analytic geometry, projecting a line-segment from point $q_1 = (x_1, y_1)$ to $q_2 = (x_2, y_2)$ on the x and y-axis will yield two intervals (line-segments on the axes), that are two legs of a triangle. The distance is also called the length of the line-segment,

Length or Distance   In 2D Euclidean space, distance from $q_1 = (x_1, y_1)$ to $q_2 = (x_2, y_2)$ is

$$d(q_1, q_2) = \sqrt{(x_2 - x_1)^2 + (y_2 - y_1)^2}$$

This is the first metric of geometry. It satisfies all conditions of the definition of metrics discussed in Chap. 3. The proof of the formula was given in various ways in Euclidean space. This Euclidean metric is not valid for spheres.

For a 2D polygonal shape, the perimeter is the total length of the edges. Therefore, for a rectangle with length a and width b, the perimeter is $2(a + b)$.

For a circle, the circumference C is $\pi \cdot$ d where d is the diameter of the circle. Even in ancient times people already found that $\frac{C}{d}$ is a constant so they called the ratio of the circumference π.

Area   The formula for the area of a triangle can be derived from the formula for the area of a rectangle: $A = a \times b$. The area of a triangle is $1/2a \times h$.

The area of a circle is $\pi \cdot r^2$, where r is the radius and $r = \text{d}/2$.

Volume In terms of volume, the volume of a cube is a^3 where a is the length of an edge. The formula for the volume of a sphere is $4/3 \cdot \pi r^3$, where r is the radius.

The general method for computing length, area, and volume for any given curve, surface, and solid object requires knowledge of differentiation and integration from calculus. Here we just present a brief description.

10.2.2   Curve Length

For an arbitrary curve in E_2, 2D Euclidean space, $s(t) = (x(t), y(t))$, where $t \in [0, 1]$. It is not difficult to know that at time t, $ds = \Delta s = s(t+\Delta t) - s(t) = \sqrt{(\Delta x^2 + \Delta y^2)}$. Therefore,

$$(ds)^2 = (dx)^2 + (dy)^2. \tag{10.1}$$

The length of the curve (arc length) is the summation of all small changes on s, Δs or ds. Therefore, the formula would be

$$\int_0^1 ds = \int_0^1 \frac{ds}{dt} dt = \int_0^1 \sqrt{\left(\frac{dx}{dt}\right)^2 + \left(\frac{dy}{dt}\right)^2} \, dt$$

In fact, the curve length (10.1) can be generalized by

$$ds^2 = E \, du^2 + 2F \, dudv + G \, dv^2$$

If the value is positive, such a curve is called the Riemann curve (or Riemann manifolds). In fact, Euclidean space is good enough for the study of any type of Riemann manifold since every Ricmann manifold can be isometrically embedded into Euclidean space in higher dimensions.

10.2.3   Surface Area

If the surface is parameterized using u and v, then the general formula for surface area $S = S(u, v)$ is the summation of all small rectangles on the surface with length ΔS_u and width ΔS_v, i.e. $T_u = \frac{dS}{du}$ and $T_v = \frac{dS}{dv}$. Therefore, the formula applies for where T_u and T_v are tangent vectors and $a \times b$ is the cross product.

$$Area = \int_S |T_u \times T_v| dudv. \tag{10.2}$$

If $z = f(x, y)$ is defined over a region D, then

$$Area = \int \int_{(D)} \sqrt{((\partial z)/(\partial x))^2 + ((\partial z)/(\partial y))^2 + 1} \, dxdy.$$

10.2.4   Solid Volume

Most of the time, we are only interested in the volume of 3D objects in 3D. In calculus, let $f(x, y) \geq 0$ be a continuous function. The volume of the solid that lies above a region R and below the $z = f(x, y)$ is given by the following formula,

$$Volume = \int \int_R f(x, y) dx dy$$

In fact, if we let E be a solid, then its solid volume is

$$Volume = \int \int \int_E dx dy dz. \tag{10.3}$$

Such volume calculation is simple. However, a related formula called Greens theorem or divergence theorem can be used to calculate the surface area S through the volume that it bounds.

$$\int_E \nabla \cdot \mathbf{F} d\Omega = \oint_S \mathbf{F} \cdot \hat{n} dS.$$

$\hat{n}$ is normal of dS.

10.2.5   Curvatures

The curvature of a curve measures the degree of curviness. It is defined as the magnitude of the acceleration of the curve.

$$\kappa(t) = \|s''(t)\|. \tag{10.4}$$

The radius of the curvature $R(t)$ is the reciprocal of the radius of curvature:

$$R(t) = \frac{1}{\kappa(t)}.$$

Therefore, for a plane curve $s(t) = (x(t), y(t))$, the curvature is

$$\kappa = \frac{|x'y'' - y'x''|}{(x'^2 + y'^2)^{3/2}} \tag{10.5}$$

The curvature for surfaces are measured by two principal curves with a local maximum curvature and local minimum curvature. We have two curvatures κ_1 and κ_2, where $\kappa_1 \kappa_2$ is called the Gaussian curvature and $(\kappa_1 + \kappa_2)/2$ is called the mean curvature.

A fundamental theorem in differential geometry, called the Gauss-Bonnet theorem, related to Gaussian curvatures states: The integral of the Gaussian curvature over the whole (closed) surface is 2π times the Euler characteristic.

The mean curvature is directly related to the minimum surfaces where the mean curvature is zero at every point. More discussion of the uses of curvatures can be found in Chaps. 12 and 14, also in [16].

10.3   Discrete Measurements for Polygons and Polyhedrons

Measurements for polygons are essential to engineering since most real world applications can be approximated by piecewise linear decompositions, i.e. by polygon type of shapes.

Length of a polygon can be measured as the summation of all edges.

10.3.1   Polygon

Area of a polygon can also be calculated by a simple formula: Assume (x_1, y_1), $\cdots, (x_n, y_n)$ are n vertices in polygon P. Its area is

$$A = \frac{1}{2}(\sum_{i=1}^{n}(x_i y_{(i+1)mod(n)} - x_{(i+1)mod(n)} y_i). \tag{10.6}$$

Proving this formula is simple since we already know the formula for calculating the area of a triangle. We split P into a triangle and another smaller polygon using just one line segment. Using mathematical induction, we get two polygons that meet the formula so the summation of the two formulae will be equal to the above formula since the direction of the new line segment is opposite of the two polygons. They will cancel each other in the final formula.

10.3.2   Polyhedron

Volume of an (Orientable) Polyhedron

Let E be the region enclosed by a polyhedron, The faces of a polyhedron are planar and each face is a polygon. The faces have piecewise constant normals. Therefore,

$$Volume = \frac{1}{3}\sum_{face_i} \mathbf{x}_i \cdot \hat{n}_i A_i \tag{10.7}$$

where $\mathbf{x}_i$ is an point (any point) on the $face_i$, $\hat{n}_i$ is the normal vector, and A_i is area of the face.

Since enumerating the faces may not be a simple task, we may need a special algorithm to count this. In later chapters we will discuss data structures for this type of computation.

Fig. 10.1 The metric
differences when p changes
in L_p spaces

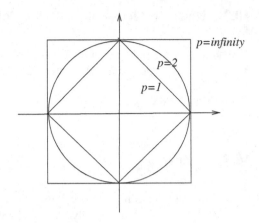

10.4   Metrics in L_p Spaces and Digital Spaces*

We have defined p-norm in Chap. 3. In fact, this is related to a class of important
spaces called L_p spaces in functional analysis, and the metrics are closely related to
direct adjacency and indirect adjacency in digital space.

In general, we define a metric for (R^2, d_p) as

$$d_p(q_1, q_2) = (|x_2 - x_1|^p + |y_2 - y_1|^p))^{\frac{1}{p}} \qquad (10.8)$$

(R^2, d_p) is called a L_p space. The p-norm $\|X\|_p = d(X, (0, 0))$ in Chap. 3.

For $R^n, X = (x_1, \cdots, x_n)$, when $p \to \infty$, $\|X\|_p \to \max\{|x_1|, |x_2|, \cdots, |x_n|\}$.
This is equivalent to:

$$\|X\|_\infty = \max\{|x_1|, |x_2|, \cdots, |x_n|\}$$

This is why we say that indirect adjacency in 2D has the metric of d_∞.

Figure 10.1 shows the characteristics when p changes around the unit circle of
each p.

The following definition and example give some essential information while we
choose a metric for a space.

Definition 10.1 A metric d on X is said to be intrinsic if for any two points x and y
in X, we can find a curve whose length is arbitrarily close to $d(x, y)$. In other words,
there is a curve that links these two points with (almost) the same length.

Example 10.1 We can examine that all d_p, $p = 1, 2, \cdot, \infty$ are intrinsic metrics.
However, for digital space, Σ_m, d_2 is not an intrinsic metric. In other words, d_1 and
d_∞ are two intrinsic metrics in the digital plane. d_2 is not an intrinsic metric in digital
space since we cannot usually find a path with the distance between two random
points x and y. For instance, $x = (i, j)$ and $y = (i + 1, j + 2)$ in (Σ_2, d_2). This
is because the shortest path from x to y has a distance of 3 (direct adjacency) or
2 (indirect adjacency or $\sqrt{(2)} + 1$ if embedded into E_2), but $d_2(x, y) = \sqrt{(2)^2 + 1}$.
Again, d_2 is an intrinsic metric if we embed a digital space into Euclidean space.

10.4.1 Geometric Measures*

Also, note that the concept of measure is different from metric. Measure is a concept usually used in probability. It is defined on a set S, where S_i is a subset of S and U is a collection of S_i.

A measure is a function μ from $\{S_i\}$ to a real number with the following properties:

(1) For all $S_i \in U$, $\mu(S_i) \geq 0$.
(2) $\mu(\emptyset) = 0$.
(3) If U is a countable set, $U = \{S_i | i \in N\}$, and U is a pairwise disjoint set, i.e. $S_i \cap S_j = \emptyset$ in U, then

$$\mu \left(\bigcup_{i \in N} S_i \right) = \sum_{i \in N} \mu(S_i).$$

To understand what a measure is, we can think about length, area, and volumes. Those are measures. Therefore, a measure defined on a set A is a nonnegative real function. The measure of a empty set would be zero. And, if $U = A \cup B$ and A and B are disjoint, then $\mu(U) = \sum \mu(A) + \mu(B)$.

Probability is also a measure. For a geometric entity, defining a measure is important.

10.5 Algorithmic Geometry

Algorithmic geometry is usually called computational geometry. It uses the algorithm technique to process the geometric problem. To find efficient and fast algorithms is the goal of this research area. Note that for the general public, computational geometry may have several different meanings. In this book, we use the term algorithmic geometry in the discussion.

10.5.1 Convex Hulls of Discrete Points

Given a set of discrete points on a plane, find the minimum convex that contains all the points. This convex is called the convex hull of the point set.

The easiest way to understand the convex hull is to use a rubber band to bind a bunch of needles that are attached to a wood plane. We can also use a long string to wrap around these points. See Fig. 10.2.

This idea can be implemented as a simple algorithm called the gift-wrapping method.

Problem of the Convex Hull Given a set of points in the plane, find the minimum convex that holds all the points. See Fig. 10.2b.

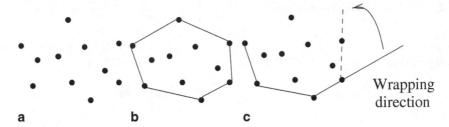

Fig. 10.2 The convex hull of points: **a** A collection of points, **b** the convex hull of these points, and **c** the gift-wrapping algorithm

Algorithm 10.1 The gift-wrapping algorithm for the convex hull. This algorithm is similar to wrap a gift.

Step 1   Find the left most point, which is also the lowest point. This point must be the corner point of the convex), denoted P.

Step 2   Draw lines from P to all the other points. Find the line with the biggest angle with respect to y-axis. This line must be on the edge of the convex.

Step 3   Take the point at the end of the line that is not P, and use this point as the new starting point P.

Step 4   If P was selected before, stop the algorithm. Otherwise, go back to Step 2.

We have presented the meaning of the algorithm in Fig. 10.2c.

Theorem 10.1 *The time complexity of the gift-wrapping algorithm is $O(nh)$, where n is the number of points in the set and h is the number of points on the boundary of the convex hull.*

Proof Algorithm analysis: Step 2 uses $O(n)$ time for calculations and comparisons. There are h points on the corners (the boundary of the convex). Therefore, the time complexity is $O(nh)$. Since h can be as many as n, this algorithm has $O(n^2)$ time complexity in the worst case scenario.   □

Another popular algorithm for convex hulls is called the Graham's scan algorithm [4]. This algorithm first uses a fast sorting algorithm to sort the angles to a reference point. Then, find the corner points of the convex hull along with the sorted order.

Algorithm 10.2 Graham's scan algorithm for the convex hull.

Step 1   Find the lowest point, which is also the rightmost point denoted by P.

Step 2   Draw dotted lines to all other points from point P. This takes $O(n)$ time.

Step 3   Sort the angles of all dotted lines in ascending order, which takes $O(n \log n)$ time if we use the merge-sort algorithm.

Step 4   Draw an edge from P to the point that is the end point of the dotted line with the smallest angle (the edge will be on the boundary of the convex hull). The new point will be set to P which uses $O(1)$ time.

Step 5   Start at P and draw the line to the end point of the next line in the sorting sequence that uses $O(1)$ time.

Step 6   If the new corner angle that is smaller than 180° (the curve would be concave
         otherwise), then back track and skip the point that makes the angle smaller
         than 180°. Continue until each point has been selected.

Theorem 10.2   *The time complexity of Graham's scan algorithm is $O(n \log n)$.*

Proof   We just give a brief algorithm analysis here. The Step 2 uses $O(n)$ time. The
Step 3 uses $O(n \log n)$ time since we can use a merge-sort algorithm. The Step 4 uses
$O(1)$ time, and the Step 5 also uses $O(1)$ time. Finally, the Step 6 uses $O(n)$ time.

When we skip points for concave cases, we do not repeat the points skipped each
time. Therefore, the total time of the algorithm is $O(n \log n)$.   □

The optimum algorithm for convex hull finding is $O(n \log h)$ that is slightly better
than Graham's scan algorithm. See [17].

10.5.2   Algorithms for Delaunay Triangulations and Voronoi Diagrams

Given n points (called sites) on a plane, for a new point x, find the closest site to x. This
problem is called the nearest neighbor problem, one of the most popular problems
in the real world (this problem predates the establishment of artificial intelligence).
Here is the example: An elementary school student transfers to the Washington DC
area from New York, and his parents want to find a new school for him. The best guess
for a school is whichever is closest to their new home. The second most important
factor is a similar problem in mathematics to finding the nearest grocery store.

This problem is a variation of the Voronoi diagram problem that involves partitioning a rectangle into smaller regions such that each site is contained within a
region. Any point in the region is closer to the site in its own region. This type of
region is called the Voronoi region.

Therefore, if we have a new point x and try to find the closest site to the new
point, we only need to decide which Voronoi region contains x.

In 1986, Fortune found an algorithm for Voronoi diagrams with time complexity
$O(n \log n)$. This is an optimal algorithm. However, it is difficult to implement [10].

Here, we introduce a technique for constructing the dual diagram of the Voronoi
diagram—Delaunay triangulation. There is a relatively simple and fast algorithm called the Bowyer-Watson algorithm to solve the Delaunay triangulation
algorithmically.

10.5.2.1   Basic Definitions and Symbols

Let P_i, $i = 1, \cdots, n$ be n points, called sites. A partitioned area, called a Voronoi
region, is bounded by edges and vertices. The vertex of the region is called the
Voronoi point. An edge of the Voronoi region must have the property that each point

on the edge must be equal distance to the two sites. A Voronoi point must have the property of being equidistant to the three sites. This means there must be a circle containing these three sites centered at each Voronoi point, the circumcircles of the triangle. Such a triangle is the Delaunay triangle.

Another definition of Delaunay triangulation is as follows: no circumcircle of any triangle contains a site. The following Bowyer-Watson algorithm is designed based on this fact.

The Bowyer-Watson algorithm is one of the most commonly used algorithms for this problem. This method can be used for computing the Delaunay triangulation of a finite set of points in any number of dimensions. After the Delaunay triangulation is completed, we can obtain a Voronoi diagram of these points by getting the dual graph of the Delaunay triangles.

Algorithm 10.3 The Bowyer-Watson algorithm is incremental in that the algorithm works by adding one point at a time to a valid Delaunay triangulation of a subset of the desired points. Then, it works in the new subset, adding points to reconstruct the new Delaunay triangulation.

Step 1   Start with three points of the set. We make the first triangle by linking three points with three edges.

Step 2   Insert a new point P. Draw the circumcircle of each existing triangle. If any of those circumcircles contains the new point P, the triangle will be marked as invalid.

Step 3   Remove all invalid triangles. This process will leave a convex polygon hole that contains the new point.

Step 4   Link the new point to each corner point of the convex polygon to form a new triangulation.

The time complexity of this algorithm is $O(n\sqrt{(n)})$. A simple procedure can be applied to get the Voronoi diagram from the Delaunay triangulation. The key of the procedure is to link the centers of the circumcircles such that two corresponding triangles share an edge. In other words, these new linking lines for the centers of the two corresponding circumcircles form a Voronoi diagram. For more details of this algorithm, see [1, 23, 24].

Another popular algorithm for this problem is called the Fortune's Algorithm. It is for Voronoi Diagrams. Fortune's algorithm is based on the fact that in a parabolic curve, the distance from the focus point to the curve point is equal to the distance from the curve point to the reference line, or the directrix line. On the other hand, the distance from the focus point to the directrix line determines the parabolic curve uniquely.

Therefore, if two sites are treated as two focus points and they share a directrix line, then the intersection point of the two parabolic curves will be equidistant to the two sites. Therefore, such an intersection point must be on the sharing edge of the two Voronoi regions.

Fortune's algorithm and a divide-and-conquer based algorithm are both in $O(n \log n)$. The detailed steps of these algorithms can be found in [10, 11, 17].

10.5.3 Closest Pair Problem

The problem of computing the closest pair of a collection of points in 2D plane is a classical problem in geometric computing. This problem states: Given n points in a plane or in R^m, find a pair of points that are the minimum distance apart [4]. We can calculate the distance for every pair to get the one with the minimum distance. However, this algorithm will take $O(n^2)$ time since we have total of $O(n(n-1)/2)$ pairs.

This problem has an $O(n \log n)$ time algorithm when we use the divide-and-conquer method for algorithm design.

Algorithm 10.4 The divide-and-conquer algorithm for the closest pair problem.

Step 1   Sort the points according to their values on the x-axis.

Step 2   Split the set of points into two (almost) equal-sized subsets. The splitting line at the x-axis is denoted by $x = x_{mid}$.

Step 3   Solve the problem in the left and right subsets. We have two minimum distances D_{left} and D_{right}.

Step 4   Let $d = \min\{D_{left}, D_{right}\}$. Make two lines on both sides of line $x = x_{mid}$ with an offset of d. The closest pair will be in two strips or only in one of either the left or right set.

Step 5   Split the strips into $d \times d$ squares. Select a point in a square in the left strip. Check all the points in the three other squares in the right strip. This will take constant time since each square only contains one or at most three points (otherwise, d is not smaller or equal to D_{right}). Checking all the points in the left strip will require $O(n)$ time. Therefore, we will have the minimum distance.

Theorem 10.3 *The time complexity for the closest pair problem is $O(n \log n)$.*

10.6   Remarks

Geometric measurement is based on the the metric that is a distance measure. Even though Euclid's Elements had the original contribution [7], but basics of measurement is now based on calculus [21]. The measurement on surfaces are even more complex, we will discuss it in Chap. 13 where we deal with differential geometry [13]. Geometric measurement is important to computer graphics [9].

Geometric computing deals with the fast method of obtaining measurements. It is mainly covered in computational geometry. In this chapter, we reviewed some algorithmic technologies that will be used in this book [6, 17].

Measurements in discrete geometry and digital geometry are different. Digital geometry for image processing and computer vision sometime requires different form of length and area computations. Digital geometry may not only use Euclidean metric [2, 12].

Another distance measurement called the Hausdorff distance could be useful too. Let S and S' be two shapes of a metric space. Their Hausdorff distance is the measure of the longest distance they can move to (not the shortest pair).

$$d_{\mathrm{H}}(S, S') = \max\{ \sup_{x \in S} \inf_{y \in S'} d(x, y), \ \sup_{y \in S'} \inf_{x \in S} d(x, y) \},$$

Hausdorff distance is the deformation distance [5, 11].

The main difference between geometry and topology is in the measurements. Strict speaking, geometry requires a metric. In other words, it must consider the distance between two points. On the other hand, topology does not focus on distance measure. However, the common ground is that they are all dealing with neighborhoods, continuity, and differentiability.

In next chapter, we will discuss numerical solutions of geometric problems [3, 19]. The combined methods today are important to data sciences. We will discuss them in Chap. 12. Some machine learning, cognitive sciences, and image processing problems are concerned [14, 18]. Some concepts in discrete mathematics of this chapter can be found in [20].

References

1. A. Bowyer, Computing Dirichlet tessellations, The Computer Journal, 24(2):162–166. 1981.
2. L. Chen, *Discrete Surfaces and Manifolds: A theory of digital-discrete geometry and topology*, 2004. SP Computing.
3. L. Chen, Digital Functions and Data Reconstruction, Springer, NY, 2013.
4. T. H. Cormen, C.E. Leiserson, and R. L. Rivest, Introduction to Algorithms, MIT Press, 1993.
5. H.S.M. Coxeter, Introduction to geometry, John Wiley, 1961.
6. de Berg, M., VAN KREVELD M., OVERMARS, M., and SCHWARZKOPF, O. Computational Geometry: Algorithms and Applications. Springer-Verlag, New York, 1997.
7. Euclid's Elements, Green Lion Press, 2002.
8. H. Federer, Geometric Measure Theory, Springer-Verlag, New York, 1983.
9. James D. Foley, Andries Van Dam, Steven K. Feiner and John F. Hughes, Computer Graphics: Principles and Practice. Addison-Wesley. 1995.
10. S. Fortune, Voronoi diagrams and Delaunay triangulations, in D.Z. Du and F. K. Hwang ed, Computing in Euclidean Geometry, World Scientific Publishing Co, 1992.
11. J.E. Goodman, J. O'Rourke, Handbook of Discrete and Computational Geometry, CRC, 1997.
12. R. Klette and A. Rosenfeld, Digital Geometry, Geometric Methods for Digital Picture Analysis, series in computer graphics and geometric modeling. Morgan Kaufmann, 2004.
13. E. Kreyszig, Differential Geometry, University of Toronto Press, 1959.
14. M. Minsky and S. Papert, Perceptrons: An Introduction to Computational Geometry, The MIT Press, Cambridge MA, 1969.
15. F. Morgan, Geometric Measure Theory, Acad. Press, INC, 1987.
16. J.-M. Morvan, Generalized Curvatures. Geometry and Computing, Vol 2. Springer Verlag, 2008.
17. D. Mount, Lecture notes on Computational Geometry, University of Maryland, 2002.
18. T. Pavlidis, Algorithms for Graphics and Image Processing, Computer Science Press, Rockville, MD, 1982.
19. W. H. Press, et al. *Numerical Recipes in C: The Art of Scientific Computing*, 2nd Ed., Cambridge Univ Press, 1993.

20. K. H. Rosen, Discrete Mathematics and Its Applications, McGraw-Hill Higher Education, Jan 2007.
21. J. Stewart, Calculus, Brooks/Cole Publishing Company, Pacific Grove, CA, 4th ed, 1999.
22. G. B. Thomas, Jr., Calculus and Analytic Geometry, 4th ed. Addison-Wesley, Reading, Mass., 1969.
23. D. F. Watson (1981). Computing the n-dimensional tessellation with application to Voronoi polytopes, The Computer Journal, 24(2):167–172.
24. H. Zimmer, Voronoi and Delaunay Techniques, lecture notes, Computer Sciences VIII, RWTH Aachen, 30 July 2005.

Chapter 11
Digital Functions, Data Reconstruction, and Numerical Geometry

Abstract This chapter contains three parts. First, we discuss data reconstruction including curve and surface fitting. Second, we cover principal component analysis, one of the most important geometric data analysis methods. Third, we present mathematical transformations for data analysis. This chapter is highly related to concurrent data sciences from theoretical perspectives. We focus on the practical methods of geometric data processing in the next chapter.

Data fitting and reconstruction are the inverse treatments of decomposition. Based on discretely sampled data points, we want to build a continuous curve, surface, or solid data volume. Even though numerical methods such as Bezier polynomials and B-Splines are the most popular in computer graphics, we first introduce discrete functions and its applications to digital function interpolation here due to the nature of this book. In the section involving principal component analysis, we briefly discuss the principle in statistics and its solution using linear algebra. Lastly, we introduce the three most important mathematical transforms including the Fourier transform, Radon transform, and wavelet transform. These transforms are fundamental to digital image processing including analysis and compression.

Keywords Data reconstruction · Fitting · Bezier polynomial · B-spline · Gradually varied function · Principle component · Mathematical transformation

11.1 Digital Functions and Data Interpolation

In this section, we introduce a discrete surface reconstruction method called gradually varied surface fitting. In 1986, Rosenfeld introduced a concept called the "continuous" function in digital space [15]. A digital function means that its values are integers on digital points. The word "continuous" allows only small changes in a neighborhood. In other words, if x and y are two adjacent points in a digital space, then $|f(x) - f(y)| \le 1$.

A gradually varied function, introduced by Chen in 1989 [1], is more general than a digital continuous function. A gradually varied function is a function from a digital space Σ to $\{A_1, A_2, \cdots, A_m\}$ where $A_1 < A_2 < \cdots < A_m$ and A_i are rational or real numbers [1]. This function possesses the following property: If x and y are two adjacent points in Σ, assume $f(x) = A_i$, then $f(y) = A_i$, $f(x) = A_{i+1}$, or A_{i-1}.

© Springer International Publishing Switzerland 2014
L. M. Chen, *Digital and Discrete Geometry*, DOI 10.1007/978-3-319-12099-7_11

11.1.1 Gradually Varied Functions

The gradually varied function is used to interpolate a discrete surface when sample points satisfy certain conditions, which we present later. We now start to define a gradually varied function.

Assume that $A_1, A_2, \cdots, A_m$ are m rational or real numbers where $A_1 < A_2 < \cdots < A_m$.

Definition 11.1 Assume $f : \Sigma_2 \to \{A_1, A_2, \cdots, A_m\}$ is a function. For two points $p, q \in \Sigma_2$, if $f(p) = A_i$ and $f(q) = A_j$, then the *level-difference* between $f(p)$ and $f(q)$ is $|i - j|$.

Definition 11.2 Let p, q be two adjacent points in Σ_2. f is said to be gradually varied on p and q if $f(p) = A_i$ implies $f(q) = A_{i-1}, A_i$, or A_{i+1}. f is said to be gradually varied if f is gradually varied on any pair of adjacent points p, q in Σ_2.

In general, we can define the above concept of gradual variation on a graph $G = (V, E)$ [2, 3]. For simplicity, we only discuss the case on Σ_2 here.

The problem of gradually varied interpolation is: Let D be a connected subset in Σ_2 and $J \subset D$. If given $f_J : J \to \{A_1, A_2, ..., A_m\}$, is there an extension of f_J, $f_D : D \to \{A_1, A_2, ..., A_m\}$ such that for all $p \in J$, $f_J(p) = f_D(p)$?

The following theorem is for the necessary and sufficient conditions of the existence of gradually varied function. It was proven by Chen in 1989 [1, 3].

Theorem 11.1 *There exists the gradually varied interpolation if and only if for any two points p and q in J, the length of the shortest path between p and q in D is not less than the level-difference between $f(p)$ and $f(q)$.*

Proof We give a constructive proof here. The basic idea of the construction is to assign a value to a point q that has not been assigned a value, but has a neighbor p, which is a sample point or has already been assigned a value.

In order to make the proof clear, we define $LD(p, p')$ as the level difference between $f(p)$ and $f(p')$: Let $f(p) = A_i$ and $f(p') = A_j$, $LD(p, p') = |j - j|$.

As usual, $d(p, p')$ denotes the length of the shortest path between p and p' in D.

(1) First, we prove the necessary condition. Suppose f is the gradually varied function on D, then f is gradually varied on every path in D. The path may be the shortest path between two points p and p' in J. Hence, the length of the shortest path is not less than the difference between the gray level of p and p', i.e. $d(p, p') \geq LD(p, p')$.

(2) Second, we prove the sufficient condition. Suppose we have $f_J : J \to \{A_1, A_2, ..., A_m\}$ and for all $p, p' \in J$, $d(p, p') \geq LD(p, p')$ (in D).

First let $f_D(p) = f_J(p)$ if $p \in J$ and $f_D(p') = \theta$ if $p' \in D - J$. Define

$$D_0 = \{p | f_D(p) \neq \theta, p \in D\}$$

Now, $D_0 = J$.

The constructive proof is the following: If $D_0 \neq D$, we can find a vertex (or point) $r \in D_0$ so that r has an adjacent point x not in D_0, i.e. $x \in D - D_0$. We can assume $f_D(r) = A_i$.

Then, let $f_D(x) = f_D(r) = A_i$, and denote

$$m(x) = \{p \,|\, f_D(p) < A_i, p \in D_0\};$$

$$M(x) = \{p \,|\, f_D(p) > A_i, p \in D_0\}.$$

There will be three cases:

(i) If there is a $p \in m(x)$, such that $d(x, p) < LD(x, p)$, we know $d(r, p) \leq d(r, x) + d(x, p)$ and $d(r, x) = 1$ (r and x are adjacent points). Therefore, $d(r, p) \leq 1 + d(x, p)$; we have $d(x, p) \geq d(x, p) - 1$. We also know $d(r, p) \geq LD(r, p)$ since $r, p \in D_0$. Thus, $d(x, p) \geq d(r, p) - 1 \geq LD(r, p) - 1$. In addition, $LD(r, p) = LD(x, p)$ since $f_D(x) = f_D(r) = A_i$. Therefore, $d(x, p) \geq LD(r, p) - 1 \geq LD(x, p) - 1$.

According to the assumption $d(x, p) < LD(x, p)$, so $d(x, p) = LD(x, p) - 1$. For any $q \in M(x)$, we have $f_D(p) < f_D(x) < f_D(q)$, and therefore, $LD(p, q) = LD(p, x) + LD(x, q)$. Again, $d(p, x) + d(x, q) \geq d(p, q) \geq LD(p, q)$, then

$$d(p, x) + d(x, q) \geq LD(p, x) + LD(x, q) \geq LD(p, x) + 1 + LD(x, q)$$

Hence,

$$d(x, q) \geq 1 + LD(x, q).$$

Therefore, we have proven that if there is a $p \in m(x)$ such that $d(x, p) < LD(x, p)$, then

$$\forall p \in m(x)(d(x, p) \geq LD(x, p) - 1) \text{ and } \forall q \in M(x)(d(x, p) \geq LD(x, q) + 1).$$

Thus, we can modify the value of x, from $f_D(x) = A_i$ to $f_D(x) = A_{i-1}$. Obviously, for the new value at x, we have,

$$d(y, x) \geq LD(y, x), \text{ if } y \in D_0 \,\&\, f_D(y) \neq A_i,$$

and

$$d(y, x) \geq LD(y, x) = 1, \text{ if } y \in D_0 \,\&\, f_D(y) = A_i.$$

(ii) If there exists a $q \in M(x)$ such that $d(x, q) < LD(x, q)$, similar to (i), we reassign $f_D(x) = A_{i+1}$. We obtain $\forall y \in D_0(d(x, y) \geq LD(x, y))$.

(iii) If (i) and (ii) are not satisfied, then $f_D(x) = A_i$ would be required.
(c) Let $D_0 \leftarrow D_0 \cup \{x\}$, then for every pair p and p' in D_0, we have

$$d(p, p') \geq LD(p, p').$$

We repeat processes (a) to (c) until $D_0 = D$. After all the points in $D - J$ are valued, we can see that f_D is gradually varied. This is because if x, y are adjacent, then $d(x, y) = 1 \geq LD(x, y)$. So $LD(x, y)$ must be 0 or 1. That is, f_D is gradually varied on every pair of adjacent points. Therefore, f_D is gradually varied on D. $\square$

We then call the following condition the gradually varied condition : For any two points p and q in J, the length of the shortest path between p and q in D is not less than the level-difference between $f(p)$ and $f(q)$.

11.1.2 The Construction Algorithms of Gradually Varied Functions

In the proof of Theorem 11.1, we have given a constructive procedure for how we assign a new value for an unknown data point assuming $p \in J$ and its neighbor $q \notin J$. First let $f(q) = f(p)$, then check whether or not $J \cup \{q\}$ with the new value satisfies the gradually varied condition. If it does, then we keep the value of $f(q)$. Otherwise, we subtract or add a level to $f(q)$. Repeating the above process will fill in all unknown value points in D.

Example 11.1 Let us look at an example. Assume that we have an array with direct adjacency. We have three sample points and we want to fill the rest of six numbers in the locations marked as "-."

$$A = \begin{pmatrix} - & 2 & - \\ - & - & 4 \\ - & - & 5 \end{pmatrix}$$

We can see that it satisfied the gradually varied condition. According to the algorithm we used in the proof. We let $A_{22} = 2$.

$$\begin{pmatrix} - & 2 & - \\ - & 2 & 4 \\ - & - & 5 \end{pmatrix}$$

But this assignment broken the gradually varied condition. There must be a value in $\{1, 3\}$ that satisfies the gradually varied condition. We have the following matrix.

$$\begin{pmatrix} - & 2 & - \\ - & 3 & 4 \\ - & - & 5 \end{pmatrix}$$

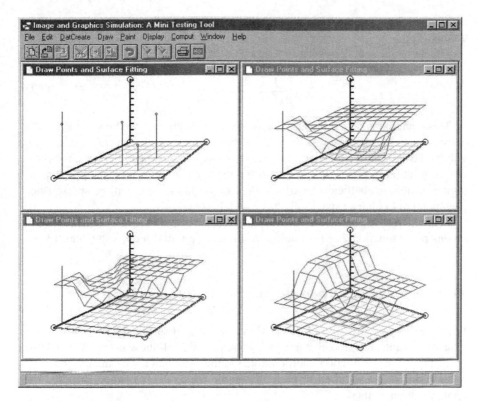

Fig. 11.1 Fitting using digital functions called gradually varied surface fitting

This new array satisfy the gradually varied condition. We can repeat the above procedure to fill all numbers.

$$
\begin{pmatrix} 2 & 2 & - \\ - & 3 & 4 \\ - & - & 5 \end{pmatrix}
\begin{pmatrix} 2 & 2 & 3 \\ - & 3 & 4 \\ - & - & 5 \end{pmatrix}
\begin{pmatrix} 2 & 2 & 3 \\ 2 & 3 & 4 \\ 2 & 4 & 5 \end{pmatrix}
$$

This case study showed the easiness of this method, we also implemented procedure in program C++ code (Fig. 11.1). For detailed proofs and time complexities, refer to [2, 3].

11.1.3 Other Discrete Surface Reconstruction

Given a boundary function f of a region, if f is continuous, then there is a smooth function for the region. This problem is called the Dirichlet problem. A classic result is that the solution of the Dirichlet problem is a harmonic function and that result is unique.

The harmonic function is the function $f(x, y)$ that has the property

$$\frac{\partial^2 f}{\partial x^2} + \frac{\partial^2 f}{\partial y^2} = 0. \tag{11.1}$$

This equation is called the Laplacian equation. There are two ways to solve the above equation discretely [3]. First, the solution that satisfies the value of a point will be the numerical average of its neighbors. We can design an iterated process to get the solution. Second, we can establish a linear equation to solve the problem. In digital plane, the coefficient matrix, called, Laplacian matrix, will be sparse since each point in Euclidean space only has four adjacent points.

In addition to use harmonic equations, we can also apply the minimum surface to reconstruct a function. For the minimum surface, we will discuss it in Chap. 13.

11.2   Numerical Curve and Surface Fitting

In this section, we give a brief description of curve and surface Fitting. Given a set of sample points, if a smooth curve that passes through all these points. We call this type of curve reconstruction interpolation. On the other hand, we can also construct a curve that is not required to pass all sample points. This method is called the approximation method.

11.2.1   Curve Interpolation and Approximation

In theory, Lagrange interpolations is the simplest method for curve interpolation. This method uses a polynomial called the Lagrange polynomial to fit all given guiding points. In other words, Lagrange interpolation provides a general method for the existence of a smooth function that always exists for any number of guiding points.

For a set of $m + 1$ data points $(x_0, y_0), \ldots, (x_i, y_i), \ldots, (x_m, y_m)$ where $x_0 < x_1 < \ldots, < x_m$, the Lagrange interpolation polynomial is given by:

$$P(x) = \sum_{i=0}^{m} y_i \cdot P_i(x) \tag{11.2}$$

where

$$P_i(x) = \prod_{0 \le j \le m;\ i \ne j} \frac{x - x_j}{x_i - x_j} \tag{11.3}$$

is called the Lagrange basis polynomial.

We can examine $P(x_i) = y_i$, which shows that this method interpolates the function exactly.

The disadvantage is that this polynomial has a very high degree, equal to the number of samples n. Such a high degree of smoothness may be unnecessary: any small error might result in a big change in fitted curves.

11.2.1.1   Bezier Polynomials

Bezier polynomials are often used in computer aided geometric design, such as in the automobile industry. The characteristic of the Bezier curve is: (1) It is an approximation method, and (2) the fitted curve is in the convex of the sample points (called control points). Bezier polynomials use the control points to manage the shape of the curve. It usually goes through two end points and does not pass through other control points.

Given a set of $m + 1$ data points $P_0 = (x_0, y_0)$, $P_1 = (x_1, y_1), \dots, P_m = (x_m, y_m)$, Bezier polynomials will be calculated by parametric curve, e.g. $P(t) = (x(t), y(t))$, which is in vector form. For two points, the Bezier polynomial is given by:

$$P(t) = (1 - t)P_0 + t P_1, t \in [0, 1] \tag{11.4}$$

In general, let $P_{\{0...k\}}$ denote the fitted function for control points $P_0, \dots, P_k$.

$$P(t) = P_{\{0...m\}}(t) = (1 - t) \cdot P_{\{0...m-1\}}(t) + t \cdot P_{\{1...m\}}(t), t \in [0, 1] \tag{11.5}$$

We can see that this formula is very similar to the formula for the linear interpolation of two points, but the reference "points" are the two lower order Bezier polynomials. In other words, Bezier polynomial fitting is defined as linear approximation recursively.

For the explicit form, we can expand the above equation to be

$$P(t) = \sum_{i=0}^{n} \binom{n}{i} (1 - t)^{n-i} t^i \cdot P_i, \tag{11.6}$$

where $\binom{n}{i} = \frac{n!}{i! \cdot (n-i)!}$ is a binomial coefficient. A Bezier polynomial is also called a Bernstein polynomial of degree n. It can also be represented using a so called Bernstein basis polynomial of degree n, which is defined as

$$\beta_{i,n}(t) = \binom{n}{i} t^i (1 - t)^{n-i}, \quad i = 0, \dots, n. \tag{11.7}$$

Therefore,

$$P(t) = \sum_{i=0}^{n} \beta_{i,n}(t) \cdot P_i \tag{11.8}$$

11.2.1.2   B-splines*

B-spline is an advanced method for curve approximation. The principle behind this method can be found in [3, 4, 6]. We present the formula for B-spline as follows:

Let the knot sequence be defined $\{t_0, t_1, ..., t_m\}$, where $t_0 \leq t_1 \leq, ..., \leq t_m$, $t_i \in [0, 1]$, and the control points are $P_0, ..., P_n$. Define the degree as $d = m - n - 1$. For instance, $d = 3$ is the cubic spline. (The knots $t_(d + 1), ..., t_(m - d - 1)$ refer to internal knots. The fitted curve among these knots will maintain the properties desired.)

Define the basis functions as

$$B_{i,0}(t) := \begin{cases} 1 & \text{if } t_i \leq t < t_{i+1} \\ 0 & \text{otherwise} \end{cases} , i = 0, \dots, m - 2 \qquad (11.9)$$

$$B_{i,j}(t) := \frac{t - t_i}{t_{i+j} - t_i} B_{i,j-1}(t) + \frac{t_{i+j+1} - t}{t_{i+j+1} - t_{i+1}} B_{i+1,j-1}(t) \qquad (11.10)$$

where $i = 0, \dots, m - j - 2$. $1 \leq j \leq d$ indicates the degree where i is the interval segment index or knots. B-spline is a linear combination of basis B-splines $B_{i,d}$ with coefficients P_i

$$P(t) = \sum_{i=0}^{m-d-2} P_i B_{i,d}(t), t \in [t_d, t_{m-d-1}].$$

We can see that this has a linear interpolation basis where the function is linearly accumulated p times.

The cubic B-spline provides a good estimate of derivatives up to the third order. The cubic B-spline formula can be represented in a matrix format:

$$P_i(t) = \frac{1}{6} \begin{pmatrix} 1 & t & t^2 & t^3 \end{pmatrix} \begin{pmatrix} 1 & 4 & 1 & 0 \\ -3 & 0 & 3 & 0 \\ 3 & -6 & 3 & 0 \\ -1 & 3 & -3 & 1 \end{pmatrix} \begin{pmatrix} f_{i-1} \\ f_i \\ f_{i+1} \\ f_{i+2} \end{pmatrix} \qquad (11.11)$$

for $t \in [0, 1]$, and $i = 2, \cdots, n - 2$. t can either represent the x-axis or y-axis (Fig. 11.2).

11.2.2   Numerical Surface Fitting

At the beginning of this chapter, we discussed the gradually varied method for continuous surface fitting. In fact, another simple way of surface fitting is to use the

Fig. 11.2 Curve fitting using piecewise linear curve a, B-spline curve b, and Bezier curve c. (by L. Chen)

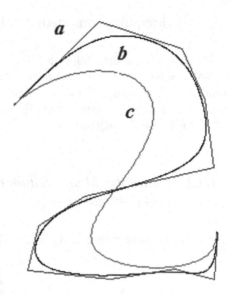

triangulation method. It projects the sample data points onto the XY-plane, makes a triangulation on the domain, and then lifts the points back to the original position. This will generate a triangulated surface in 3D. This method is a special case of meshing of a closed continuous surfaces, which we discussed in Chap. 10.

Mathematically, we can fit a surface based on a second-order surface in 3D space:

$$x^2 + a_1 y^2 + a_2 z^2 + a_3 yz + a_4 zx + a_5 xy + a_6 x + a_7 y + a_8 z + a_9 = 0. \quad (11.12)$$

This equation needs nine sample points to solve for nine coefficients. If there are more than nine sample points, then we need to use the so called least squares method [3] to get a best approximation. However, the best way today is to use the tensor product of two B-spline curves. B-spline surfaces are constructed by two B-spline curves to make a surface-area or patch. It is used in computer graphics and engineering [8, 14].

$$S(u, v) = \sum_{i=0,n} \sum_{j=0,m} B_{i,n}(u) B_{j,m}(v) P_{i,j}, \, u, v \in [0, 1]. \quad (11.13)$$

where $P_{i,j}$ are $(n + 1)(n + 1)$ control points (sample points) and $B_{i,n}(u)$ and $B_{j,m}(v)$ are B-spline basis functions discussed in the above sections.

Some advanced surface reconstruction techniques are also based on tensor products. For instance, non-uniform rational B-splines (NURBS) are very common today in computer graphics in controlling flexible shapes. NURBS are made for geometric design since getting the value of the weights w_{ij} is a huge problem. If we use a uniform rational and the weights w_{ij}'s are the same, then NURBS will be similar to B-spline [4].

11.3   Principal Component Analysis

For a set of sample data points in 2D, the method of principal component analysis (PCA) can be used to find the major direction of the data. It is essential to data processing. The method is also related to regression and other methods in numerical analysis and statistical computing [10]. Even though, there are many ways to explain PCA, the best is through statistics.

11.3.1   Concepts: Mean, Standard Deviation, Variance, and Covariance

For a set of sample points, $x_1, x_2 \cdots, x_n$, the mean is the average value of x_is:

$$\bar{x} = \frac{1}{n} \sum_{i=1}^{n} x_i,$$

so we may use a random variable X to represent this data set.

Variance is the measure of how data points differ from each other in a whole. The sample variance of the data set or the random variable X is defined as

$$Var(X) = \frac{1}{n} \sum_{i=1}^{n} (x_i - \bar{x})^2,$$

The square root of the variance is called standard deviation σ_X, i.e. $Var(X) = \sigma_X^2$. Note that people sometimes use $n - 1$ instead of n in the formula.

For simplicity, let $X = \{x_1, x_2 \cdots, x_n\}$ and $Y = \{x_1, x_2 \cdots, x_n\}$. The covariance of X and Y is defined as

$$Cov(X, Y) = \frac{1}{n} \sum_{i=1}^{n} ((x_i - \bar{x})(y_i - \bar{y}))$$

It is important to note that the correlation of X and Y is defined by

$$Cor(X, Y) = \frac{Cov(X, Y)}{\sigma_X \cdot \sigma_Y}.$$

We can see that if $X = Y$, then $Cov(X, Y) = Var(X) = \sigma_X^2$ and $Cor(X, Y) = 1$.

11.3.2   Covariance Matrix and Principal Components

Given a set of 2D points, $(x_1, y_1), \cdots, (x_n, y_n)$, one can treat X as a random variable of the first component of the 2D vector and Y as the second component. We define the covariance matrix to be:

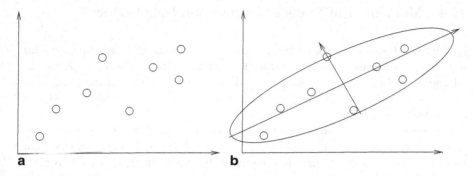

Fig. 11.3 Principal component analysis: **a** Original data points, and **b** the eigenvectors of the covariance matrix

$$M(X, Y) = \begin{pmatrix} Cov(X, X) & Cov(X, Y) \\ Cov(X, Y) & Cov(Y, Y) \end{pmatrix} \tag{11.14}$$

Geometric direction of a set of vector data points can be determined through principal component analysis (PCA). The largest eigenvalue of the matrix indicates the major axis by calculating the corresponding eigenvector.

The application of PCA is to find the major axis of the data in 2D. See Fig. 11.3.

In general, this method can be extended to analyze data in m dimensional space. Assume we have m random variables in X_i and n samples for each variable. In other works, we have N sample points $p_i = (x_1, x_2, \cdots, x_m)$, $i = 1, ..., N$. We can extend the covariance matrix to be $m \times m$,

$$M(X_1, \cdots, X_m) = \begin{pmatrix} Cov(X_1, X_1) & Cov(X_1, X_2) & \cdots & Cov(X_1, X_m) \\ Cov(X_2, X_1) & Cov(X_2, X_2) & \cdots & Cov(X_1, X_m) \\ \cdots & \cdots & \cdots & \cdots \\ Cov(X_m, X_1) & Cov(X_m, X_2) & \cdots & Cov(X_m, X_m) \end{pmatrix}$$

We will have m eigenvalues, $\lambda_1, \ldots, \lambda_m$. We assume that $\lambda_1 \leq, ..., \leq \lambda_m$. The eigenvector V_i of λ_i indicates the i-th principal component. This means that most of the sample data is along the side of vector V_i, compared to vector V_j, if $j > i$.

For data storage, we can determine that most of the data is covered by the first few principal components. We can use linear transformation to attach this to the original data and store the transformed data in a much lower dimension to save space. This is why PCA is one of the most effective methods in BigData and data science today. It was discovered many times by different researchers [9].

11.4   Mathematical Transformations and Data Science*

Mathematical transformations are the basic tools for data analysis, especially for image processing and computer vision. Fourier transform is the most fundamental approach. In this section, we mainly discuss the principles of the three most widely used mathematical transformations: Fourier transform, Radon transform, and wavelet transform.

The basic idea of transforms are to decompose the signal, image, or information in some way. Then, we extract the most important information and remove the least important information. We then put the majority of the information back to get a clear or filtered data. Putting the decomposed information back is called inversion. Therefore, the entire process of mathematical transformations has two steps: forward transformation and inversion.

11.4.1   Fourier Transform and Image Filtering

Let $f(t)$ be a periodic wave function that has an amplitude (intensity) at time t. Also $e^{(xj)} = cos(x) + jsin(x)$ is a complex number where $j = \sqrt{(-1)}$ is the imaginary unit. The Fourier transform of $f(t)$ represents the function on frequency domain U.

$$F(u) = \int_{-\infty}^{\infty} f(x)e^{-j2\pi ut} dt. \tag{11.15}$$

u represents frequency. The inverse Fourier transform is the following:

$$f(t) = F^{-1}(t) = \int_{-\infty}^{\infty} F(u) \, e^{j2\pi tu} \, du \tag{11.16}$$

Scientists use part of the frequency such as $u \in [-B, B]$ to reconstruct $f(t)$. When we think about the original $f(t)$, it contains much noise beyond $[-B, B]$ in its frequency domain.

In image processing, the Fourier transform can also be used to filter an image such as isolated noise. It is done by convolution as defined below:

$$(f * g)(t) = \int_{-\infty}^{\infty} f(\tau)g(t - \tau)d\tau$$

Understanding this formula in signal processing is not hard. Thinking about the integral requires summation, and $g(t)$ is a digitized function that only contains three consecutive 1's. If $g = 00 \cdots 01110 \cdots 00$, then $(f*g)(t) = f(t-1)+f(t)+f(t+1)$ is just the accumulation of three consecutive values added together. If we divide the value by 3, then we get the average value of f on an interval. For 2D image f, the process is similar. We add nine pixels together and divide by 9. The convolution is equivalent to the average of the surrounding pixels.

11.4.2 The Radon Transform and Data Reconstruction by Projection

The Radon transform is the mathematical foundation of computerized tomography (CT). Assume that we have a partial transplanted object (like a jade stone) and we want to know the internal structure of the object. One thing we can do is to use a flash light to light the object at a point. We can check the opposite sides of the lighted points and other places. The light passing through the object would give some information. For instance, if there is a small metal inside of the jade stone, the light passing through the metal would be very weak. If the jade is an evenly distributed substance, then the light would be smooth at all places on the edge.

The intensity of how the light passes is called projection. That is a function of the summation (integral) of the line that goes through the object. The Radon transform guarantees that if we know all the projected values then we will know the intensity of each point of original object. This is called reconstruction based on projection.

The line equation for the projection is $\rho = x \cos \vartheta + y \sin \vartheta$. $T = \rho$ is the distance from the origin to the line.

The Radon transform $T \in [0, \infty)$ and $\vartheta \in [0, 2\pi]$, now the function f Radon transformed with variables (ϑ, T):

$$\mathbf{R}f(T, \vartheta) = \int_{-\infty}^{\infty} \int_{-\infty}^{\infty} f(x, y)\delta(T - x \cos \vartheta - y \sin \vartheta)dx\,dy \qquad (11.17)$$

where δ function keeps the values on the straight line $T - x \cos \vartheta - y \sin \vartheta = 0$, which has a positive value and the rest of the location of the δ function will be (almost) zero.

The complete theory behind Radon transforms is beyond the scope of this book. The key idea of the inversion (to get the original density function of objects) is the following: (1) Use δ function to represent the line integration as a type of Fourier transform, then use the formula of inversion of the Fourier transform to get the final formula [13].

11.4.3 Wavelet Transform and Image Compression

The wavelet transform is similar to the Fourier transform in terms of decomposing the signal. However, the Fourier transform only decomposes the signal into sines and cosines, i.e. in frequency domain. The wavelet transform has two advantages: (1) The wavelet transform can use many types of basic functions (basis) called wavelets. (2) Wavelet transform can perform simultaneous localization in both time and frequency domains [5, 9, 14].

We only introduce the discrete wavelet in this section. A function $\psi(x)$ is a wavelet. $\{\phi_{sk} : s, k \in I\}$ is called a basis of ψ, if for integer s (scale factor) and k (time factor)

$$\phi_{sk}(x) = 2^{\frac{s}{2}} \psi(2^s x - k),$$

Let $\phi_{sk}(x)$ be basis functions where s is the scale and k is time. Let $f(x)$ be a function.

The integral wavelet transform is the integral transform defined as

$$W_\phi(\alpha, \beta) = \frac{1}{\sqrt{|\alpha|}} \int_{-\infty}^{\infty} \bar{\psi}(\frac{x - \beta}{\beta}) f(x) dx \qquad (11.18)$$

where $\bar{g}(x)$ is the complex conjugate of $g(x)$. (For the complex number $z = a + ib$, its conjugate is $\bar{z} = a - ib$. $\bar{g} = g$ for real functions.)

We can represent a function $f(x)$ as

$$\mathbf{f}(x) = \sum_{s,k=-\infty}^{\infty} c_{sk}\phi_{sk}(x)$$

where

$$c_{sk} = W_\phi(2^{-s}, k2^{-s}).$$

The scale factor always decreases by 2. This method can only be used for multi-resolution or compression. People can select any indexes for the special purposes of compression. The simplest form of the wavelets is called Haar wavelets.

An example of using Haar wavelets is as follows. For instance, in Haar wavelets, $H(t)$ can be described as

$$H(t) = \begin{cases} 1 & 0 \leq t < 1/2, \\ -1 & 1/2 \leq t < 1, \\ 0 & \text{otherwise.} \end{cases}$$

We have used Haar wavelets in meteorological data analysis for outlier tracking [11].

11.5   Remark: Data Reconstruction in Science and Engineering

In this Chapter, we briefly discussed the methods of data reconstruction in science and engineering. We began with the discrete fitting method use digital functions [3]. This method is called gradually varied functions. It is related to Rosenfeld's applications to image segmentation [15]. It is also related to a classic problem called Whitney's problem in mathematics [7, 12, 16]. We gave an extensive coverage in [3].

We also introduce some of the most important methods in data fitting including Bezier polynomials and B-Splines. They are the most popular.

This chapter also presented principal component analysis, and it will be used in Chap. 12 when we discuss the application in Google-search and other data science methods. The method containing eigenvalues and eigenvectors has essential importance in many areas in applied sciences. Principal component analysis is a statistical method [10].

This chapter also introduces the three most important mathematical transforms: the Fourier transform, Radon transform, and wavelet transform. Due to the fact that they are not the central topics of this book, we only provide the short discussion.

References

1. L. Chen, The necessary and sufficient condition and the efficient algorithms for gradually varied fill, Chinese Science Bulletin, 35:10(1990).
2. L. Chen, *Discrete Surfaces and Manifolds: A theory of digital-discrete geometry and topology*, 2004. SP Computing.
3. L. Chen, Digital Functions and Data Reconstruction, Springer, NY, 2013.
4. E. Cohen, R. Riesenfeld, G. Elber, Geometric Modeling with Splines. An Introduction, A.K. Peters Ltd., Wellesley, MA, 2001.
5. I. Daubechies, Ten Lectures on Wavelets, Society for Industrial and Applied Mathematics, 1992
6. C. de Boor A Practical Guide to Splines, Springer-Verlag, 1978.
7. C. Fefferman, Whitney's extension problems and interpolation of data, Bull. Amer. Math. Soc. 46 (2009), 207–220.
8. James D. Foley, Andries Van Dam, Steven K. Feiner and John F. Hughes, Computer Graphics: Principles and Practice. Addison-Wesley. 1995.
9. R. C. Gonzalez, and R. Wood, *Digital Image Processing*, Addison-Wesley, Reading, MA, 1993.
10. Johnson, R. A., and Wichern, D. W., Applied Multivariate Statistical Analysis, 6th ed. Prentice–Hall, Upper Saddle River, NJ. 2007.
11. C. T. Lu, Y. Kou, J. Zhao, and L. Chen, Detecting and Tracking Region Outliers in Meteorological Data, Information Sciences, Vol 177, pp1609–1632, 2007.
12. E. J. McShane, Extension of range of functions, Edward James McShane, Bull. Amer. Math. Soc., 40:837–842, 1934.
13. T. Pavilidis, Algorithms for Graphics and Image Processing, Computer Science Press, Rockville, MD, 1982.
14. W. H. Press, et al. *Numerical Recipes in C: The Art of Scientific Computing*, 2nd Ed., Cambridge Univ Press, 1993.
15. A. Rosenfeld, Continuous' functions on digital pictures, Pattern Recognition Letters 4 (1986), 177–184.
16. H. Whitney, Analytic extensions of functions defined in closed sets, *Transactions of the American Mathematical Society* **36**: 63–89, 1934.

Chapter 12
Geometric Search and Geometric Processing

Abstract In this chapter, we focus on cutting edge problems in geometric data processing. These problems have common properties and usually can be summarized as generally as: Given a set of n data points $x_1, ..., x_n$ in m-dimensional space, R^m, how do we find the geometric structures of the sets or how do we use the geometric properties in real data processing? Geometric data representation, image segmentation, and object thinning are some of the most successful applications of discrete and digital geometry. Along with the fast development of wireless networking, geometric search, especially R-tree technology, has become a central method for quickly identifying and retrieving a geometric location. 3D thinning is one of the best applications developed through digital geometry by preserving topological structures while reducing pixels or voxels. The classic methods of geometric pattern recognition such as the k-means and k-nearest neighbor algorithms are also included. The newest topic in BigData and data science is concerned with these methods.

Keywords Graph search algorithm · Quad-tree and R-tree · PageRank · Graph-cut · Classification and segmentation · Manifold learning · Persistent analysis · Cloud data · Incomplete data classification

12.1 Geometric Searching and Matching

Searching and matching are two basic tasks. Search is used to find some object in a set or space. For instance, searching for a number in an integer set, we know the binary search method is the fastest for a sorted array of integers.

If the set is the random set, then the search will be trivial since we can only compare the elements from the set one by one. When the set contains a structure, such as an order, the search may use properties of the structure. In space, the set could be a topological structure such as components, a geometric structure such as distance metrics, or an algebraic relation such as rules.

Matching usually means searching for an object that may contain several elements in a set. The set may hold a structure, meaning that the elements in the set have relations and connections. Finding a substring in a DNA sequence is a good example. More profound research in finding a protein structure is a geometric problem.

© Springer International Publishing Switzerland 2014

L. M. Chen, *Digital and Discrete Geometry,* DOI 10.1007/978-3-319-12099-7_12

Matching could also mean that we can find a partial or best match. For instance, finger print matching.

Recent research shows great interest in high dimensional search. This search is usually related to the nearest neighbor method. The fast algorithm is based on prepro-duced data structure. To find a new data point in a set of points, usually in the form of an n-dimensional vector, we want to find the point that is closest to the new point. R-tree data structure was previously implemented in order to perform a fast search. This application is important in wireless networking. For instance, we have a number of wireless tower stations that cover cellular phone communication. When a new phone joins in the area, we need to locate the new phone and appoint a station to commu-nicate with it. To search the cell phone without calculating its distance from all the stations, we will need a special data structure. The most popular today is the R-tree.

Another hot research topic due to the need of BigData applications is called subspace recovery, an interesting question that has received much attention lately:

Thinking about a simple problem, there are 100 sample points on a 2D plane. We want to find a line that contains most of these sample points. Is this problem NP hard? Its decision problem can be described as follows: Is there a line that contains at least 50 points? What about a curve instead of a line? This problem is highly related to manifold learning where we try to determine a cloud point set that represents a manifold such as a curve. The data points are not only 50–100, and they are 50 –100 GB, how do we use cloud computing technology to solve this problem?

The general version of this problem called robust subspace recovery, relates to dimension reduction [16, 23]: Given a collection of m points in R^n, if many but not necessarily all of these points are contained in a d-dimensional subspace T, can we find m? The points contained in T are called inliers and the remaining points are outliers. This problem has received considerable attention in computer science and statistics. However, efficient algorithms from computer science are not robust to adversarial outliers, and the estimators from robust statistics are difficult to compute in higher dimensions.

The problem is finding a T dimensional space that contains most of the points or a given ratio of inner points. How do we determine T in the fastest way? Does T have some sort of boundary?

Geometric search, especially high dimensional search, is a hot topic related to data science, along with incomplete data search, which is related to artificial intelligence.

12.2   Searches in Graphs

The very basic geometric search method is the search method for general graphs. The best search method was first made by Tarjan [14] and first used in computer graphics and computer vision by Pavlidis [32]. These techniques are called breadth first search and depth first search.

The data structures that refer to these two techniques are queues and stacks made by adjacency lists.

In this section, we explain how the basic algorithms of graph theory are translated into geometric search. These algorithms are: (1) The breadth first search algorithm, (2) The shortest path algorithm, and (3) The minimum spanning tree algorithm.

The maximum flow and minimum cut algorithms are also related to this topic. We give a brief introduction.

12.2.1   Breadth First Search and Depth First Search

Breadth first search is a fast search approach to get a connected component on a graph. It begins at a vertex p and searches for all adjacent vertices. Then, it "inserts" all adjacent points (neighbors) into a queue. "Removing" a vertex from the queue, the algorithm calls the point p and then repeatedly finds p's neighbors until the queue is empty. Marking all the vertices we visited, the marked vertices form a connected component. This technique was introduced by Tarjan [14].

Algorithm 12.1 Breadth-first-search technique for all point-connected components.

Step 1:   Let p_0 be a node in G. Set
          $L(p_0) \leftarrow *$ and $QUEUE \leftarrow QUEUE \cup p_0$
          i.e., labeling p_0 and p_0 is sent a queue $QUEUE$.
Step 2:   If QUEUE is empty, go to Step 4; otherwise,
          $p_0 \leftarrow QUEUE$ (top of $QUEUE$). Then,
          $L(p_0) \leftarrow 0$.
Step 3:   For each p with an edge linking to p_0,
          do
          $QUEUE \leftarrow QUEUE \cup p$ and $L(p) \leftarrow *$. Then, go to Step 2.
Step 4: $S = \{p : L(p) = 0\}$ is a connected part.
Step 5:   If p is un-visited vertex, Let $p_0 = p$, Repeat Step 1.
          Otherwise Stop.

Breadth first search only visit a new node twice: insert to the queue and remove from the queue. So this algorithm is a linear time algorithm. It is a fast search approach. Example 4.1 in Chap. 4 gives the detailed example for this process.

Depth first search is similar to breadth first search. Depth first search will find a new node continuously until no more new node can be found. When the algorithm run in the journey, it saves all visited nodes in a stack. Since we only go with a path try to find the "deepest" node, there are other branches we might missed at the first try. So the algorithm return to each visited node in the order pushed in stack. It was first in last out. We check other branches from a pop-up node, after all nodes are popped out, this algorithm will find all nodes in a component. This algorithm will visit all edges twice at most. Therefore, it is also a linear algorithm on edges.

12.2.2 Dijkstra's Algorithm for the Shortest Path

Finding the shortest path in the weighted graphs is a classic topic in graph theory. We have presented an algorithm in Chap. 2. Here we introduce another algorithm for the shortest path. Dijkstra's algorithm [14] for shortest paths in weighted graphs can be modified and used to solve this problem. The drawback of Dijkstra's algorithm is that it cannot take negative edge. Dijkstra's algorithm is faster that Bellman-Ford Algorithm presented in Chap. 2. We can also involve a special data structure to make Dijkstra's algorithm even more fast.

The idea of this shortest path algorithms is the following: Start at the vertex that is the departing node. Then, we record the distance from the departing node to all of its neighbors. The record may not be the shortest path involving those notes. We update the record and put the updated value on the nodes, a process called relaxation.

After that, we extend the notes by one more edge, do a relaxation, and continue to repeat these two steps until we have reached every node in the graph.

The algorithmic technique used in this problem is called dynamic programming. Even though from a vertex to another vertex, there may be exponential number of paths, however if we only interested in the shortest path, we only need to care about the length we travel, we do not have pass a certain vertex. There are only $n(n-1)/2$ pairs of vertices. Update the minimum distance while we calculate. This is called dynamic programming.

Algorithm 12.2 The modified Dijkstra's algorithm can be used to find the connectivity from a source point to all other points in the graph.

Step 1: Let $T = V$. Choose the source point a
$L(a) = 0; L(x) = \infty$ for all $x \in T - \{a\}$.
Step 2: find all neighbors v of a node u with L value $L(v) = L(u) + w(u, v)$.
$T \leftarrow T - \{v\}$
Step 3: For each x adjacent to v do
$L(x) = max\{L(x), min\{L(v), L(x) + w(x, v)\}\}$
Step 4: Repeat steps 2-3 until T is empty.

12.2.3 Minimum Spanning Tree

A spanning tree of a graph G is a tree which contains all vertices of G. A graph G has a spanning tree if and only if G is connected. The problem of finding a minimum spanning tree is to find a tree such that the total weights reach the minimum for a weighted graph.

We have discussed the minimum spanning tree in Chap. 2. We did give an algorithm, the Kruskal's algorithm, and an example. Here we want to present the methodology of designing the algorithm. This methodology is called the greedy algorithm.

In the Kruskal's algorithm, the tree T initially contains all vertices but no edges. Because of this, it starts an iterative process, adding an edge to T under the condition

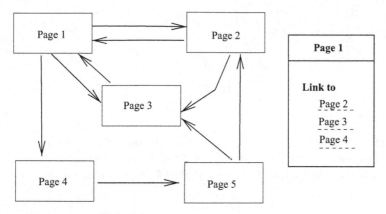

Fig. 12.1 Example of a web link graph

that the edge has the minimum weight. We continue to add a minimum edge to T as long as not allowing a cycle appearance in T. When T has $|G| - 1$ edges, the process stops.

The principle where we always made a minimum or maximum selection is called the greedy method. In artificial intelligence, the greedy method constantly applies to applications. In most of cases, it may not obtain an optimal solution rather an approximation. However, for the minimum spanning tree, the greedy method reaches the optimum.

Here we want to give another algorithm called The Prim's algorithm [14].

Algorithm 12.3 The Prim's algorithm. It is to find a minimum spanning tree T of a weighted graph $G = (V, E)$.

Step 1: Select any point p in V, let $T = \{p\}$.
Step 2: Repeat steps 3–4 until T has $|V|$ vertices.
Step 3: Select a new vertice p not in T, choose the minimum edge linking to T. It means that p link to any vertex in T as long as the new edge will reach reach the minimum.
Step 4: Let $T \leftarrow T \cup \{p\}$. Go to Step 3.

12.2.4 Online Search, Google Search, and PageRank

The principle of the Google search method, PageRank, is simple and elegant [4, 31]. It is to establish a link graph, then calculate the importance of each web node (page). To explain this method, we start with the adjacency matrix of the page link graph, a directed graph. This matrix for Fig. 12.1 is the following:

$$
M = \begin{bmatrix}
0 & 1 & 1 & 1 & 0 \\
1 & 0 & 1 & 0 & 0 \\
1 & 0 & 0 & 0 & 0 \\
0 & 0 & 0 & 0 & 1 \\
0 & 1 & 1 & 0 & 0
\end{bmatrix}
$$

Its weight graph with distribute the average contribution to each of outgoing nodes is shown below. For instance, the first page "Page 1" gives $1/3$ contribution to each of its pointed neighbors: Page2, Page3, and Page4. We will have the weight matrix:

$$
W = \begin{bmatrix}
0 & \frac{1}{3} & \frac{1}{3} & \frac{1}{3} & 0 \\
\frac{1}{2} & 0 & \frac{1}{2} & 0 & 0 \\
1 & 0 & 0 & 0 & 0 \\
0 & 0 & 0 & 0 & 1 \\
0 & \frac{1}{2} & \frac{1}{2} & 0 & 0
\end{bmatrix}
$$

The transpose of W will be the matrix we are interested, denoted as M_{PR} Such a representation is very intuitive.

On the other hand, Page and Brin have used the following formula to rank the importance of each web page [4] called the PageRank of a page A:

$$
PR(A) = \frac{(1-d)}{N} + d(PR(T_1)/C(T_1) + ... + PR(T_n)/C(T_n)) \tag{12.1}
$$

where N is the total number of pages considered, $PR(T_i)$ is the PageRank of page T_i which link to page A, $C(T_i)$ is the number of links going out of page T_i and $d \in [0, 1]$ is a damping factor which is usually set to 0.85.

A simple algorithm runs the above formula on an iterating manner. It stops as an error limit will meet.

It is astonish when we check the relationship between the matrix M_{PR} and $PR(A)$. Let vector $x = (1/N, \cdots, 1/N)$. x^T is the transpose of x. we will get another vector $M_{PR}x^T$, and so on $M_{PR}^k x^T$. We know that $M_{PR}^k x^T$ will converge to a vector when k is big enough. Then

$$
M_{PR}^{k+1} x^T = M_{PR}[(M_{PR})^k x^T] = (M_{PR})^k x^T,
$$

so in such a case, $y = (M_{PR})^k x^T$ is the eigenvector of M_{PR}. After we add the dumping factor to the matrix, we have

$$
G = \begin{bmatrix}
\frac{(1-d)}{N} \\
\cdots \\
\frac{(1-d)}{N}
\end{bmatrix} + d \cdot M_{PR} \tag{12.2}
$$

the eigenvector of G will be approximately

$$\begin{bmatrix} PR(A_1) \\ \cdots \\ PR(A_N) \end{bmatrix}.$$

Now, G is called the *Google* matrix. (Some good examples can be found at *pr.efactory.de/e-pagerank-algorithm.shtml*.)

12.3   Graph Cut and Its Applications to Image Segmentation

Let graph $G = (V, E)$. If we partition V into two disjoint sets A and B, i.e. $V = A \cup B$ and $A \cap B = \emptyset$, then the edges from A to B are called a cut. If G is a weighted graph, then $w(e) = w(u, v)$ is the weight on edge $e = (u, v)$. We need to then find a partition for A and B such that

$$\min\{cut(A, B) = \Sigma_{u \in A, v \in B} w(u, v)\}$$

is a minimum. This is called a minimum cut problem. It is proven that a minimum cut gives the maximum flow [14, 22].

If we compare the graph cut to image segmentation (we have defined the image segmentation in Sect. 4.5 in Chap. 4.), we can see that there are some similarities. Let us assign high values on the connection (edges) to the nearby pixels if they have similar intensity (brightness), and assign small values on the connection to the nearby pixels if they have large differences in terms of intensity. Therefore, the minimum cut would offer us a relatively good segmentation.

However, there are so many data points in a picture and some neighborhoods have exactly the same values if we use the weight graph for all the points in the pictures. For instance in $|V| = 1024 \times 1024 = 2^{20}$, the graph for E and w may be $|E| = |V|^2 = 2^{40}$ in size. This may require Bigdata or cloud computing techniques to process.

Thus, a technique of solving this issue begins with safely partitioning the picture into smaller regions, specifically near convex regions for building an easier new graph, where each node of the graph represents a homogeneous region similar that is treated as one big pixel. Therefore, the minimum cut for big pixels will be much faster.

A method called normalized cut was invented so that the procedure almost always finds the part that contains fewer vertices. Normalized cut will balance the size factor of the cut [38]. We present a brief introduction to this technology, which is very popular today. The normalized cut defines a measure as follows:

$$N\_cut(A, B) = \frac{cut(A, B)}{Assoc(A, V)} + \frac{cut(B, A)}{Assoc(B, V)}, \tag{12.3}$$

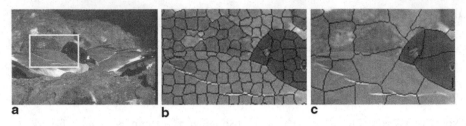

Fig. 12.2 Improved N-cut example by Levinshtein. **a** Original image, **b** partitioned by bigger pixels, and **c** the final segmentation displayed by the partition curves

where

$$Assoc(S, V) = \Sigma_{u \in S, v \in V} w(u, v)$$

is the total "association" between S and V. $cut(A, B)$ and $cut(B, A)$ may differ if we consider direct graphs. It is proven that finding the smallest $N - cut(A, B)$ is an NP-hard problem. We can only find its approximate solution in a reasonable time. This method involves an excellent graph theory tool called the Laplacian matrix along with the eigenvector to obtain this approximation of the solution.

Let D be a diagonal matrix with $d_{ii} = \Sigma_j w_{ij}$ and $d_{ij} = 0$ if $i \neq j$ and $W = [w_{ij}]$. Shi and Malik showed that X is the binary membership function of a set S with 1 value for the component-index number in S and -1 value if the component-index number not in S [38]. If $cut(S, V - S)$ is the normalized cut then

$$\frac{X^T(D - A)X}{X^T D X}$$

is minimized with the condition $X^T D \mathbf{1} = 0$. Therefore, solving the following eigenvalue equation will provide a solution for this problem.

$$(D - W)X = \lambda D X$$

The second smallest λ value (the smallest is zero) indicates its eigenvector X as the smallest segment when X is translated into a binary vector.

If D is diagonal matrix with $d_{ii} = \Sigma j w_{ij}$, then $d_{ij} = 0$ if $i \neq j$ and $W = [w_{ij}]$. An improved method of normalized cut was presented in [29]. In Levinshtein's PhD thesis, He used some rules by simply removing the edges connecting the two parts. See Fig. 12.2c.

Mathematically, this is a beautiful method. However, in practice, it is still quite slow in performance. Ren and Malik improved this algorithm using a method similar to the method of finding the similarity of bigpixels [34]. Scientists in Canon Inc observed that it is related to the λ-connected method by Chen [7, 8].[1] We will discuss λ-connectedness at the last section of this chapter. In 3D or other massive

---

[1] VerticalNews reported a patent application by Canon Inc with the description on this. *http://www.spclab.com/research/lambda/VerticalNewsReportsRelatedChen91a.pdf*.

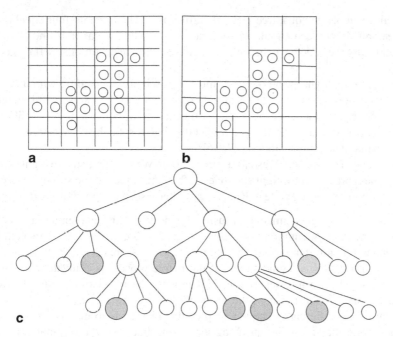

Fig. 12.3 Quad trees

search, when the data points are not fit into a single computer, the bigdata technology can be used with the big-pixels or the big surface cells.

12.4   Spatial Data Structures: Quadtrees, Octrees, and R-trees

Geometric processing must build on some efficient data structures. Even though graph structures can be used to represent all geometric data sets, to represent graphs in a computer is very costly in terms of space. In many applications, we develop effective data structures. The most popular ones for geometric problems are quadtrees, octrees, and R-trees. We specifically discuss these advanced data structures in this section.

We also provide simple examples to describe the necessities of each data structure.

12.4.1   Quadtrees and Octrees

The quadtree is for 2D data sets or objects. The root of a quadtree contains a 4-subtree. If we partition a 2D square into 4 identical smaller subsquares, then each subtree will represent a subsquare. When storing a binary image, we can save space when a subsquare contains no information (it is blank) ([20, 32, 35]; Fig. 12.3)

In this example, the quadtree method partitions a 2D space into four equal quadrants, subquadrants, and so on. We stop at a node called a leaf if all the elements represented by the node have the same value. We continue decomposing if this is not the case.

The depth or height of the tree is at most $\log_2 n$ if we are dealing with an $n \times n$ region. We can save space since a leaf may represent a lot of pixels in real images.

The octree uses the same principle as the quadtree, but it is used for 3D data storage. This is because we can split a 3D cube into 8 subcubes. Therefore, octrees are for spatial data representation [27, 36].

Each node in an octree represents a cubic region. We subdivide it into eight octants that will be represented by eight subnodes. When the values of the subregion are the same. The corresponding node will be a leaf. Otherwise, we keep dividing the region.

Example 12.1 Image segmentation using quadtrees: This segmentation is called Split-and-merge segmentation. The key is to give an order of merge. This is not an equivalence relation that is why the order of merge is important. Pavlidis and his PhD student found this segmentation. If the merge process is equivalence relation, it is not needed for a tree separation. If it is a similarity relation, as the Uniform rule of distance to the mean of the grouped sub-segments does, the order of the grouping makes difference. The key step of this method is to merge two segments together when they are adjacent along a splitting line. For the original split-and-merge segmentation, two segments (can be viewed as two big-pixels) can be merged if the maximum distance of the value of the merged segment to the mean value of the merged segment is smaller or equal to ϵ. Split-and-merge segmentation is based on a quadtree partition of an image and hence is sometimes called quadtree segmentation.

12.4.2 R-tree and Wireless Networking

The idea of the R-tree comes from the B-tree in database systems. Using a rectangle, as small as possible, we cover all data points in a 2D space and then use smaller subrectangles to cover the points inside the parent rectangle. The subrectangles in the same level can overlap each other. See Fig. 12.4.

Since the dimensions of each rectangle are not fixed, this allows for the most flexibility in the partition of data in a space. The "R" in R-tree stands for rectangle. This method is a very popular technique in querying and search.

A popular problem is: Given a point location, find all of the rectangles or leaves that contain a chosen point [15, 43]. The idea of R-trees is to use the bounding rectangle boxes to decide whether or not we need to search the inside a subtree. Therefore, most of the nodes in the tree are never read during a search. The idea of R-tree is from B-trees in data structure [14], that have variable number of subtrees (son nodes). Not like a quad tree. It makes R-trees suitable for large data sets and databases,

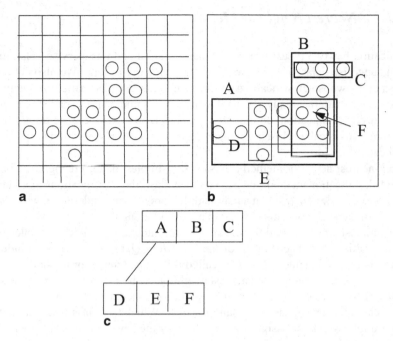

Fig. 12.4 R trees

There are many applications using *R*-trees in wireless networking. For instance for queries in wireless broadcast system, we need to quickly find a group of users in a heavy rain area to tell them that there might be a flooding possibility. This is especially useful in social network systems. Social networking categorizes individuals into specific groups. Another example for R-tree to be used, a social network such as *facebook* does not usually contain the location parameter for users. However, when a local social event needs to hurry some local attendees, they can ask *facebook* to send message to the members around. R-tree is a best way to identify the people and send them message.

12.5   Classification and Clustering: Distance Related Methods

In this section, we introduce the most popular classification methods that all use geometric distance as the measuring standard. In this section, we assume the samples are in vector format, i.e. each sample will be represented as a vector (called a feature vector).

12.5.1 k-Nearest Neighbor Method

Classification has two types: supervised classification and unsupervised classification. In supervised classification, we know the category of data. We also collected some samples whose classifications are known. This is called pattern recognition. We know some pattern already and want to put a new sample into its pattern category. Unsupervised classification, on the other hand, is a process where we do not know the pattern. This is called clustering, meaning that we can only partition the data into several categories [40].

The k-nearest neighbor method (kNN) is the simplest pattern recognition or supervised classification method. It means that we know the classifications of a sample data set $S = \{Q_1, ..., Q_m\}$. When a new sample comes for consideration, we want to know which data category this new sample would belong to.

The simplest way is to find the distance of the sample P to all the samples with known classifications. The shortest distance of $d(P, Q_i)$ $i = 1, ..., m$ will indicate the class that should include P. This is called the nearest neighbor method.

kNN is a generalization of the nearest neighbor method. If we assume a class (or category) contains multiple samples in S and the classified data contains some noise, then we can consider the closest k samples in S. Whichever class that can contain these k samples would be assigned as the class for the new sample. That is why this is called the kNN method [1, 40].

We can also add a machine learning component into kNN. In machine learning [30], we split S into two sets, one set would be the training set S_T and the other would be the testing set S_R. We know the classification results for both sets. For each element R_i in S_R, if we run kNN, then we get the result for S_R. The correct ratio represents the accuracy of the training set to the testing set.

12.5.2 k-Means Method

The k-means method is an advanced unsupervised classification method. Given a set of vectors, if we want to classify the data into two categories, then what is the best we can do? The k-means algorithm would provide a very reasonable way to do the partition. In Fig. 12.5, we first select two initial vectors as the "centers"(called sites) of the two categories. Second, for each elements P in the set, we put it into one of the categories and recording the distance to the center. Third, update the two center locations such that the total summation of the distance gets smaller. The algorithm will halt if there is no improvement when we move the two centers.

The idea of the diagram can be used in the k-means algorithm: Based on the initial k sites, partition the space using the Voronoi diagram. Then, we can move the center to the geometric centroid of the new partition before we recalculate the result. This is an iterated process. The only problem is with the local minimum meaning that this algorithm may converge at the local best result. In other words, the algorithm

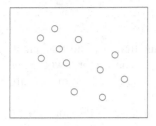 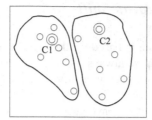 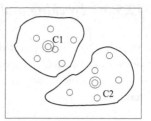

Fig. 12.5 The k mean algorithm

will stop after a local minimum is attained. The idea of the k-means algorithm first started in 1957 by H Steinhaus [40].

One algorithm was designed by Lloyd in 1982 that used the partitioned area and not the discrete data points in the new center calculation. Incorporating the Voronoi diagram partition only speeds up some portions of the algorithm. In the global sense of optimization, we can prove that this method is NP-hard. A local search algorithm was implemented by D.Mount et al. [26].

A simple version of the k-mean algorithm is as follows:

Algorithm 12.4 The k-Means Algorithm.

Input: A set of data points $X = \{x_1, \cdots, x_n\}$, each x_i is a m dimensional real vector.

Output: k-means tries to find $P = \{c_1, c_2, \cdots, c_k\}$, where each c_i is a vector. Then, all the data points will be partitioned into k subset of X, S_i associating with c_i, satisfying:

$$\min \sum_{i=1}^{k} \sum_{x_j \in S_i} \|\mathbf{x}_j - c_i\|^2. \tag{12.4}$$

Step 1   Randomly select k vectors $c_1, c_2, \cdots, c_k$ as the initial centers.

Step 2   For each data point x_i, calculate the distance $d(x_i, c_j)$, setting x_i as S_t if $d(x_i, c_t)$ is the smallest.

Step 3   Calculate the new geometric center of S_j by

$$c_j = \frac{1}{|S_j|} \sum_{x_i \in S_i} x_i$$

Step 4   Repeat Step 2 until the total square of the distances in (12.4) can no longer be improved.

Since there only exists a finite number of such partitions, the algorithm will converge locally. More theoretical results were found in recent years.

12.5.3 Tracking and Mean Shift

Tracking the set of point data is based on the density function. To find the center point of each small window, we always move the current mean vector towards the direction of the maximum increase in density. This method is called the mean shift method .

The procedure guarantees the convergence of the point towards the points with the highest density in a designed window [18]. Note that the gradient of the density function is zero at this point since the derivative of the highest point is zero. This process can start at a randomly selected point in the search space, this method can be used to find the backbone or framework of the data set. Mean shift can also be used to find the thinning of an object statistically.

Mathematically, mean shift can be used to find the centric trice of a cloud point data. We first need to assume a statistical model of density function, such as the Gaussian kernel: $Kernel(x - x_0) = e^{-c(x-x_0)^2}$ where x_0 is a initial estimated center where c is a predefined constant. Then the next mean will be determined by

$$m(x_0) = \frac{\sum_{x \in N(x_0)} Kernel(x - x_0)x}{\sum_{x \in N(x)} Kernel(x - x_0)} \tag{12.5}$$

where $N(x_0)$ is the neighborhood window of x_0, a set of points for which $Kernel(x_0) \neq 0$.

Then we will set the new center x_0 to be $m(x_0)$, and repeat the above calculation to get a trice of x_0 as the track of the mean shift.

12.5.4 Support Vector Machine Algorithms

The support vector machine algorithm (SVM) is another machine learning model. It is an advanced method in machine learning algorithms that combines both classification and regression analysis. This method is to find a best separation line between two classes (a middle line to both classes) [40].

In Fig. 12.6, for an original data set (a), we can find three reasonable lines to separate them (b). However, line a will give the largest gap. So a will be selected by SVM. The data points A, B, and C will be called the supporting vectors.

12.6 Case Study: Cloud Data Computing

Given a set of data points in m-dimensional space, in this section, we will ask several questions and give primary answers to these questions.

This section is highly related to the current BigData and data science topics. We mainly cover two topics: Manifold learning and persistent homology analysis.

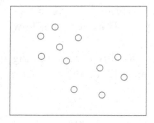 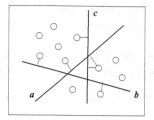 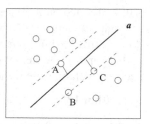

Fig. 12.6 The support vector machine algorithm

All of the major pattern recognition methods were presented in this chapter. Now, we will discuss when will use them in this section. After that, for the questions that cannot be answered using existing methods reviewed before. We will try to use the methods to be introduce in the rest of the chapter.

A data set is called a cloud if its data points are randomly arranged, but they are dense and usually have a lot of data points. Cloud computing as well as cloud data computing is highly related to networking.

Let M be a cloud data set. We can ask the following questions:

Q1. How do we find the orientation of the data set? We can use the principle component analysis to get the primary eigenvectors.

Q2. How do we classify M into different classes? The best method is k-means if there is no time limit. We can use use SVM to separate the data set.

Q3. How do we find the partition of the space using the data set? The Voronoi diagram is the best choice.

Q4. If some categorization is given, some instance, we know the data classification of a subset of M, we can use k-NN to find the classification for unknown data points.

Q5. How do we save the data points? And later, we like to retrieve the data points. We can use quadtree or R-tree to save the data. So we can save a lot of space [36].

Q6. If most of the data are not occupying the entire space, can we find a subspace to hold the data not waste the other space? This problem is related to dimension reduction [23] and manifold learning [39, 42].

Q7. How do we find the topological structure of the data sets? A technology called persistent homology analysis was proposed to solve this problem. [5, 19]

Q8. When the cloud data is dense enough to fill entire space, we will call M the image. We will use image processing method to solve the problem related [35].

12.6.1 Manifold Learning on Cloud Data

Manifold learning was originally proposed for nonlinear dimensionality reduction [3, 9, 23]. The main task is to identify a surface or manifold where the most data will be located on it. The example always used in manifold learning is a 2D Swiss roll that is sampled randomly and it is embedded in a 3D space. The question is how do we extract the information?

Isomap and the Kernel principle component analysis are usually used in finding such a structure. Here we just present the algorithm for Isomap [2, 6, 39].

Algorithm 12. 5 The Isomap algorithm. M is a cloud data set.

Step 1   For each point $x \in M$, determine the neighbors of each point. we can K-NN or MST to get the information.

Step 2   Construct the graph with the neighbors found in Step 1. the edge will be weighted by the Euclidean distance.

Step 3   Calculate the shortest path between two vertices using Dijkstra's algorithm

Step 4   Multidimensional scaling makes edges between two vertices. Cut off the data points beyond the clip level. Determine local dimension of the data. This step is called lower-dimensional embedding.

Step 5   Compare the results in Step 4. make decision on the dimensions for all local neighborhood, For instance, the most of local neighborhood are 2D. We make the 2D out put.

In Step 5, one can use principle component analysis to help us to find the local dimensions [42]. In fast algorithm design, there may be some geometric data structures that can assist us to find new faster algorithms. See Fig. 1.5a.

12.6.2 Persistent Homology and Data Analysis

Cloud data usually does not have a topological structure except they are located individually. Each point is an independent component. However, to interpret that each data point is a sampling, meaning that each point represent an area or volume, but we do not know how big the area or volume is? Then this problem will relate to so called persistent homology analysis [5, 19].

In topology, homology usually indicates the number of holes in each dimension. We will introduce the formal concept of homology in Chap. 13. Here we only treat this concept as the number of holes. It is called the Betti number of the manifold.

In the method of persistent homology, we make a sample point grows in its volume with a radius r. When r changes from 0 to a big number, the data will change from individual data points to the large volume until fills entire space.

The persistent homology method will calculate the homology groups (number of holes in each dimension) for each r. It would make some sense that the same topology (homology groups) that covers most of r will be the primary topological structure of M, the data set. See Fig. 1.5b.

In Chap. 14, we will specifically discuss how we calculate homology groups [12]. In terms of making a data shape to grow, mathematical morphology may provide some nice operators [20].

12.7 Topological Image Processing: Thinning Algorithms

Image thinning is to provide a simplest structure to a complex image for human or computer to interpret or recognize. Therefore, thinning usually will extract or keep the middle curves of the image components. In addition, thinning maintains the topological structure of an image while deleting as many pixels as possible.

One of the most popular methods is called Zhang-Suen thinning for 2D images [41].

Let $S \in \Sigma_2$ be a connected component and S be represented as set of "1s" with background pixels that are marked "0"

N_8 is the 8-Neighbourhood of the center pixel P

$$\begin{bmatrix} P4 & P3 & P2 \\ P5 & P & P1 \\ P6 & P7 & P8 \end{bmatrix}.$$

where $c(P)$ = number of neighbors with value 1, i.e. $c(P) = P1 + P2 + \cdots + P8$ and $b(P)$ = number of boundary components, i.e. number of 0–1 transitions in sequence (P1,P2, ... ,P8,P1).

The pixel $P \in S$ can be deleted if P is a boundary point, i.e. $c(P) \neq 8$. If $c(P) = 7$, then deleting P may cause a hole if $P2$, $P4$, $P6$, or $P8$ is 0. At the same time, P as a center point will be kept if $P1$, $P3$, $P5$, or $P7$ is 0.

Therefore, P can only be deleted if $(2 \leq c(P) \leq 6)$ and P is a "corner" point in S.

Algorithm 12.6 The Zhang-Suen thinning algorithm is as follows:

Step 1: Mark pixel $P \in S$ ($P = 1$) if $(2 \leq c(P) \leq 6)\hat{(}b(P) = 1)\hat{(}P1 \cdot P3 \cdot P5 = 0)\hat{(}P3 \cdot P5 \cdot P7 = 0)$.
Step 2: Delete all marked pixels.
Step 3: Mark pixel $P \in S$ ($P = 1$) if $(2 \leq c(P) \leq 6)\hat{(}b(P) = 1)\hat{(}P1 \cdot P3 \cdot P7 = 0)\hat{(}P1 \cdot P5 \cdot P7 = 0)$.
Step 4: Delete all marked pixels.
Step 5: Repeat Steps 1–4 until no pixel can be marked.

For 3D thinning, one of the best algorithms was designed by Lee et al [28]. This algorithm specifically use digital topology to define the problems. The digital surface points (Chap. 5) are used to identify the data point that can be deleted. They also use the Euler characteristic to maintain the topology unchanged when deleting a point. Homann had implemented this algorithm in a relatively simple way [24]. We present this algorithm as below:

Algorithm 12.7 3D thinning algorithm. Let M be a connected component in 3D.

Step 1: Pick a new point (or pixel) x in M.
Step 2: If x is not a boundary surface pixel, go to Step 1. Then, consider one the six possible direct directions in 3D at a time (using $mod(6)$ to pick direction

once at a time.) to keep the thinning symmetrically. In other word, keep the centrelines are not shifted.

Step 3: Test if the deletion of x will change the Euler characteristic. If it does, go to Step 1. We want to make sure that no new hole is created or no hole will be removed. In [28] There is a look-up table that can be used.

Step 4: Marked x and delete x.

Step 5: Repeat Steps 1–4 until no pixel can be deleted.

We will discuss how to calculate of the Euler characteristic in Chap. 14. We need to treat a 3D pixel as a 3D-cell. So we can use the algorithm for 3D manifold calculation for thinning.

12.8 Connectedness: Geometric Search Using Incomplete Information

Connectivity is a basic measure in many research areas of mathematical science and social sciences: (1) In the discrete case, two vertices are said to be connected if there is a path between them; (2) in the continuous case, two points, a and b, are connected in space, S, if there is a continuous function, $f : [0, 1] \to S$, such that $f(0) = a$ and $f(1) = b$; (3) In social science, we also can say that two persons in an institution are connected if one person is under the supervision of the other. However, these connectivity relations only describe either full connection or no connection.

In this section, we introduce a systematic approach called the λ-connectedness method to the problems in digital and discrete geometry [7–9, 11]. It can be applied to image segmentation or classification, searching, and data reconstruction.

λ-connectedness can be used to measure incomplete relations between two vertices, points, human beings, etc. We start with the definition of λ-connectedness, then discuss some examples.

λ-connectedness is a measure for partial connectivity among data points, a set of grouped data, or a collected objects. This method is based on a graph, $G = (V, E)$, and an associate function, ρ, on the vertices of the graph, where ρ is called the potential function.

A metric, $C_\rho(x, y)$, is defined for the λ-connectedness on the vertices, $x, y \in G$, with respect to ρ. This metric is not only based on the distance but also relates to the value on the vertices.

12.8.1 Connectedness and Segmentation

Let (Σ_2, f) be a digital image. If p and q are adjacent, we define a measure called "neighbor-connectivity" below:

$$\alpha_f(p, q) = \begin{cases} 1 - \frac{\|f(p) - f(q)\|}{H} & \text{if } p \text{ and } q \text{ are adjacent} \\ 0 & \text{otherwise} \end{cases} \tag{12.6}$$

where $H = \max\{f(x)|x \in \Sigma_2\}$.

Let $x_1, x_2, ..., x_{n-1}, x_n$ be a simple path. The path-connectivity β of a path $\pi = \pi(x_1, x_n) = \{x_1, x_2, ..., x_n\}$ is defined as

$$\beta_f(\pi(x_1, x_n)) = \min\{\alpha_f(x_i, x_{i+1})|i = 1, ..., n-1\} \qquad (12.7)$$

or

$$\beta_f(\pi(x_1, x_n)) = \prod\{\alpha_f(x_i, x_{i+1})|i = 1, ..., n-1\} \qquad (12.8)$$

Finally, the degree of connectedness (connectivity) of two vertices x, y with respect to ρ is defined as:

$$C_f(x, y) = \max\{\beta_f(\pi(x, y))|\pi \text{ is a (simple) path.}\} \qquad (12.9)$$

For a given $\lambda \in [0, 1]$, point $p = (x, f(x))$ and $q = (y, f(y))$ are called λ-connected if $C_f(x, y) \geq \lambda$.

If Eq. (12.7) is used, λ-connectedness is reflexive, symmetric, and transitive. Thus, it is an equivalence relation. If Eq. (12.8) is used, λ-connectedness is reflexive and symmetric. Therefore, it is a similarity relation. Generalized λ-Connectedness can be found in [9].

λ-connectedness was proposed to describe the phenomenon of gradual change, specifically in geophysical and geological layer search [7, 9]. In other words, these layers exhibit gradual or progressive changes in a layer, but sudden changes frequently occur between two layers. A λ-connected component can be viewed as a layer, and two layers should be separated by the λ-connected search.

We know that a digital image can be represented by a function: $f : \Sigma_2 \rightarrow [0, 1]$. So, if p, q are adjacent and there are only a "little" difference between $f(p)$ and $f(q)$, then pixel $(p, f(p))$ and $(q, f(q))$ are said to be λ-adjacent. If there is a point r that is adjacent to q and $(q, f(q)), (r, f(r))$ are λ-adjacent, then $(p, f(p)), (r, f(r))$ are said to be λ-connected. Similarly, we can define the λ-connected on a path of pixels.

Assume that there is a 4×4 small image, all pixels $p_{i,j}, i, j = 1, 2, 3, 4$, are given in the following array [8].

$$\begin{bmatrix} .2 & .1 & 1. & .8 \\ .1 & .8 & .9 & .5 \\ .7 & .8 & .4 & .6 \\ .7 & .4 & .6 & .8 \end{bmatrix} \qquad (12.10)$$

Assume the "little" difference is set to be 0.2. Two pixels are λ-adjacent if the difference between two adjacent elements are not greater than 0.2. It is easy to see

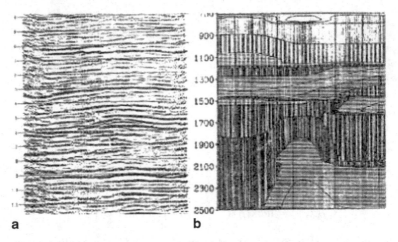

Fig. 12.7 Examples of waveform data and gray-scale seismic data. (**a**) A waveform seismic section, and (**b**) a velocity section

that there are three λ-connected components in the image. The second component is shown below:

$$
\begin{bmatrix}
 & & 10. & 8.0 \\
 & 8.0 & 9.0 & \\
7.0 & 8.0 & & \\
7.0 & & &
\end{bmatrix}. \tag{12.11}
$$

The other two components are located at the up-left corner and the low-right corner of the original matrix (1), respectively.

In seismic data processing, researchers usually deal with two types of layers: the geological layer and the seismic layer. The geological layer is formed by sedimentary rocks from one geological time period to another. On the other hand, a seismic layer is distinguished by geophysical properties of sedimentary rocks represented in the form of seismic data in the stratum, the most popular in use are velocity and porosity. Since the geological layer cannot be identified just by using seismic data, the work described in this chapter is only the seismic layer search.

Seismic data has two forms in general: waveform seismic data, and gray-scale seismic data. Both waveform data and gray-scale seismic data can be viewed as digital images (See Fig. 12.7). A digital image can be represented as a function from a 2D/3D array to an interval in R, the real number set. Mathematically, such an array can be regarded as a graph if we define: (1) each element in the array is a vertex of the graph, (2) the edge set consists of all pairs of elements a, b, denoted by (a, b), where a and b are adjacent in the array.

To determine a layer in the stratum is to find a subset of vertices which have similar properties (or values) in the gray scale image. λ-connectedness is defined to describe the relationship among pixels in the image.

12.8.1.1 λ-connected Region Growing Segmentation

In a gray scale image, intensity is the uniformity measure. A region (or segment) in an image may be viewed as a connected group of pixels, all with similar brightness. The region growing method begins with a single pixel, and then by examining its neighbors to find a maximum sized connected region of similar pixels. In this manner, regions grow from single pixels. One can also use a region or grouped set of pixels as a seed instead of a single pixel. In this case, after selecting the partition (group of pixels), a uniformity test is applied to the region to see if it qualifies as a partition. If the test fails, the region is subdivided into smaller regions. This process is repeated until all regions are uniform. (The major advantage of using small regions rather than single pixels is that it reduces the sensitivity to noise.)

Region growing forms an equivalence relation to partition the image. λ-connected segmentation is used to partition the image by searching each λ-connected component in the image. The fast algorithm design technique such as depth-first-search or breadth-first-search can be used for implementation [14].

Algorithm 12.8 The breadth-first-search technique for λ-connected segmentation. This algorithm is very similar to standard breadth-first-search.

Step 1:   Let p_0 be a point in Σ. Set
$L(p_0) \leftarrow *$ and $Queue \leftarrow Queue \cup p_0$
i.e., labeling p_0 and p_0 is sent a Queue $Queue$.

Step 2:   If STACK is empty, go to step4; otherwise,
$p_0 \leftarrow Queue$ (top of $Queue$). Then,
$L(p_0) \leftarrow 0$.

Step 3:   For each p linking p_0, if
$L(p) \neq 0, L(p) \neq *$, and $C(p, p_0) \geq \lambda$, then
$Queue \leftarrow Queue \cup p$ and $L(p) \leftarrow *$. Then, goto Step2.

Step 4:   Stop. $S = \{p : L(p) = 0\}$ is one λ-connected part.

Algorithm 12.8 is time optimal, and the time complexity of it is $O(|n \cdot \Sigma_N|)$.

12.8.1.2 λ-connected Split-and-Merge Segmentation

We give an example of split-and-merge segmentation in Sect. 12.4. They key step of this method is to merge two segments together when they are adjacent along a splitting line. For the original split-and-merge segmentation, two segments (can be viewed as two big-pixels) can be merged if the maximum distance of the value of the merged segment to the mean value of the merged segment is smaller or equal to ϵ.

The idea of this method is as follows: One starts with tree nodes (representing square regions of the image) at some intermediate level of the quad-tree. If a square is found to be non-uniform, then it is replaced by its four son-squares (split). Conversely, if four son-squares are segmented and a region in a son-square can merge with a region in another son-square in terms of adjacency and uniformity, they will be merged until no more pair of regions that can be merged. These four son-squares are replaced by a single square(merge), called a segmented square. This process continues recursively until no further splits or merges are possible.

In traditional split-and-merge segmentation, the uniformity is measured by the mean of the merged region. This is a statistical measure. In λ-connected split-and-merge segmentation, we merge two regions into one if the merged region is λ-connected for any possible path in the region. Such a region is called a normal λ-connected set [7]. Split-and-merge segmentation only preserves reflexivity and symmetry and is not a mathematical partition or equivalence classification.

Algorithm 12.9   The split-and-merge technique for λ-normal-connectedness.

Step 1:   Let Σ_2 be a $(2^l \times 2^l)$ array. $L \leftarrow \Sigma_2$. And let L is a set.
Step 2:   If region L is already homogeneous, i.e. L is λ-normal-connected set,
           $NOMERGE \leftarrow NOMERGE \cup L$ and go to step 3; otherwise,
           divide L into four subsquares, L_{11}, L_{12}, L_{21}, and L_{22};
           then $NOSPLIT \leftarrow NOSPLIT \cup \{L_{11}, L_{12}, L_{21}, L_{22}\}$.
Step 3:   If there exists four regions which are four son of some region Ls
           in $NOMERGE$, namely L_{11}, L_{12}, L_{21}, and L_{22},
           they will be conquered by merging approach. It tests any two
           λ-normal-connected subset which are neighbor,
           and belongs to different regions in L_{11}, L_{12}, L_{21}, and L_{22}, merge
           these subset if their union is a λ-normal-connected subset of L.
           $NOMERGE \leftarrow NOMERGE - \{L_{11}, L_{12}, L_{21}, L_{22}\}$. Otherwise,
           if there are no such four regions; then stop when $NOSPLIT$ is empty,
           choose a region from $NOSPLIT$, send it to L, and goto Step2 else.

We can prove the following result: There is an $O(n|\Sigma_2| + |\Sigma_2|log|\Sigma_2|)$ time split-and-merge algorithm for normal λ-connected segmentation [7]. For more information about the λ-connected method, refer to [9].

12.8.2 λ-connected Segmentation for Big-pixels and BigData Related Sets

As we discuss in Sect. 12.3, the normalized cut used big-pixel technique to reduce the size of a graph. In fact, λ-connected split-and-merge segmentation is just a big-pixel related technique. λ-connectedness can be just defined on a general graph not only rectangle domain [9]. So use λ-connected search, we can find segments for big-pixels.

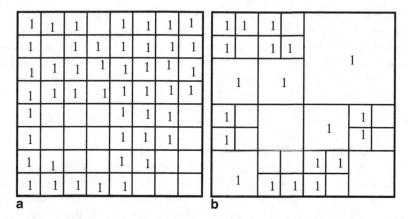

Fig. 12.8 The image and its quadtree representation. **a** an image, and **b** the quadtree representation of this image

When an image is already compressed in a quadtree format, we can even perform a segmentation without decode the data to a real image. In such a way, we can deal with BigData sets in a very fast manner.

In typical image segmentation applications, the domain is a rectangular region. Quadtree and octree representations are commonly used in medical imaging and spatial databases to compress data [20, 36].

If an image is stored or compressed by a quadtree, then the algorithm presented in this section provides a method that does not require restoration or decoding of the quadtree code before the image can be used. In other words, the quadtree partition is directly used to build a graph, and then a λ-connected segmentation is performed on the new graph. The advantage of using such a strategy is to significantly increase the segmentation speed.

A compressed image represented in the quadtree shall have a leaf index with value [13]. In Fig. 12.8, we represent an image into a quadtree format where "1" means that there is non-zero data points in the region with small variations. So each square marked as "1" is a big-pixel.

An image is split into four quadrants, namely Q_0, Q_1, Q_2 and Q_3, which represent the upper-left, upper-right, bottom-right and bottom-left quadrants, respectively. Specific formats are used to describe the structure of the compressed image in the quadtree representation. For example, $(Null, 0)$ means that the entire image is filled by "0," $(< 3 >, 128)$ means that the bottom-left quadrant is filled by "128," and $(< 2 >< 1 >, 255)$ means that the upper-right quadrant and the bottom-right quadrant of the image is filled by 255. In this example, the leaf size may be computed by: $\frac{n}{2^2}$ where n is the length of the image.

Typical image segmentation must go through each point so the time complexity must be at least $O(n^2)$ [14], where n is the length of the image and we assume $n = 2^k$. In the quadtree technique, a leaf $(< 2 >< 1 >, 255)$ will represent $n/2^2 \times n/2^2$ pixels.

Assume that the number of quadtree leaves is N. Then the segmentation algorithm can be described by first defining the adjacency graph $G_Q = (V_Q, E_Q)$ for the quadtree stored image where each leaf is a node in G_Q. If u and v are two adjacent leaves in V_Q, then $(u, v) \in E_Q$. In a 2D image, there are two types of neighborhood systems: the 4-neighbor system where each point has only four neighbors and the 8-neighbor system where each point has eight neighbors.

Lemma 12.1 G_Q *has at most* $3N$ *edges for a 4-neighbor system. And* G_Q *has at most* $4N$ *edges for an 8-neighbor system.*

For the detailed proof of this lemma, see [9, 10, 13].

How much time is needed to build G_Q? It depends on how the quadtree code is stored. Basically, there are two ways. In the first way, the quadtree code is stored in the "depth-first" mode and we do the following: recursively store the first quadrant, Q_0 and its off-springs, then store Q_1 and its off-springs, and so on. In this mode, we can build G_Q quickly since we only need to compare the neighbors in the quadtree code sequence to see if they are adjacent. According to Lemma 3.1, the time complexity of the algorithm is linear. (How we store an image in quadtree format is not the focus of this book. We can recursively generate the quadtree-code for $Q_0, ..., Q_3$. In this method, we only store the code if it is a leaf, which would only take $O(N)$ time.) In the second way, the quadtree code is stored in the "breadth-first" mode: we sequentially store the quadtree index codes for the largest blocks, then second largest blocks, and so on. The time to get G_Q may be longer, since we need to check if the current block is adjacent to any previous block. The time complexity could be $O(N \cdot N)$.

Lemma 12.1 provides us another advantage. We only need an $O(|V_Q|)$ time algorithm to perform the segmentation using λ-connectedness. The value of $|V_Q|$ is usually much smaller than $n \cdot n$, the original image size. Even through $|V_Q|$ is dependent on the actual image, it is very reasonable to say that the average is $O(n)$. Therefore,

Theorem 12.1 *There is an* $O(|V_Q|)$ *time algorithm to perform segmentation using* λ-*connectedness.*

Without decoding the quadtree code in the original image, we cannot perform a statistical mean-based segmentation since it is not a mathematical classification. A leaf (or a block) added to a segment probably does not satisfy the requirement of $|p - mean| < \delta$ since the mean may change. We may need to break a leaf to get a more precise segmentation. Developing this idea can lead to another algorithm: (1) separate the leaf into 4 sub-blocks, (2) if one sub-block can merge into the segment, repeat this step, (3) insert the rest of the sub-blocks into the quadtree code sequence, then repeat. This algorithm is faster than restoring the whole image, but is slower than λ-connected quadtree segmentation [10, 13].

This method of building graph G_Q has essential significance to direct calculate the homology groups without decode the image into real Euclidean space. The idea is this, quadtree representation does not change inward and outward points in the formula related to hole counting. we can get the exact number of holes just check the leaves of the quadtree and its neighbors to determine the number of inward and outward points. So we can use it for the calculation of genus directly. See Chaps. 6 and 14. for details including the definition of homology groups.

Storing images always generates the BigData sets. The algorithms discussed in this section will direct relate to BigData and data science for recent trend of studies. We will talk more about this in Chap. 15.

12.9 Remarks

In this chapter, we overview the geometric processing for many applications. Beginning with geometric search, we reviewed some of main algorithms in graphs. We use the famous the Google search algorithm (PageRank) as example to describe the importance of graph search algorithms and its matrix representation. Then we discuss another important application to image segmentation such as the normalized cut.

Secondly, we introduce spatial data structures especially quadtrees and R-trees. And their applications to Wireless network applications.

Thirdly, we review the major classification methods for geometric problems such as kNN and k-mean et al. We also introduce one of the most important applications in digital geometry—Thinning algorithms.

At last, we focus on the discrete method for incomplete data analysis using geometric method. We call it as λ-connected method.

For λ-connectedness, in fact, partial relations have been studied in other aspects. Random graph theory allows one to assign a probability to each edge of a graph [3]. This method assumes, in most cases, each edge has the same probability. On the other hand, Bayesian networks, are often used for inference and analysis when relationships between each pair of states/events, denoted by vertices, are known. These relationships are usually represented by conditional probabilities among these vertices and are usually obtained from outside of the system [25]. We can see that λ-connectedness is different from those techniques.

In fact, the gradually varied function introduced in Chap. 11 is the special case of λ-connected sets [10]. λ-connectedness can also be applied to numerical fitting problems [10, 11]. The relationship between a continuous function and a λ-connected function was investigated in [8–10]. The main material of the book relating to λ-connectedness is from [8, 11]. Graphics and numerical methods can also be used in geometric data visualization, see [17, 33].

References

1. S. Arya, T. Malamatos, and D. M. Mount. Space-time tradeoffs for approximate nearest neighbor searching. J. Assoc. Comput. Mach., 57:1–54, 2009.
2. M. Belkin, P. Niyogi, Laplacian Eigenmaps for Dimensionality Reduction and Data Representation, Neural Computation, June 2003; 15 (6):1373–1396.

3. B. Bollobas, *Random Graphs*, Academic Press. 1985.
4. S. Brin and L. Page, The anatomy of a large-scale hypertextual Web search engine, Computer Networks and ISDN Systems 30: 107–117. 1998.
5. G. Carlsson and A. Zomorodian, Theory of multidimensional persistence, Discrete and Computational Geometry, Volume 42, Number 1, July, 2009.
6. L. Cayton, Algorithms for manifold learning. Technical Report CS2008-0923, UCSD, 2005.
7. L. Chen, The λ-connected segmentation and the optimal algorithm for split-and-merge segmentation, Chinese J. Computers, 14(2), pp 321–331, 1991.
8. L. Chen, λ-Connectedness and Its Application to Image Segmentation, Recognition, and Reconstruction, University of Bedfordshire, U.K, July, 2001. (http://ethos.bl.uk/OrderDetails.do?uin=uk.bl.ethos.427595)
9. L. Chen, *Discrete Surfaces and Manifolds: A theory of digital-discrete geometry and topology*, 2004. SP Computing.
10. L. Chen, Digital Functions and Data Reconstruction, Springer, NY, 2013.
11. L. Chen, and O. Adjei. lambda-Connected Segmentation and Fitting, Proceedings of IEEE conference on System, Man, and Cybernetics 2004. 3500–3506.
12. L. Chen, and Y. Rong, Digital topological method for computing genus and the Betti numbers, Topology and its Applications, Volume 157, Issue 12, 2010, Pages 1931–1936.
13. L Chen, H. Zhu and W. Cui, Very Fast Region-Connected Segmentation for Spatial Data: Case Study, IEEE conference on System, Man, and Cybernetics, 2006.
14. T. H. Cormen, C. E. Leiserson, and R. L. Rivest, Introduction to Algorithms, MIT Press, 1993.
15. M. Demirbas, H. Ferhatosmanoglu, Peer-to-peer spatial queries in sensor networks, in 3rd IEEE Int. Conf. on Peer-to-Peer Computing, Linkoping, Sweden, Sept. 2003.
16. D. L. Donoho and C. Grimes. Hessian Eigenmaps: new locally linear embedding techniques for high-dimensional data. Technical Report TR-2003–08, Department of Statistics, Stanford University, 2003.
17. James D. Foley, Andries Van Dam, Steven K. Feiner and John F. Hughes, Computer Graphics: Principles and Practice. Addison-Wesley. 1995.
18. K. Fukunaga and L. D. Hostetler, The Estimation of the Gradient of a Density Function, with Applications in Pattern Recognition. IEEE Transactions on Information Theory, 21 (1): 32–40, 1975.
19. R. Ghrist, Barcodes: the persistent topology of data, Bull. Amer. Math. Soc., 45(1), 61–75, 2008.
20. R. C. Gonzalez, and R. Wood, *Digital Image Processing*, Addison-Wesley, Reading, MA, 1993.
21. J. Goodman, J. O'Rourke, Handbook of Discrete and Computational Geometry, CRC, 1997.
22. F. Harary, Graph theory, Addison-Wesley, Reading, Mass., 1969.
23. M. Hardt and A. Moitra. Algorithms and hardness for robust subspace recovery. In COLT, pages 354–375, 2013.
24. H. Homann, Implementation of a 3D thinning algorithm. Oxford University, Wolf- son Medical Vision Lab. 2007.
25. F. V. Jensen, Bayesian Networks and Decision Graphs, New York: Springer, 2001.
26. T. Kanungo, D. M. Mount, N. Netanyahu, C. Piatko, R. Silverman, and A. Y. Wu, A Local Search Approximation Algorithm for k-Means Clustering, Computational Geometry: Theory and Applications, 28 (2004), 89–112.
27. R. Klette and A. Rosenfeld, Digital Geometry, Geometric Methods for Digital Picture Analysis, series in computer graphics and geometric modeling. Morgan Kaufmann, 2004.
28. T. C. Lee, R. L. Kashyap, and C. N. Chu. Building skeleton models via 3-D medial surface/axis thinning algorithms. Computer Vision, Graphics, and Image Processing, 56(6):462–478, 1994.
29. A. Levinshtein, Low and Mid-level Shape Priors For Image Segmentation, PhD Thesis, Department of Computer Science University of Toronto, 2010.
30. T. M. Mitchell, Machine Learning, McGraw Hill, 1997.

31. L. Page, S. Brin, R. Motwani, and T. Winograd, The PageRank Citation Ranking: Bringing Order to the Web. Technical Report. Stanford InfoLab. 1999.
32. T. Pavilidis, Algorithms for Graphics and Image Processing, Computer Science Press, Rockville, MD, 1982.
33. W. H. Press, et al. *Numerical Recipes in C: The Art of Scientific Computing*, 2nd Ed., Cambridge Univ Press, 1993.
34. X. Ren, J. Malik, Learning a classification model for segmentation, Proc. IEEE International Conference on Computer Vision, pp. 10–17, 2003.
35. A. Rosenfeld and A. C. Kak, *Digital Picture Processing*, 2nd ed., Academic Press, New York, 1982.
36. H. Samet, The Design and Analysis of Spatial Data Structures. Addison Wesley, Reading, MA, 1990.
37. L. K. Saul and S. T. Roweis. Think Globally, Fit Locally: Unsupervised Learning of Low Dimensional Manifolds. Journal of Machine Learning Research, v4, pp. 119–155, 2003.
38. J. Shi and J. Malik, Normalized cuts and image segmentation, IEEE Transactions on pattern analysis and machine intelligence, pp 888–905, Vol. 22, No. 8, 2000.
39. J. B. Tenenbaum, V. de Silva, J. C. Langford, A Global Geometric Framework for Nonlinear Dimensionality Reduction, Science 290, (2000), 2319–2323.
40. S. Theodoridis and K. Koutroumbas, Pattern Recognition, Academic Press, FL, 2003.
41. T. Y. Zhang, C. Y. Suen, A fast parallel algorithm for thinning digital patterns, Communications of the ACM, v. 27 n. 3, p. 236–239, 1984.
42. Z. Zhang and H. Zha, Principal Manifolds and Nonlinear Dimension Reduction via Local Tangent Space Alignment, SIAM Journal on Scientific Computing 26 (1) (2005), 313–338.
43. B. Zheng, W.-C. Lee, and D. L. Lee. Spatial Queries in Wireless Broadcast Systems. Wireless Networks, 10(6):723–736, 2004.

Part V
Advanced Topics

Chapter 13
Discrete Methods in Differential Geometry

Abstract Nowadays, differential geometry is not only still one of the most profound research areas of mathematics after having had great influence in physics for more than a century, but it has also recently begun to play a very important role in computer graphics and image processing. The Poincare conjecture was believed to be proven by Perelman in 2004. However, other mathematicians are still looking into the details of the proof where not all parts are constructive. Researchers in digital topology have already started to explore the possibility of proving this conjecture in digital or discrete cases. This would also be exciting since a pure digital proof must be able to be implemented in terms of algorithms and would be constructive. In this chapter, we introduce the basic knowledge of differential geometry and some practical topics in its applications to computer graphics and computer vision. Due to the fact that differential geometry has a close relationship to variational analysis and harmonic functions, we also include a brief review of the principle of variational analysis. This chapter emphasizes some important topics of the discrete methods in differential geometry including circle packing, curvature flow, and minimum surface calculations.

Keywords Differential geometry · Discrete geometry · Riemannian metric · Curvature · Fundamental forms of surface · Gaussian-bonnet theorem · Discrete conforming mapping · Circle packing · Curvature flow · Minimum surface

13.1 Basics of Differential Geometry

Originally, differential geometry was the study of the geometric properties of curves and surfaces using differentials and integrals. The simplest example calculates the length of a smooth curve.

Later, people began studying the geometry of differential manifolds. The concept of differential manifolds is a very general one in that it is not defined in Euclidean space. Each point in such a manifold is contained in a neighborhood that is homeomorphic to a Euclidean space. Therefore, the distance from one point to another on a differential manifold must be defined in a dynamic way when the small Euclidean-like coordination frame moves from one to another in parallel. In this section, we mainly introduce differential geometry of surfaces.

© Springer International Publishing Switzerland 2014

L. M. Chen, *Digital and Discrete Geometry*, DOI 10.1007/978-3-319-12099-7_13

13.1.1   The Riemannian Metric

In Euclidean space, the distance between two points is measured by the Pythagorean theorem. Its differential form is

$$ds^2 = du^2 + dv^2$$

for a curve s in the u, v plane. Extending this formula to a more general form, we have

$$ds^2 = E du^2 + F_1 du\, dv + F_2 dv\, du + G dv^2$$

where A, F_1, F_2, and G are functions of u and v. Such a distance measure is called the Riemannian metric. When we require that the coefficients of $dv\, du$ and $du\, dv$ be the same, we will have

$$ds^2 = E du^2 + 2F du\, dv + G dv^2 \qquad (13.1)$$

Differential geometry using the Riemannian metric is called Riemannian geometry. A differentiable manifold with Riemannian metric is called a Riemannian manifold. We can see that when $F = 0$ and E and G are 1, this metric is the Euclidean metric.

Example 13.1 We gave the arc length of a sphere in Chap. 3, Formula (3.10). Using the Riemannian metric form, we can prove that

$$ds^2 = r^2 du^2 + r^2 \cdot sin^2(u) dv^2 \qquad (13.2)$$

where r is a constant radius and u, v are two angles in spherical coordinates. In this case, $E = r^2$, $F = 0$, and $G = r^2 \cdot sin^2(u)$.   □

   In general, for n variables, a Riemannian metric in a local coordinate system can be written as

$$g = \Sigma_{i,j}^n g_{ij}(x) dx_i dx_j \qquad (13.3)$$

where $g_{ij}(x) = g_{ji}(x)$ and the matrix $[g_{ji}(x)]_{n \times n}$, called Riemannian metric tensor, is a positive definite matrix. This means that g is always positive when dx_i is not all zero.

13.1.2   Curvatures of Surfaces

We defined the curvature of a curve $C(t)$ in Chap. 10 as the second derivative to t, $\frac{d^2C}{dt^2}$. See Eq. (10.4).
   To understand the meaning of the curvature in $C(t)$ on a plane, we can think of it as a measure of how sensitive the tangent line is when its base point moves other

Fig. 13.1 A point $p \in S$ and its tangent plane T_p. Two planes A and B containing **n**, the normal line at p. There are two normal curvatures $\mathbf{k_n}(a)$ and $\mathbf{k_n}(b)$ of curve a in A and curve b in B, respectively

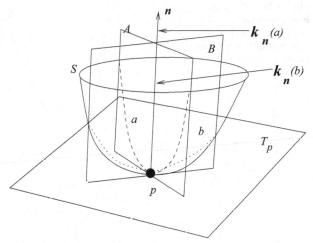

nearby points. If $C(t)$ is a straight line, then the curvature would be zero. Therefore, the curvature of a large circle would be comparatively smaller than that of a small circle. Thus the curvature of a circle is the reciprocal of its radius r: $\kappa = \frac{1}{r}$.

Curvatures on surfaces are much more complex, and they have many types. For the different types of curvatures on a surface S, we first calculate the maximum value of the curvature and the minimum value of the curvature around a point $p \in S$. The curves we select on S from the plane are perpendicular to the tangent plane T_p at point p. See Fig. 13.1. The curvatures of curves $a = S \cap A$ and $b = S \cap B$ are called normal curvatures.

It follows that the maximum and minimum values of the normal curvatures at a point are called the principal curvatures, κ_1 and κ_2. Based on the principal curvatures, two types of important curvatures can be defined. These are the Gaussian curvature and the mean curvature.

$$K_G = \kappa_1 \cdot \kappa_2, \tag{13.4}$$

and,

$$H = \frac{\kappa_1 + \kappa_2}{2}. \tag{13.5}$$

More general definitions of curvatures are illustrated in Fig. 13.2. (This part may be difficult to understand.) Again, let T_p be the tangent plane at point p on the surface. **n** is the normal line at p on T. n is perpendicular to T_p. Let C be an arbitrary curve on a surface S and K be the curvature of curve C. The tangent vector at point p of curve C will have a normal vector N with respect to C. The projection of k to **n** is called the normal vector, and the projection of **k** to T_p is called the geodesic curvature $\mathbf{k_g}$. Therefore,

$$\mathbf{k} = \mathbf{k_n} + \mathbf{k_g}$$

Fig. 13.2 Curvatures:
Relationship among
curvatures on a curve and
their normal curvature and
geodesic curvature

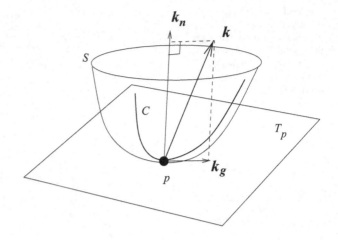

where $\mathbf{k}$, $\mathbf{k_n}$, and $\mathbf{k_g}$ are vectors. We can also write $\mathbf{k} = k \cdot N$. Their scalar values are
denoted as (geodesic curvature) k_g, and (normal curvature) k_n.

Note that two different curves passing through p on S that have different curvatures
($\mathbf{k}$) may have the same normal curvature.

13.1.3   Fundamental Forms of Surfaces

To describe the characteristics of a surface, two facts are the most important: the
length between two points and the length of the curvatures on the surface. Therefore,
there are two fundamental forms for surfaces that describe these properties. They are
called the first fundamental form and the second fundamental form.

These two forms are also related since the second fundamental form is the derivative of the first fundamental form (at the normal vector of the tangent plane of a point
on the surface).

The first fundamental form is just the metric form (13.1):

$$I = E\,du^2 + 2F\,du\,dv + G\,dv^2.$$

To understand the first fundamental form, we can rewrite the differential arc length
formula of a curve $s(t) = (u(t), v(t))$ on a surface $S = r(u, v)$ in Chap. 10, we have

$$ds^2 = r_u \cdot r_u\,du^2 + 2r_u \cdot r_v\,du\,dv + r_v \cdot r_v\,dv^2.$$

Therefore, the first fundamental form is just the generalization of the above formula. The Riemann metric tensor for this case is: $g_{11} = (r_u)^2$; $g_{12} = g_{21} = r_u \cdot r_v$;
$g_{22} = r_v \cdot r_v$.

The second fundamental form describes how curviness of a surface, what we call
curvature. Thinking about the tangent plane at a point on the surface, with respect

to the tangent plane, we rotate our space to make this tangent plane our (local) domain plane. For example, if the tangent plane is parallel to the xy-plane, so the first derivative of the surface would be zero (see Fig. 13.2).

Therefore, the second derivative determines the curvature of the surface. If we use Taylor expansion, we have,

$$r(u + du, v + dv) = r(u, v) + r_u du + r_v dv + 1/2(r_{uu} du^2 + 2r_{uv} du\, dv + r_{vv} dv^2)$$
$$+ \cdots . \tag{13.6}$$

thus,

$$dr = r(u + du, v + dv) - r(u, v) = 1/2(r_{uu} du^2 + 2r_{uv} du\, dv + r_{vv} dv^2). \tag{13.7}$$

This is the differentiation of the direction of a curve on a surface at point $(u + du, v + dv)$. This is very similar to the Hessian matrix that determines the extreme points of a function $f(x, y)$. In general, we use tangent planes to replace the xy-plane, which is the vector perpendicular to the tangent plane at (u, v). Therefore, this local coordinate system is a moving frame (or tangent plane in this case).

The second fundamental form is the generalization of (13.7)

$$II = L\, du^2 + 2M\, du\, dv + N\, dv^2. \tag{13.8}$$

So, the second fundamental form determines the curvature of a surface (in the local sense).

In calculus, we know that the determinate of the Hessian matrix can determine the type of extreme points. $LM - N^2$ will also have the same functionality as $f_{xx} f_{yy} - (f_{xy})^2$ in the Hessian matrix.

To understand the function f on an xy-plane is equivalent to rotating any point on the Ricmann manifold such that its tangent plane is the xy-plane in the local sense. That is why the second fundamental form is the generalized Hessian matrix. Moving the frame along with the normal vector of the tangent plane of points on a curve in the manifold is called parallel transport. We discuss the important concept of connection in the later sections of this chapter. (This is also related to the Gaussian map.) In higher dimensional space, the supersurface, an $(n - 1)$-manifold, can also have the generalization of the second fundamental form.

We define the normal curvature of the general Riemannian metric as :

$$k_n = \frac{II}{I} = \frac{L du^2 + 2M du\, dv + N dv^2}{E du^2 + 2F du\, dv + G dv^2}$$

The Gaussian curvature $K_G = \kappa_1 \cdot \kappa_2$ of a surface is given by

$$K_G = \frac{LN - M^2}{EG - F^2} \tag{13.9}$$

The mean curvature is given by

$$H = \frac{LG - 2MF + NE}{2(EG - F^2)} \tag{13.10}$$

Modern Riemannian geometry mainly studies global properties. Since any differential manifold can be embedded into a Euclidean space isomorphically proved by Nash, this means that the shape can be retained without deformation. On the other hand, any differential manifold can be decomposed piecewisely. Therefore, in this book, what we discuss regarding discrete geometry of the differential manifold in Euclidean space is mathematically valid. For details, refer to Thurston's book [27].

13.1.4 Gauss-Bonnet Theorem

The Gauss-Bonnet Theorem is not only one of the fundamental theorems in differential geometry, but it is also one of the most important results in the entire field of mathematics. This single theorem connects very well different branches of modern mathematics, including topology, geometry, and algebra. Its generalization, the Atiyah-Singer index theorem has many applications in theoretical physics. We first present a simpler form of the Gauss-Bonnet theorem and then explain its meaning.

Theorem 13.1 *Suppose S is a closed surface or compact two-dimensional Riemannian manifold,*

$$\int_S K_G \, dA = 2\pi \chi(S), \tag{13.11}$$

where K_G is the Gauss curvature and $\chi(S)$ is the Euler characteristic of S.

We know that the Euler characteristic of S is the number related to genus in from Chap. 9, Eq. (9.2). We also know that

$$\chi(S) = 2 - 2g$$

for a closed surface. If the surface is simply connected, without any holes, then $g = 0$ and

$$\int_S K_G \, ds = 4\pi,$$

The integral of the Gauss curvature on a closed surface is a constant. This theorem has significant applications in 3D image processing where the goal is to find the number of holes in a object, which we discuss in Chap. 14. It states: If M is a closed 2-dimensional digital manifold, the genus g is

$$g = 1 + (M_5 + 2M_6 - M_3)/8,$$

where M_i indicates the number of surface-points, each of which has i adjacent points on the surface. This is the simplest formula for the Gauss-Bonnet theorem in 3D digital space [4].

Consider the boundary ∂S. Then the Gauss-Bonnet theorem will be,

$$\int_S K_G \, dA + \int_{\partial S} k_g \, ds = 2\pi \chi(S),$$

where k_g is the geodesic curvature of ∂S [14].

13.2   Variational Principle and Harmonic Analysis

In this section, we introduce the variational principle and harmonic functions. The variational principle came from the solution of the Dirichlet problem: given a continuous function f on boundary ∂D of D, is there a differentiable extension F of f on D?

The solution resulted in the discovery of the minimum energy function for an integral:

$$E[f(x)] = \int_{\Omega} (f_x^2 + f_y^2) dx dy. \qquad (13.12)$$

Usually $|\nabla f|^2$ replaces $(f_x^2 + f_y^2)$ for a more general representation. ∇f is the gradient in the vector field with unit vectors $\mathbf{e}_i$:

$$\nabla f = \frac{\partial f}{\partial x_1} \mathbf{e}_1 + \cdots + \frac{\partial f}{\partial x_n} \mathbf{e}_n$$

The Dirichlet problem is also important in mathematical physics. The solution f to this problem is unique and satisfies the following equation

$$\frac{\partial^2 f}{\partial x^2} + \frac{\partial^2 f}{\partial y^2} = 0 \qquad (13.13)$$

This is called a harmonic function.

The method used in finding the solution is called the variational principle, which laid out the foundation for functional analysis.

Harmonic functions have important applications in science and engineering. The linear function is always harmonic.

Harmonic functions also play an important role in solving the famous minimal surface problem [7]. We discuss its algorithm in the later sections of this chapter.

Harmonic functions have some important properties: (1) The maximum and minimum values must be on the boundary of the domain (if f is not a constant), and (2) The average of the values in a neighborhood circle equals the value at the center of the circle.

For a simply connected 2D region D and its boundary J, we have the following:

Theorem 13.2 *For a bounded region D and its boundary J, if f on J is continuous, then there is a unique harmonic extension F of f such that the extension is harmonic in D − J.*

More generally, the solution for the Dirichlet problem can be extended to a connected with genus greater than 0 [7].

This theorem plays a significant role in existence of conformal mapping and circle packing we will discuss next. Especially, the algorithm we present next in circle packing.

Variational principle is an important tool for geometric problems in data processing. A functional $E[f, B]$ was proposed by Mumford and Shah for image

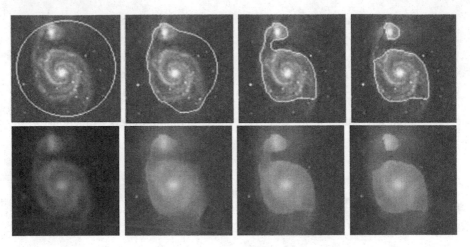

Fig. 13.3 Image segmentation using modified a modified Mumford-Shah functional

segmentation [10]. (We introduced the concept of image segmentation in Chaps. 3 and 12.) This has become a trendy research field in discrete geometry.

For an image F in a domain D, let f be the model (a reconstructed image) and let B be the boundaries of each segmented component of f. The Mumford-Shah functional is defined as [18]

$$E[f, B] = \alpha \int_D (F - f)^2 dA + \beta \int_{D-B} |\nabla f|^2 dA + \gamma \int_B ds$$

where $\alpha + \beta + \gamma = 1$ are weights. This formula means that the difference between the original image and fitted image should be small because the first term is $\int_D (F - f)^2 dA$. The internal variation of f should be small because of the second term (the standard segmentation has the same value in a segmented component). The total boundary length should also be small in many cases. In Fig. 13.3, Vese and Chan obtained a nice segmentation using a modified Mumford-Shah functional. The first row is the original data F and the second row is the fitted data f. The second column is the intermediate iteration result and the last column is the final result of the segmentation [28].

13.3   Discrete Conforming Geometry

The conforming mapping came from complex analysis for angle-preserving transformation in 2D complex plane. A simply connected 2D region can always conformal map to a disk. This is the famous Riemann conforming mapping theorem. Riemann used a simple construction of some complex functions to prove this theorem. This theorem is also valid for any bounded 2D surface due to the Dirichlet principle [7].

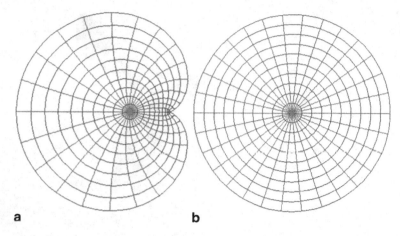

Fig. 13.4 Conforming mapping with angle-preserving: **a** a deformed disk, and **b** a disk

Fig. 13.5 Conforming mapping for geometric design by Y. Yang et al.: **a** planar design, and **b** the 3D realization

We can have an example that shows the conforming mapping from a deformed disk back to a disk. See Fig. 13.4.[1]

Researchers in computer graphics sometimes use conforming mapping to make 3D pictures or geometric design of architectures. See Fig. 13.5 [2, 29]

The question remaining here is how we actually make such a conformal mapping when an actual 2D region is given? In other words, how do we design an algorithm to construct such a mapping for an arbitrary boundary?

One of the solutions was suggested by W. Thurston in 1985. His idea is to use so called circle packing. We will next introduce the circle packing for obtaining the conforming mapping, and then discuss the direct harmonic function for discrete conforming mapping.

---

[1] The example was made by the Java applet by Michael A. Lee and Kevin E. Schmidt on conforming mapping at $http://fermi.la.asu.edu/ccli/applets/confmap/index.html$.

Fig. 13.6 Circle packing on
3D surface by Yang

13.3.1   Circle Packing

Circle packing is to use circles to cover a given surface such that all circles touch another without overlapping. The circles can be equal or vary in sizes. If two circles are tangent, we can link a line segment from two centers of the circles. These line-segments will form a partition of the surface. Therefore, circle packing is a method of decomposition of a surface. However, since circles cannot fill the total space of a surface, we use terminology of density to describe the ratio. Figure 13.6 gives an example of circle packing on 3D [29].

Researchers in computational geometry or algorithmic geometry are interested in how the best density circle packing can be [11]. In this section, we focus on its relationship to conforming mapping.

For an 2D region D, Thurston observed that circle packing could be used to approximate conformal mappings: (1) First, pack small circles with radius r in a hexagonal tessellation of the plane, (a circle will tangent to six circles within region D), (2) constructs a planar graph G from the intersection graph of the circles, and (3) add an extra vertex that is adjacent to all the circles on the boundary of D. We just present the main idea below:

Let C be a 2D disk having the boundary circle with radius R. The planar graph G can be represented by a circle packing of C. The circles from the packing of D has a one-to-one mapping to the circles in C.

Thurston's conjecture of the circle packing: Using the Mobius transformation, when the radius r approaches zero, the functions from the packing of D to C constructed in this way would approach the conformal function given by the Riemann mapping theorem. Thurston's conjecture was proven by Rodin and Sullivan [24]. Collins and Stephenson designed an algorithm for circle packings [6, 26]. This algorithm use angle relaxation for convergence (Fig. 13.7).

Fig. 13.7 Conforming
mapping using circle packing:
a Original packing, and **b** the
result of Thurston's algorithm
modified by Stephenson and
his colleagues

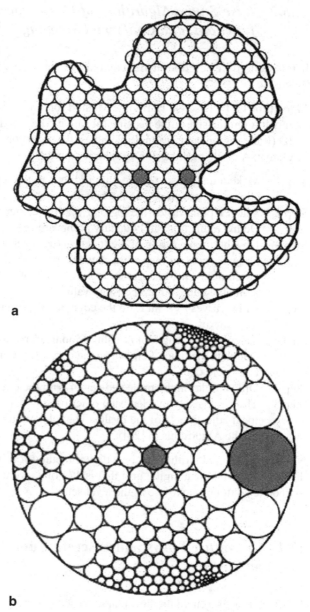

13.3.2   Case Study: Algorithms of Circle Packing, Harmonic functions, and Conformal Mapping

In this section, we describe an idea combining ideas from [26] and discrete harmonic functions.

Algorithm 13.1   Discrete Harmonic Method for circle packing.

   Input: A unique size circle packing (hexagon type) for a simply connected region in 2D (If the region contains multiple holes, the algorithm is similar).

   Output: A circle packing of a circular region C.

Step 1   Make a unique size circle with radius ϵ in the region (hexagon type). Cut all incomplete circles on the boundary out.

Step 2   Select a point inside the region c_0 (at the center of the region if possible). Define a function: all boundary points are zero and $f(c_0) = 1$.

Step 3   Do a harmonic fitting of the function (using discrete method [3]).

Step 4   Draw the contour map of the region A based on the fitted value (isosurface). (A contour line, also called isoline, is a curve where the function values along the curve have a constant value.)

Step 5   Draw curves r_θ from c_0 to the edge points with right-angles to all contour map curves.

Step 6   Calculate the length of r_θ using Mobius transformation (any one or all) to calculate the ratio of circles on the curve. Compare this to the radius of the big-circle C.

Step 7   Draw circles in the projected center with the circle's original radius.

Step 8   The circles may overlap each other or not be attached (hexagon type). Do local adjustments of the ovals. (The relaxation method can be used based on the Algorithms in [26]).

Now the detailed algorithm can be a digital method for faster performance if using digital grids. Making a digital grid that is $1/4$ of ϵ, we use digital harmonic fitting to find the contour map to use as the level set.

13.4   Curvature Flows and Discrete Curvature Flows on Surfaces

The Ricci flow is a curvature flow that was used to attempt to solve the Poincare conjecture by Hamilton and Perelman. Most of mathematicians believed that Perelman and others have completed the proof [17]. The Poincare conjecture was one of the most famous unsolved problems in mathematics. It states that every simply connected closed 3D manifold is homeomorphic to the 3D sphere S^3. Some mathematicians are still looking into the details of the proof where not all parts are constructive. A. V. Evako suggested to study the Poincare conjecture in digital space so that it would

Fig. 13.8 Riemann curvature $R(X, Y)Z$ where X, Y, and Z are vector fields

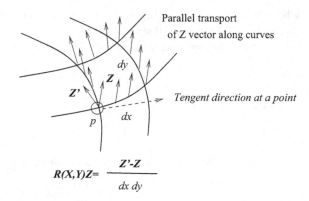

Parallel transport of Z vector along curves

Tengent direction at a point

$$R(X,Y)Z= \frac{Z'\text{-}Z}{dx\,dy}$$

give an algorithmic proof. This would also be interesting since a pure digital proof must be able to be implemented and must be constructive [21].

Hamilton did design a plan for proving the Poincare conjecture, first by putting a Riemannian metric on a simply connected closed 3-manifold and then adjusting this metric by modifying the curvature at each point. Until the manifold has a constant curvature, it must be a 3-sphere. The idea is simple in this process. Thinking about a deformed circle, in order to make it a rounded circle again, we just need to expand the inward (negative curvature) parts of the circle and contract the outward (positive curvature) parts. We use Ricci flow to make the negative curvature points positive and reduce the value of the curvature if it is too large.

This short section only provides a basic introduction to this newest development in differential geometry and its applications to discrete problems. We only intend to show some related concepts for the purpose of future uses in massive geometric data processing.

Let us first introduce the Riemann curvature tensor, which is defined on the Levi-Civita connection. The Levi-Civita connection is similar to the directional derivative in Eucldean space. It is called the covariant derivative. Let C be a curve on a manifold M, where p and q are two closed points and T_p and T_q are two tangent space at p and q, respectively. u is the tangent vector at p and v is the tangent vector at q along C. Therefore, we can calculate the directional derivative of v in the direction of u (on C), $D_u(v)$. Since $D_u(v)$ may or may not be in T_p, we make a projection perpendicular to T_p, where the projected vector is called the covariant derivative or connection and is denoted by $\nabla_u v$. In fact, in order to make the projection, we must move the vector from q to p, a move called parallel transport. This definition is valid for any vector field (and its related bundle).

Intuitively, the Riemann curvature $R(X, Y)Z$ at point p can be defined as follows: Let Z' and Z be two unit vectors where Z' is a parallel transport of Z starting at p and traveling along the boundary of a small cell and returning to point p. See Fig. 13.8. Note that $R(X, Y)Z$ is equal to $Z' - Z$ divided by the area of the small cell.

Formally, the curvature tensor can be defined using Levi-Civita connection:

$$R(u, v)w = \nabla_u \nabla_v w - \nabla_v \nabla_u w - \nabla_{[u,v]} w \qquad (13.14)$$

where $[u, v]$ is the Lie bracket for two vector fields u and v. It satisfies $[X, Y](f) = X(Y(f)) - Y(X(f))$ for all smooth functions f.

The Riemann curvature was expressed in the coordinate system,

$$R^a_{\sigma\mu\nu} = dx^a(R(\partial_\mu, \partial_\nu)\partial_\sigma)$$

Its tensor form is the following.

$$R_{a\sigma\mu\nu} = g_{ab}R^b_{\sigma\mu\nu}.$$

The Ricci curvature is a special curvature defined by the Riemann curvature:

$$Ric(u) = \sum_i R(u, e_i)e_i, \tag{13.15}$$

where $\{e_i\}$ is an orthonormal basis of the tangent space at a point in M. We have Ricci curvature tensor

$$R_{ab} = R^k{}_{akb}.$$

Then, the Ricci flow equation is given by

$$\partial_t g_{ij} = -2R_{ij}.$$

It is to say that if we have a Riemannian manifold with metric tensor g_{ij} (see (13.15)), we can compute the Ricci tensor R_{ij}. The calculation can be done by using the so called sectional curvature,

$$\kappa(u, v) = \langle R(u, v)v, u \rangle.$$

where u and v are the orthonormal vectors on the tangent space at p and $\langle \cdot, \cdot \rangle$ is the inner product. The other way of simplifying the equation is viewing the flow equation as

$$\partial_t g_{ij} = -2K_G g_{ij}.$$

where K_G can be the Gaussian curvature if M is a 2D manifold.

The Ricci flow equation does not suggest any actual algorithms for modifying a 3D manifold into a 3-sphere, since checking which point has the positive curvature on a continuous space is not possible. Chow and Luo introduced an equation called combinatorial Ricci flow for this computational purpose [5]. They used a weighted triangulation based on the circle-packing algorithm to get the initial curvature in combinatorial form. Then, they used a procedure to modify the angles of the triangles to converge to a constant curvature.

In Chow-Luo's definition, let $A^{(i)}$ be the cone type angle at the vertex v_i which is the sum of all inner angles having vertex v_i. The discrete Gaussian curvature K_i at v_i is defined to be $2\pi - A^{(i)}$. This definition is compatible with the definition we will use in Chap. 14 as well. The combinatorial Ricci flow is the following,

Fig. 13.9 The minimum surface example of a tent of a musical event

Let the curvature K_i at v_i be defined as $2\pi - \Sigma a_i$. The combinatorial Ricci flow is the following:

$$\frac{dr_i}{dt} = -K_i \cdot r_i.$$

Compare this with the normalized Ricci flow that is defined for all vertices:

$$\frac{dr_i}{dt} = -(K_i - K_{avg}) \cdot r_i.$$

Chow-Luo's method makes K_i closer to K_{avg} computationally [5]. A partial implementation of this method can be found in [12]. A more efficient discrete flow is called combinatorial Yamabe flow were studied in [9, 16].

13.5 Discrete Minimum Surfaces

The problem of minimal surfaces is related to Plateau's problem that proves the existence of a minimal area surface with a given boundary. The minimal area surface is equivalent to having a mean curvature of zero in all inner points. Physicists like to make minimal surfaces by dipping a wire frame into a soap solution, generating a soap film. Another popular example is the tent of a musical event, see Fig. 13.9.

The mathematical solution of this problem is to find a function that satisfies the follow equation:

$$(1 + f_x^2)f_{yy} - 2f_x f_y f_{xy} + (1 + f_y^2)f_{xx} = 0. \tag{13.16}$$

This equation is different from a Laplace equation as $f_{xx} + f_{yy} = 0$. This problem was solved by Douglas and Rado using Dirichlet's principle and conformal mapping in the 1930s [7].

Computationally, solving the problem is not easy. It requires an iterated process that can approach a triangulated surface with boundaries near minimal surfaces. Several algorithms were proposed and the curvature was calculated in discrete models. The mean curvature was updated to be zero or near zero in 3D. Before we introduce the algorithms for the solution of discrete minimum surfaces, we first introduce the Laplace-Beltrami operator.

13.5.1   Laplace-Beltrami Operator

The Laplace-Beltrami equation is a generalization of the Laplace equation for functions in a plane. The Laplace-Beltrami equation is for functions with a domain of the arbitrary two-dimensional Riemannian manifold. It is the divergence of the gradient in short: $\Delta f = \operatorname{divgrad} f$. For a surface with local coordinates and first fundamental form (13.1), the Laplace-Beltrami equation defined as a tensor form is as follows

$$\Delta F = g^{ij}\left(\nabla_{X_i}\nabla_{X_j} F - \nabla_{\nabla_{X_i} X_j} F\right)$$

where g^{ij} is the inversion of g_{ij} in metric tensor and $\nabla_{X_j} F$ is a connection (covariant derivative). It can also be written as:

$$\Delta f = \frac{\partial}{\partial x}\left(\frac{(F f_y - G f_x)}{\sqrt{EG - F^2}}\right) + \frac{\partial}{\partial y}\left(\frac{(F f_x - G f_x)}{\sqrt{EG - F^2}}\right) \qquad (13.17)$$

For $E = G = 1$ and $F = 0$, Δf becomes the Laplace operator. $\Delta f = 0$ is a solution of the minimum of the Dirichlet integral on a surface with the metric form Eq. (13.1)

13.5.2   Algorithms of Discrete Minimum Surfaces

According to [22, 23] the numerical solution of mean curvature H_p, at discrete point p and surrounded by triangles with q_i as neighbors, can be represented as:

$$H_p = \frac{\frac{1}{2}\Sigma_{q_i}(cota_i + cotb_i)\cdot(p - q_i)}{A(V(p))}$$

where $A(V(p))$ is the area of the Voronoi cell $V(p)$. A digitized mean curvature will be derived based on this principle in Chap. 14.

The formula of $\frac{1}{2}\Sigma_{q_i}(cot(a_i) + cot(b_i))\cdot(p - q_i)$ is derived from the Laplace-Beltrami operator for triangulated surfaces. This is obtained by minimizing the Dirichlet energy over the triangulation. In general, the Laplace-Beltrami operator is for Riemann manifolds, which is the generalization of the Laplace operator in Euclidean space. In other words, the Dirichlet principle is valid for Riemann manifold domains, but its solution is the Laplace-Beltrami equation [20, 25].

To get a minimal surface requires designing an iterated procedure to make H_p zero for all inner points p in the triangulation. The algorithm according to Pinkall and Polthier is briefed as follows [22]:

Algorithm 13.2   The discrete minimal surface.

Step 1   Take the initial surface M_0 with boundary ∂M (the polygonal representation of ∂M) as the first approximation of M.

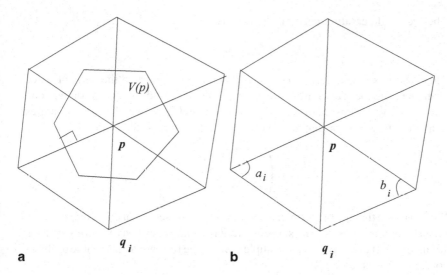

Fig. 13.10 The discrete mean curvature calculation: **a** the Voronoi area around vertex p, and **b** angles opposite to the edge (p, q_i)

Step 2   Compute the next surface M_{i+1} by solving the linear Dirichlet problem based M_i on with the minimal condition based on H_p.

Step 3   Set $i \leftarrow i + 1$ and continue to repeat Step 2.

Another way is to adjust each point p such that H_p approaches zero (or a constant for constant mean curvature surfaces). A spring model of the solution was proposed by Jiang et al. [13].

13.6   Remarks: From Gaussian Curvatures to Sectional Curvatures

Let M be a manifold [1, 8, 19], especially the hypersurface of a Riemann manifold. We can observe the curvature on M in each 2D tangent plane in the tangent space T_p at a point p of the manifold. This curvature is called sectional curvature. If u and v are two linearly independent tangent (unit) vectors at point p, then

$$K(u, v) = \frac{\langle R(u, v)v, u \rangle}{\langle u, u \rangle \langle v, v \rangle - \langle u, v \rangle^2}$$

is called the sectional curvature with regards to a plane containing u and v. We know R is the curvature tensor. If u and v are orthonormal, knowing that $\langle u, u \rangle \langle v, v \rangle - \langle u, v \rangle^2$ is the square of the area of parallelogram u,v, then

$$K(u, v) = \langle R(u, v)v, u \rangle.$$

The sectional curvature tensor is denoted as

$$[K(u_i, u_j)]_{n \times n}$$

This is an $n \times n$ matrix. $u_1, \cdots u_n$ are orthonormal vectors in the tangent space. We can use principal curvatures as examples to understand the nature of sectional curvatures. As we discussed in the second fundamental form, which is for curvatures, the curvature tensor regarding to two orthonormal vector variables are

$$\begin{pmatrix} L & M \\ M & N \end{pmatrix}.$$

We can prove that the two principal curvatures of this sectional curvature matrix is just the two eigenvalues. A Gaussian curvature is a multiplication of two eigenvalues of this matrix. Let us use this as an example to solve the following eigenvalue problem. According to formula (3.23)

$$Det(A - \lambda I) = 0$$

will be the solution of eigenvalues of matrix A. Therefore,

$$\begin{vmatrix} L - \lambda & M \\ M & N - \lambda \end{vmatrix}.$$

we have,

$$(L - \lambda)(N - \lambda) - M^2 = 0$$

then,

$$(\lambda^2 - (L + N))\lambda + (LN - M^2) = 0$$

Assuming that λ_1 and λ_2 are two roots of the equation, then $\lambda_1 \cdot \lambda_2 = (LN - M^2)$. In addition, $\frac{\lambda_1 + \lambda_2}{2} = \frac{L+N}{2}$. These are exactly the same value as K_G in Formula (13. 9) and H in Formula (13.10) when $E = G = 1$ and $F = 0$, respectively.

In general, for a manifold M in Euclidean space, the principal curvatures are the eigenvalues of its second fundamental form. If $\kappa_1 = \lambda_1, ..., \kappa_n = \lambda_n$ are the n principal curvatures at a point $p \in M$ and $x_1, ..., x_n$ are orthonormal eigenvectors in principal directions, then the sectional curvature of M at p is given by

$$K(x_i, x_j) = \kappa_i \kappa_j$$

This is the Gaussian curvature regarding to the tangent plane containing x_i and x_j. The general Gaussian curvature is $Det(K) = \kappa_1 \cdot \kappa_2 \cdots \kappa_n$. Lee's book has extensive discussions on this topic [15].

It is a fantastic relation between curvatures and eigenvalues. We have discussed that the principal components in Chap. 11 are the eigenvalues of the covariant matrix of random variables. We also discussed the eigenvalues of the PageRanking matrix is the indicator in Google search. Now, we have shown Riemann's idea that the curviness of a general manifold can be measured by looking at all the Gaussian curvatures of a 2D subsurface, which is also the multiplication of eigenvalues. (Good Mathematics are Always "Connected!")

In addition to the current trends of research on manifold learning, Riemann manifold learning was also studied. The topic is related to geometric processing that was covered in Chap. 12.

Acknowledgement The author would like to thank Dr. Feng Luo's comments on curvatures. Many thanks to Dr. Steffen Rohde for his comments and help on the algorithm about circle packing. Dr. K. Polthier provided some references that are very helpful as well.

References

1. P. S. Alexandrov, Combinatorial Topology, New York: Dover, 1998.
2. A. Bobenko, Y. U. Suris, Discrete Differential Geometry: Integrable Structure, AMS, 2008.
3. L. Chen, Digital Functions and Data Reconstruction, Springer, NY, 2013.
4. L. Chen, and Y. Rong, Digital topological method for computing genus and the Betti numbers, Topology and its Applications, Volume 157, Issue 12, 2010, Pages 1931–1936.
5. B. Chow, F. Luo, Combinatorial Ricci flows on surfaces. J. Differential Geom. 63 (2003), no. 1, 97–129.
6. C. Collins and K. Stephenson, A circle packing algorithm, Computational Geometry. Theory and Applications 25 (3): 233–256, 2003.
7. R. Courant, Dirichlet's principle, conformal mapping, and minimal surfaces, The Dover Publications, Inc., 2005.
8. H. S. M. Coxeter, Introduction to geometry, John Wiley, 1961.
9. D. Glickenstein, A combinatorial Yamabe flow in three dimensions. Topology 44 (2005), no. 4, 791–808.
10. R. C. Gonzalez, and R. Wood, *Digital Image Processing*, Addison-Wesley, Reading, MA, 1993.
11. J. Goodman, J. O'Rourke, Handbook of Discrete and Computational Geometry, CRC, 1997.
12. A. Henniges, T. Williams, M. Wilson, Combinatorial Ricci flows, University of Arizona Undergraduate Research Program, 2008 (Supervisor: Dr. David Glickenstein) http://math.arizona.edu/ura-reports/083/Wilson.Mitch/Midterm.pdf.
13. Y. Jiang, L. Chen, Q. Chen, Q. Peng, J. X. Chen, Computing discrete minimal surfaces using a nonlinear spring model, IEEE Computing in Science and Engineering, Vol 12m No. 6, 2010, pp 74–79.
14. E. Kreyszig, Differential Geometry, University of Toronto Press, 1959.
15. J. M. Lee Riemannian Manifolds: An Introduction to Curvature, Springer, 1997.
16. F. Luo, Combinatorial Yamabe flow on surfaces. Commun. Contemp. Math. 6 (2004), no. 5, 765–780.
17. J. Morgan, G. Tian, Ricci flow and the Poincare conjecture, Clay Mathematics Monographs, Cambridge, MA, 2007.
18. D. Mumford and J. Shah, "Optimal Approximations by Piecewise Smooth Functions and Associated Variational Problems", Communications on Pure and Applied Mathematics XLII (5): 577–685, 1989.

19. M. Newman, Elements of the Topology of Plane Sets of Points, Cambridge, London, 1954.
20. R. Osserman, A Survey of Minimal Surfaces. Dover, New York, 1986. in, http://www.polthier.info/articles/habil/polthier-habil2002.pdf
21. Personal communication with A. V. Evako on the Poincare conjecture in digital spaces.
22. U. Pinkall, K. Polthier. Computing discrete minimal surface and their conjugates. Exp. Math. 2(1): pp. 15–36, 1993.
23. K. Polthier, Polyhedral Surfaces of Constant Mean Curvature, Habilitationsschrift, Technische Universität Berlin, http://www.polthier.info/articles/habil/polthier-habil2002.pdf (2002)
24. Rodin, Burton; Sullivan, Dennis (1987), "The convergence of circle packings to the Riemann mapping", Journal of Differential Geometry 26 (2): 349–360.
25. M. Schiffer, D. C. Spencer, "Functionals of finite Riemann surfaces", Princeton Univ. Press (1954). also see (Laplace-Beltrami equation. Encyclopedia of Mathematics. URL: http://www.encyclopediaofmath.org/index.php?title=Laplace-Beltrami_equation&oldid=22707)
26. Ken Stephenson, "Introduction to Circle Packing: the Theory of Discrete Analytic functions", Cambridge University Press, 2005,
27. W. Thurston, Three-Dimensional Geometry and Topology, Volume 1, Princeton University Press, 1997.
28. L. A. Vese and T. F. Chan, A multiphase level set framework for image segmentation using the Mumford and Shah model, International Journal of Computer Vision 50(3), 271–293, 2002.
29. Y. L. Yang, Y. J. Yang, H. Pottmann, and N. J. Mitra, Shape space exploration of constrained meshes. ACM Trans. on Graph (Proc. of SIGGRAPH Asia) 30, 124:1–12., 2011.

Chapter 14
Advanced Digital Topology and Applications

Abstract In this chapter, we present advanced topics in digital geometry and topology including digital curvatures and their applications. First we give a brief overview of current development of computational topology that overlaps digital topology. Second, we introduce digital Gaussian curvatures and prove the digital form of the Gauss-Bonnet theorem. The new formula that calculates genus is $g = 1 + (|M_5| + 2 \cdot |M_6| - |M_3|)/8$ where M_i indicates the set of surface-points, each of which has i adjacent points on the surface. This formula provides the types of topological invariants such as genus and homology groups for 3D image processing. We also design a linear time algorithm that determines such invariants for digital spaces in 3D. Such computations could have applications in medical imaging as they can be used to identify patterns in 3D imaging. We then discuss the implementation of the method. After that, we introduce digital mean curvatures and its applications to 3D image classifications.

Keywords Digital topology · Digital homology · Image processing · Gaussian curvature · Mean curvature · Digital gaussian–bonnet theorem · Pattern recognition · Algorithm design

14.1 Topology and Computing

The most famous problem in topological computing is how to decide whether a simply connected 3D manifold is homeomorphic to a 3D sphere. This problem is called the Poincare conjecture. Even though this problem was believed to be solved in 2004. Some mathematicians are not convinced that the current proof of the theorem is totally constructive.

In any case, the actual procedure that converts a piece-wise linear representation of a simply connected 3D manifold into a sphere has not yet been found.

A problem related to this problem is called the 3-sphere recognition algorithm, which determines whether a triangulated 3-manifold is homeomorphic to the 3-sphere. However, the best algorithm runs in exponential time and also requires exponential space [33].

© Springer International Publishing Switzerland 2014

L. M. Chen, *Digital and Discrete Geometry,* DOI 10.1007/978-3-319-12099-7_14

If we can determine that a 3D triangulated manifold is simply connected, then it is surely homeomorphic to the 3-sphere based on Perelman's proof of the Poincare conjecture.

Computational Topology is a research area in computational geometry or algorithmic geometry. It solves the topological problems using algorithms, for example, homotopy groups and homology groups of a manifold [12, 18, 31].

Computing topological invariants has been of great importance in understanding the shape of an arbitrary 2D or 3D object [17]. The most powerful invariant of these objects is the fundamental group [16]. Unfortunately, fundamental groups are highly non-commutative and therefore difficult to work with. In fact, the general problem in determining whether two given groups are isomorphic is undecidable (meaning that there is no algorithm that can solve the problem) [29]. For fundamental groups of 3D objects, this problem is decidable but no practical algorithm has yet been found.

As a result, homology groups have received the more attention because their computations are more feasible and they still provide significant information about the shape of the object [11, 18]. However, for general simplicial complexes, the problem of computing homology groups is still not completely solved [3].

Using topology to calculate the topological properties of geometric objects is currently a popular topic in science and engineering. It is a fascinating thing since topology can be used in every day life and is not esoteric enough to only remain in textbooks. People do not have to be serious mathematicians to understand these topics.

Due to the development of cloud computing, many researchers have made considerable progress in homology computations including persistent homology and dimension reduction in geometric data processing. We have discussed this part in Chap. 11.

14.2   Digital Homology and Image Processing

Images are stored in digital space. For image related applications, the simplest topological question is determining how many holes are in an image. Counting them one by one is the obvious technology. However, this is very time consuming for big data sets as most images are 1000×1000 pixels today.

The digital method can solve this topological question in a reasonable amount of time.

Another development is even more crucial and important. This area is called persistent homology, which calculates how many holes are in each scale for scattered data or cloud data. Since we may not have a solid object to use in calculating these holes, we only use some random samples.

Each sample can represent an area or an m-dimensional ball, but we do not know how big the ball is. The persistent homology will tell us all the ways for this process of finding all possible holes in each level.

This leads to a main problem addressed in this chapter: Given a 3D object in 3D Euclidean space R^3, determine the homology groups of the object in the most effective way by only analyzing the digitization of the object.

In this chapter, we design an optimal algorithm with time complexity $O(n)$ to compute the genus and homology groups in 3D digital space, where n is the size of the input data. The method used is based on cubical images with direct adjacency, also called (6,26)-connectivity images in discrete geometry. There are only six types of local surface points in such a digital surface.

The Gauss-Bonnet theorem in differential geometry will be used to determine the genus of 2-dimensional digital surfaces. The new formula derived in this section that calculates genus is $g = 1 + (|M_5| + 2 \cdot |M_6| - |M_3|)/8$ where M_i indicates the set of surface-points, each of which has i adjacent points on the surface.

On the other hand, the mean curvature was also interesting in differential geometry especially in the calculation of minimum surfaces, we have discussed it in Chap. 13. In this chapter, we introduce digital mean curvatures. We use it in 3D image classification as a case study. In this chapter, we mainly discuss the solution made by digital topology, especially the application of digital curvatures.

14.3   Digital Curvatures in 3D

In this section, we give formulas for two important curvatures in digital space: the Gaussian curvature and the mean curvatures. These are called digital curvatures.

14.3.1   Digital Gaussian Curvatures

In 3D digital space, assume K_i is the digital Gaussian curvature of elements in M_i, $i = 3,4,5,6$ shown in Fig. 5.8 where M_i is the set of points each element has i adjacent points in the surface. In other words, M_3 is the set of corner point shown as Fig. 5.8a. M_4 are shown in b and c, M_5 has just one case in (d), and M_6 has two cases shown in (e) and (f).

Since the classification is so important to the proof of the main theorem of this chapter, we redraw Fig. 5.8 again as below:

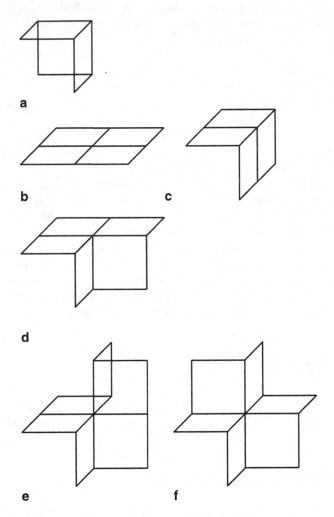

All types of simple surface points as shown in Fig. 5.8

Let us recall the Gauss-Bonnet theorem (13.11) that states if M is a closed manifold, then

$$\int_M K_G dA = 2\pi(2 - 2g) \qquad (14.1)$$

where K_G is the Gaussian curvature and g is the genus of M. Its discrete form is the following

$$\Sigma_{p \in M} K(p) = 2\pi(2 - 2g) \qquad (14.2)$$

where $K(p)$ is the Gaussian curvature at point p with omitting G in K_G without losing generality.

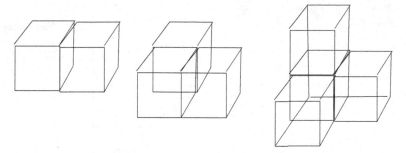

Fig. 14.1 Simple closed surface for digital Gaussian curvatures

We can prove the following lemma based on this formula.

Lemma 14.1 *Let K_i be the digital Gaussian curvature for the element in M_i, $i = 3, 4, 5, 6$.*

(1) $K_3 = \pi/2$,
(2) $K_4 = 0$, for both types of digital surface points,
(3) $K_5 = -\pi/2$, and
(4) $K_6 = -\pi$, for both types of digital surface points.

Proof This lemma can be proven by directly applying the discrete form of Gaussian curvatures in [15, 32]. We give another prove here.

(a) We can see that K_4 is always 0 since $K_G = K_1 \times K_2$ and one of the principal curvatures must be 0 for any point in M_4. We know that there exists a simply closed surface that contains only eight points of M_3 and several points within M_4. Therefore, $8 \cdot K_3 = 2\pi \cdot (2 - 0)$. See Fig. 14.1a. Therefore, $K_3 = \pi/2$. So we have proved (1) and (2) in the lemma.

(b) There is a simply closed surface that only contains ten points of the type in M_3, and two points in M_5. See Fig. 14.1b. Therefore, $K_5 = -K_3$. In Fig. 14.1c, we have 13 M_3 points and three M_5 point and one M_6 point. So, $13K_3 + 3K_5 + K_6 = 4\pi$. Therefore, $K_6 = -2\pi/2 = -\pi$.   □

Lemma 14.1 can also be calculated by the discrete Gaussian curvature theorem [32]. The curvature of the center point of the polyhedron is determined by

$$\int_M K_G dA = 2\pi - \Sigma_i \theta_i. \tag{14.3}$$

Therefore, we can obtain the same results as Lemma 14.1 using above equation. For example, in 3D digital space, the angle of one face is $\pi/2$. So, $K_5 = 2\pi - 5 \cdot \pi/2 = -\pi/2$. The proof of Lemma 14.1 is necessary if we are using the Gauss-Bonnet theorem first and then obtaining the curvature for each surface point. We include a proof here since the reference [32] does not contain such a proof.

Fig. 14.2 The Voronoi region
in digital space

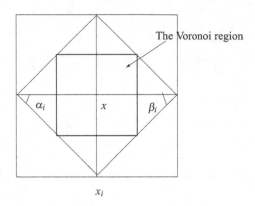

The Voronoi region

α_i x β_i

x_i

14.3.2   Digital Mean Curvatures

If H_i denotes the digital mean curvature of elements in M_i, $i = 3, 4, 5, 6$ in 3D digital space, then we have

Lemma 14.2

(1) $H_3 = \frac{4}{\sqrt{3}}$.
(2) $H_4 = 0$ for flat neighborhood, and $H_4 = \sqrt{2}$ for a bend neighborhood.
(3) $H_5 = \frac{4}{5}$.
(4) $H_6 = 0$, for both types of digital surface points.

This lemma can be proven easily based on the formula derived by Meyer et al. in 2002 [26].

A minimum surface can be defined as a surface whose mean curvature at every point is 0. There are several algorithms for obtaining minimum surfaces. Here we give a brief proof for Lemma 14.2. The formula derived in [26] for the mean curvature normal at point x is (also see Sect. 13.5.2.):

$$H(x) \cdot \mathbf{n} = \frac{1}{2 \cdot A} \Sigma_{x_i \in N(x)} (\cot \alpha_i + \cot \beta_i)(x - x_i) \qquad (14.4)$$

where N is the set of (discrete) neighbors of x. α_i and β_i are angles in two different triangles that are both opposite to line segment xx_i, which is shared by the triangles. A is called the Voronoi region of x. For digital space, the Voronoi region is easy to determine. See Fig. 14.2.

$$H(x) \cdot \mathbf{n} = \frac{1}{A} \Sigma_{x_i \in N(x)} (x - x_i) \qquad (14.5)$$

Note that $H(x) \cdot \mathbf{n}$ is a vector. $|(x - x_i)| = 1$ and $\cot \alpha_i = \cot \beta_i = 1$. in Fig. 14.2. $A = \frac{1}{4} \cdot i$ for M_i points. $|\Sigma x_i \in N(x)(x - x_i)| = \sqrt{(3)}$ for M_3; $|\Sigma x_i \in N(x)(x - x_i)| = 0$ for M_4 flat, and $\sqrt{(2)}$ for M_4 that is bend in Fig. 5.8 (3); $|\Sigma x_i \in N(x)(x - x_i)| = 1$ for M_5; $|\Sigma x_i \in N(x)(x - x_i)| = 0$ for M_6; Therefore, $H_3 = \frac{1}{3 \cdot \frac{1}{4}} \sqrt{(3)} = \frac{4}{\sqrt{(3)}}$. For bend M_4 points (Fig. 5.8c), $H_4 = \frac{1}{4\frac{1}{4}} \sqrt{(2)} = \sqrt{(2)}$. $H_5 = \frac{1}{5\frac{1}{4}} \cdot 1 = \frac{4}{5}$.

14.3.3 Digital Principal Curvatures

The principal curvatures at a point p of a surface, denoted κ_1 and κ_2, are the maximum and minimum value of curvatures of the curves on normal planes (intersecting with the surface). The relationship among the principal curvature, the Gaussian curvature, and the mean curvature were given in the formulas (13.4) and (13.5): $K = k_1 \cdot k_2$ and $H = (k_1 + k_2)/2$. Therefore [23, 26],

$$k_1 = H + \sqrt{H^2 - K}, k_2 = H - \sqrt{H^2 - K}$$

It is easy to get,

Lemma 14.3 (a) $k_1^{(3)} = \frac{4}{\sqrt{3}} + \sqrt{\frac{16}{3} - \frac{\pi}{2}} = 4.24913$, $k_2^{(3)} = \frac{4}{\sqrt{3}} - \sqrt{\frac{16}{3} - \frac{\pi}{2}} = 0.369675$.

(b) $k_1^{(4b)} = 0$ for a flat M_4 point,
$k_2^{(4b)} = 0$;
$k_1^{(4c)} = 2\sqrt{2} = 2.82843$, for a bend M_4 point
$k_2^{(4c)} = 0$.

(c) $k_1^{(5)} = \frac{4}{5} + \sqrt{\frac{16}{25} + \frac{\pi}{2}} = 2.28687$,

$k_2^{(5)} = \frac{4}{5} - \sqrt{\frac{16}{25} + \frac{\pi}{2}} = -0.686875$;

(d) $k_1^{(6)} = \sqrt{\pi} = 1.77245$,
$k_2^{(6)} = -\sqrt{\pi} = -1.77245$ for both types of M_6 digital surface points.

14.4 Gauss-Bonnet Theorem of Closed Digital Surfaces

Cubical space with direct adjacency, or (6,26)-connectivity space, has the simplest topology in 3D digital spaces. It is also believed to be sufficient for the topological property extraction of digital objects in 3D. Two points are said to be adjacent in (6,26)-connectivity space if the Euclidean distance between these two points is 1, i.e., direct adjacency.

Let S be a closed (orientable) digital surface in 3D grid space in direct adjacency. We know that there are exactly 6-types of digital surface points [5, 9].

In Chap. 9, we have proved a theorem using Euler's theorem of planar graphs. This theorem (Theorem 9.2) stated for a simply connected digital surface S, we have,

$$|M_3| = 8 + |M_5| + 2|M_6|.$$

where M_i (M_3, M_4, M_5, M_6) is a set of digital points with i neighbors [5]:

The limitation of this result is obvious since it is only for the simply connected surface. It cannot be used in calculating a surface with holes or genus. The following theorem have solved this problem completely. This theorem is the simplest form of the famous Gauss-Bonnet theorem.

Theorem 14.1 *(Chen-Rong [8]) If S is closed digital surface, we have*

$$g = 1 + (|M_5| + 2 \cdot |M_6| - |M_3|)/8. \tag{14.6}$$

Proof We use the digital form of the Gauss-Bonnet theorem (14.2) Its discrete form is

$$\Sigma_{\{p \text{ is a point in } s\}} K(p) = 2\pi \cdot (2 - 2g)$$

where g is the genus of S. So, we have

$$\Sigma_{i=3}^6 K_i \cdot |M_i| = 2\pi \cdot (2 - 2g),$$

$$\pi/2 \cdot |M_3| - \pi/2 \cdot |M_5| - \pi \cdot |M_6| = 2\pi \cdot (2 - 2g),$$

$$|M_3| - |M_5| - 2|M_6| = 4\pi \cdot (2 - 2g).$$

Therefore, $2g = 2 + (|M_5| + 2 \cdot |M_6| - |M_3|)/4$. Thus,

$$g = 1 + (|M_5| + 2 \cdot |M_6| - |M_3|)/8.$$

□

Without discussing intensively, we already know that the algorithm of calculating the genus of a digital surface in 3D is very simple. We just need to count for all points in M_i. In other words, given a closed 2D manifold, we can calculate the genus g by counting the number of points in M_3, M_5, and M_6. Then we can get the genus using Eq. (14.6).

Lemma 14.4 *There is an algorithm that can calculate the genus of S in linear time.*

Proof Scanning through all points (vertices) in S and counting the neighbors of each point, we can see that a point in M has four neighbors indicating that it is in M_4, as are M_5 and M_6. We can then put points in each category of M_i and then use formula (14.6) to calculate the genus g.   □

The two following examples show that the formula (14.6) is correct.

Example 14.1 The first example, shown in Fig. 14.3, is the easiest case. In Fig. 14.3a, there are eight points in M_3 and no points in M_5 or M_6. (To avoid conflict between the closest digital surface and the 3-cell [5], we can insert some M_4 points on the surface but not at the center point.) According to (14.6), $g = 0$.

We extend the surface to a genus 1 surface as shown in Fig. 14.3b where there are still eight M_3 points and also eight M_5 points. Thus, Fig. 14.3b satisfies Eq. (14.6).

In Fig. 14.3c, S has 16 M_5 points and 8 M_3 points. So $g = 2$. Using the same method, one can insert more handles.

Example 14.2 The second example comes from the Alexander horned sphere. See Fig. 14.4. First we show a "U" shaped base in Fig. 14.4a. It is easy to see that there are 12 M_3 points and 4 M_5 points. So $g = 0$ according to Eq. (14.6). Then, we attach

Fig. 14.3 Simple examples of closed surfaces with $g = 0, 1, 2$

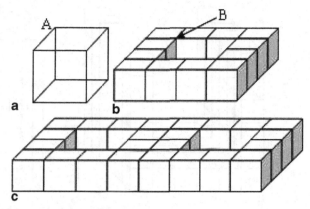

a handle to Fig. 14.4a shown in Fig. 14.4b. We have added 4 M_3 points and 12 M_5 points. $g = 1 + (|M_5| + 2 \cdot |M_6| - |M_3|)/8 = 1 + (4 + 12 - 12 - 4) = 1$. Finally, we add another handle to the other side of the "U" shape in Fig. 3a, the genus number increases by one since we still add 4 M_3 points and 12 M_5 points shown in Fig. 14.4c and we have. $g = 2$ for (c). For more complex cases like the Alexander horned sphere (the finite case), we only need to insert two smaller handles to an existing handle, so the genus will increase accordingly.

The theorem 14.1 is an amazing theorem in digital topology. It connects both differential geometry and discrete geometry in a perfect way. It is very practical since the digital space does not have an error in angle calculation. There is is no need to worry about the numerical error corrections.

The above idea can be extended to simplicial cells (triangulation) or even general CW k-cells. This is because for a closed discrete surface, we can calculate Gaussian curvature at each vertex point using formula (14.3). However, for real world problems, we have to consider the actual error accumulated when S is very big. Theoretically, we still have the following result.

Lemma 14.5 *There is an algorithm that can calculate the genus of a closed simplicial surface in $O(|E|)$ where E the set of 1-cells (edges).*

There are examples that $|E|$ is not linear to the number of vertices $|V|$.

14.5 Homology Groups of Manifolds in 3D Digital Space

We have introduced the concept of homology groups in Chap. 13. In this section, we can get complete expression of homology groups of a 3D manifold in 3D digital space. Homology groups are other invariants in topological classification. For a k-manifold, Homology group H_i, $i = 0, ..., k$ indicates the number of holes in each i-skeleton of the manifold [10, 12, 19, 22].

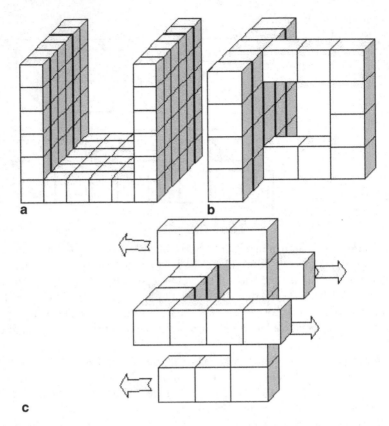

Fig. 14.4 An example that comes from the Alexander horned sphere in digital space

Consider a compact 3-dimensional manifold in R^3 whose boundary is represented by a surface. We show that its homology groups can be expressed in terms of its boundary surface (Theorem 14.2). This result follows from standard results in algebraic topology [16].

First, we recall some standard concepts and results in topology. Given a topological space M, its homology groups, $H_i(M)$, are certain measures of i-dimensional "holes" in M. For example if M is a solid torus, its first homology group, $H_1(M) \cong Z$, is generated by its Longitude, which goes around the obvious hole. For a precise definition, see [16]. Let $b_i = \mathrm{rank} H_i(M, Z)$ be the ith Betti number of M. The Euler characteristic of M is defined by

$$\chi(M) = \sum_{i \geq 0} (-1)^i b_i$$

If M is a 3-dimensional manifold, then $H_i(M) = 0$ for all $i > 3$ essentially because there are no i-dimensional holes. Therefore, $\chi(M) = b_0 - b_1 + b_2 - b_3$. Furthermore, if M is in R^3, it must have a nonempty boundary. This implies that $b_3 = 0$.

The following lemma is well known for 3-manifolds. It holds, with the same proof, for any odd dimensional manifolds. We only present the results in the section. For detailed proofs, refer to [8].

Lemma 14.6 *Let M be a compact orientable 3-manifold (which may or may not be in R^3).*

(1) If M is closed (i.e. $\partial M = \emptyset$), then $\chi(M) = 0$.
(2) In general, $\chi(M) = \frac{1}{2}\chi(\partial M)$.

Theorem 14.2 *Let M be a compact connected 3-manifold in S^3. Then*

(1) $H_0(M) \cong Z$.
(2) $H_1(M) \cong Z^{\frac{1}{2}b_1(\partial M)}$, i.e. $H_1(M)$ is torsion-free with a rank of half of rank $H_1(\partial M)$.
(3) $H_2(M) \cong Z^{n-1}$ where n is the number of components of ∂M.
(4) $H_3(M) = 0$ unless $M = S^3$.

14.6 Algorithm Design: Theory and Practical Implementation

The algorithm mentioned in Sect. 14.4 is a theoretical result. The implementation of the algorithm must consider all possible cases in practical data collection. We first need to find the boundary and then decide if the boundary is a 2D manifold [5, 25]. If the boundary data that connects voxel data sets are not purely defined digital surfaces [20, 21, 24], we will have three options: (1) we need to modify the data to meet the requirement before genus calculation, (2) if the change of the original data set is too great, we may need to stop the modification instead of outputting a result for reference, and (3) we make some limited changes, and then produce a result.

The difference between the theoretical results and practical data processing is that we may not always get the input data we expected. In our case, the boundary of a solid object should be treated as a surface. However, practically, this might not always be the case. Some researchers also consider making real data sets "well"-organized. Siqueira et al. considered making a 26-connected data set well-composed [36].

Our new algorithm and implementation will perform: (1) pathological cases detection and deletion, (2) raster space to point space (dual space) transformation, (3) linear time algorithm for boundary point classification, and (4) genus calculation.

14.6.1 A Linear Algorithm of Finding Homology Groups in 3D

Based on the results we presented in Sects. 14.4 and 14.5, we now describe a linear algorithm for computing the homology group of 3D objects in 3D digital space.

Assuming we only have a set of points in 3D. We can digitize this set into 3D digital spaces. There are two ways of doing so: (1) by treating each point as a cube-unit that is called the raster space, (2) by treating each point as a grid point, which is also called the point space. These two are dual spaces. Using the algorithm described in [5], we can determine whether the digitized set forms a 3D manifold in 3D space in direct adjacency for connectivity. The algorithm is in linear time.

Algorithm 14.1 Let us assume that we have a connected M that is a 3D digital manifold in 3D. $H_i = H_i(M)$ is the i-th homology group in this section.

Step 1. Track the boundary of M, $S = \partial M$, which is a union of several closed surfaces. This algorithm only needs to scan though all the points in M to see if the point is linked to a point outside of M. That point will be on boundary.

Step 2. Calculate the genus of each closed surface in ∂M using the method described in Sect. 2. We just need to count the number of neighbors on a surface. and put them in M_i, using the formula (14.6) to obtain g.

Step 3. Using the Theorem 14.2, we can get H_0, H_1, H_2, and H_3. H_0 is Z. For H_1, we need to get $b_1(\partial M)$, which is just the summation of the genus in all connected components in ∂M. H_2 is the number of components in ∂M. H_3 is trivial.

Lemma 14.7 *Algorithm 14.1 is a linear time algorithm.*

Proof Step 1 uses linear time. We can first track all points in the object using breadth-first-search. We assume that the points in the object are marked as "1" and the others are marked as "0." Then, we test if a point in the object is adjacent to both "0" and "1" by using 26-adjacency for linking to "0." Such a point is called a boundary point. It takes linear time because the total number of adjacent points is only 26. Another algorithm tests whether each line cell on the boundary has exactly two parallel moves on the boundary [5]. This procedure only takes linear time for the total number of boundary points in most cases.

Step 2 is also in linear time by Lemma 14.4.

Step 3 is just a simple math calculation. For H_0, H_2, and H_3, they can be computed in constant time. For H_1, the counting process is at most linear.   □

Therefore, we can use linear time algorithms to calculate g and all homology groups for digital manifolds in 3D based on Lemma 14.4 and Lemma 14.7.

Theorem 14.3 *There is a linear time algorithm to calculate all homology groups for each type of manifold in 3D.*

To some extent, researchers are also interested in space complexity that is regarded to the running space needed beyond the input data. Our algorithms do not need to store past information, and the new algorithms presented in this section are always $O(\log n)$. Here, $\log n$ is the number of bits needed to represent a number n.

14.6.2 Input Data Sets

In real application, input data formats are critical to algorithm design. We will use cubical data that is also the data format for MRI and CT data. In cubical data samples, we assume the sampling is contiguous, where each sample point is normally followed by another sample point in its neighborhood.

Random sampled points can introduce a level of uncertainty. In this case, we usually cannot calculate the genus without making an assumption. For instance, we would not be able to know where a hole is. In order to obtain simplicial decomposition (usually triangulation), we usually need to use Voronoi or Delaunay decomposition with boundary information. This means the boundary must be assumed.

A new technology, called persistent homology analysis, tells us how to find the best estimation for the location of holes as we discussed in Chap. 12. However, this method is not a precise analysis [4, 39].

Even though the method described in this section can be modified in persistent analysis, we mainly deal with the method of precise genus and homology group calculation. In other words, our assumption is that the digital object consists of cubical points (digital points, raster points). Each point is a cube, which is the smallest 3D object. The edge and point are defined with regards to the cube and an object may contain several connected components using a cube-linking path.

14.6.3 Searching Connected Components of a Cubical Data Set

Connected component search is an old task that can be done by using Tarjan's breadth-first search. Pavlidis was one of the first people to realize and use this algorithm in image processing. This problem is also known as the labeling problem. The complexity of the algorithm is $O(n)$ [30].

The problem is what connectivity is based of. In 3D, we usually have 6-, 18-, 26- connectivity. Since real data has noise, we also have to consider all of those connectivities and we must use 26-connectivity to get the connected components.

Therefore, the connected component of real processing is not a strictly 6-connected component. The topological theorem generated previously in [8] is no longer suitable. So we need to transform a 26-connected component into a 6-connected component. This should be done by a meaningful adding or deleting process since optimization on the minimum number of changes could be an NP-hard problem.

Problem of Minimum Modifications Given a set of points in 3D digital space, if this set is not a manifold, assume that the points are connected in a connectivity defined using adding or deleting processes to make the set a 3D manifold. The question becomes: is there a polynomial algorithm that makes the solution have minimum modifications where adding or deleting a data point will be counted as one modification [6]?

A similar problem was considered in [35] in which a decision problem of adding was proposed.

This problem can be extended to a general k-manifold in n-D space. Even though we have the 6-connected component, there may still be cases that contain the pathological situation, which requires special treatment. We will discuss this issue in the next subsection.

14.6.4   Pathological Case Detection and Deletion

In this section, we only deal with Jordan manifolds, meaning that a closed $(n - 1)$-manifold will separate the n-manifold into two or more components. For such a case, only direct adjacency will be allowed since indirect adjacency will not generate Jordan cases.

That is to say, if the set contains indirect adjacent voxels, we need to design an algorithm to detect the situation and delete some voxels in order to preserve the homology groups.

It is known that there are only two such cases in cubical or digital space [5]: two voxels (3-cells) share a 0-cell or a 1-cell. Therefore, we want to modify the voxel set to only contain voxels where two of these cases do not appear. Two voxels share exactly a 2-cell, or there is a local path (in the neighborhood) of voxels where two adjacent voxels share a 2-cell [5]. A special case was found in [36] that is the complement case of the case in which two voxels share a 0-cell (see Fig. 14.6a). This special case may create a tunnel or could also be filled. We will simplify it by adding a voxel in a $2 \times 2 \times 2$ cube. Such a case in point space is similar to case (a) in Fig. 14.6 since the boundaries of these cases are the same.

The problem is that many real data sets do not satisfy the above restrictions (also called well composed image). The detection is easy but deleting certain points (the minimum points deletion) to preserve the homology is a bigger issue.

The following rules (observations) are reasonable: In a neighborhood N_p that contains 8 cubes and 27 grid points then the following is true.

a) If a voxel only shares a 0-cell with a voxel then this voxel can be deleted (Fig. 14.6a).
b) If a voxel only shares a 1-cell with a voxel then this voxel can be deleted (Fig. 14.6b).
c) If a boundary voxel v shares a 0,1-cell with a voxel, assuming that v also shares a 2-cell with a voxel u, then u must share a 0,1-cell with a voxel that is not in the object M. Therefore, u is on the boundary and deleting v will not change the topological properties.
d) If in a $2 \times 2 \times 2$ cube there are six boundary voxels and the complement (two zero-valued voxels) of these voxels is the same as Fig. 14.6a, then we can add a voxel to this $2 \times 2 \times 2$ cube such that the new voxel shares as many 2-cells in the set as possible. This means that we want the additional voxels to be inside the object as much as possible Fig. 14.5.

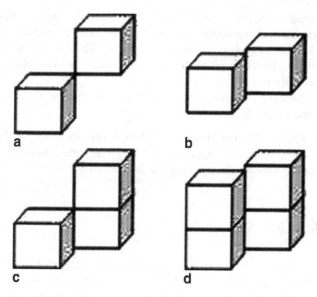

Fig. 14.5 Pathological cases

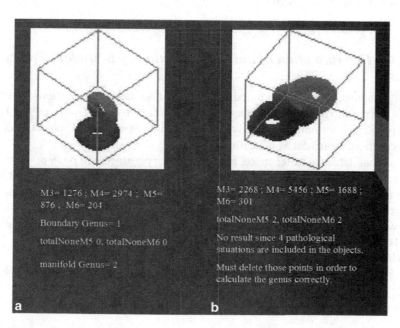

M3= 1276 ; M4= 2974 ; M5= 876 , M6= 204

Boundary Genus= 1

totalNoneM5 0, totalNoneM6 0

manifold Genus= 2

M3= 2268 ; M4= 5456 ; M5= 1688 ; M6= 301

totalNoneM5 2, totalNoneM6 2

No result since 4 pathological situations are included in the objects.

Must delete those points in order to calculate the genus correctly.

Fig. 14.6 Example of constructed data without using pathological case elimination

We have implemented or modified the above rules to fit the theoretical definition of digital surfaces. We also design an algorithm based on these rules to detect and delete some data points while preserving the topology. This is essential to calculating the genus correctly. However, when the object becomes more complex, pathological situations may still exist.

The mathematical foundation of the above process that eliminates pathological cases is still under investigation.

Mathematical Foundation of Modifying a 3D Object into a 3D Manifold: Given a set of points in 3D digital space, how would we modify the data set into a manifold without losing or changing the topology (in mathematics)?

14.6.5 Boundary Search

In general, a point is on the boundary if and only if it is adjacent to one point in the object and one point not in the object (in 26-connectivity). A simple algorithm that goes through each point and tests the neighborhood will determine whether a point is on the boundary or not. This is a linear time and $O(log(n))$ space algorithm.

The only thing special about this boundary detection is that we use 26-connectivity to determine the boundary points. This is to take all possible boundary points into consideration in the next step.

14.6.6 Determination of the Configuration of Boundary Points

When all boundary points are found, we need to find their classifications. In other words, we need to determine whether a special point is in M_3, M_4, M_5, or M_6. Here is the problem, if we only have one voxel, is it a point (0-cell) or a 3D object (3-cell)? In this section, we treat it as a 3-cell.

The input data is in raster space, but the boundary surface will be in point space. We must first make the translation. Then, for each point on the surface, we count how many neighbors exist in order to determine its configuration category. After that, we use formula (14.6) to get the genus.

If we still need to find homology groups, we can just use the simple calculations based on Theorem 14.4 to obtain them. Using the program, we get the genus = 5 for a modified real image (Figs. 14.7 and 14.8).

In Fig. 14.9, we show the result of real data processing for a modified 3D bone image.

Fig. 14.7 The processing
result after pathological case
elimination

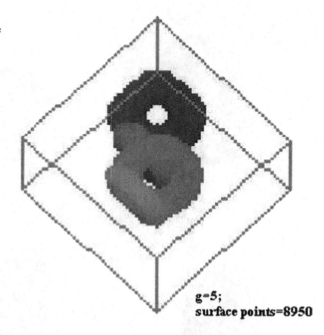

g=5;
surface points=8950

Fig. 14.8 The data processing
for a modified 3D real image

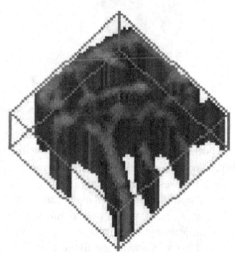

14.7 Case Study: Digital Curvatures Applied to 3D Object Analysis and Recognition

In this section, we describe a new method that uses digital curvatures for 3D object analysis and recognition. For direct adjacency in 3D, digital surface points have only six types. It is easy to determine and classify the digital curvatures of each point on

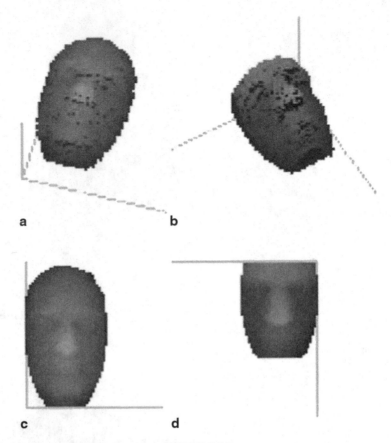

Fig. 14.9 Two original human faces from NIST-FRGC data sets

the boundary of a 3D object. This is simpler than the case of triangulation on the boundary surface of a solid; the curvature can be of any real value.

We focuses on the global properties of categorizing curvatures for small regions. We use both digital Gaussian curvatures and digital mean curvatures to characterize 3D shapes. Then, we propose a multi-scale method and a feature vector method for 3D similarity measurement.

In this experiment, we found that Gaussian curvatures mainly describe global features and average characteristics such as the five regions of a human face. On the other hand, mean curvatures can be used to find local features and extreme points such as the nose in 3D facial data.

A 3D object can be represented by one or several closed surfaces (2D-manifolds). Curvatures that describe the degree of change at a point on the surface have been used for many years in 3D image processing [1, 27].

The typical technology related to curvatures is as follows: (1) triangulation of the surface, (2) fit the digital image to a continuous surface (using B-spline), and (3) calculate the standard Gaussian and/or mean curvatures.

Worring and Smeulders showed that even in 2D, the interpretations of digitized curves may generate completely different curvatures [38]. In other words, the same image will generate different curvature maps if the triangulation of the image is different.

This section presents a preliminary study of using digital curvatures of digital surfaces as features to classifying different 3D objects. Digital curvatures can be used for 3D object matching, classification, and recognition. Since direct adjacency only has six types of digital surface points in local configurations, it is much easier to determine and classify the digital curvatures for every point on the boundary surface of a 3D object.

14.7.1 Meaning of Digital Curvatures

In 1986, Besl and Jain presented a systematic method to use curvatures in image processing [2]. Besl and Jain used the signs (positive, zero, negative) of Gaussian curvature and mean curvature to classify 3D surface points. Their technique uses triangulation for decomposition and then calculates the Gaussian curvature (K) and mean curvature (H). Eight principle shapes can be identified:

(1) Peak (surface) point $K > 0, H < 0$,
(2) Flat point $H = 0, K = 0$,
(3) Pit point $H > 0, K > 0$,
(4) Minimal point $H = 0, K < 0$,
(5) Ridge point $K = 0, H < 0$,
(6) Saddle ridge point $H < 0, K < 0$,
(7) Valley point $H > 0, K = 0$, and
(8) Saddle valley point $H > 0, K < 0$.

In fact, cases (1) and (3) are mirror images of the same shape with a change of sign in H. This also applies to cases (5) and (7) and also cases (6) and (8).

Given a set of cloud points in 3D, we assume that they are connected. Since cubical space with direct adjacency, or (6,26)-connectivity space, has the simplest topology in 3D digital spaces, we use this as the 3D image domain. This is also believed to be sufficient for the topological property extraction of digital objects in 3D.

In this space, two points are said to be adjacent in (6,26)-connectivity space if the Euclidean distance between these two points is 1. If this condition is not satisfied, we can use the a 3-ball to cover the neighboring space to make do persistent analysis as described in Chap. 12.

Let M be a closed (orientable) digital surface in 3D grid space in direct adjacency. We know that there are exactly 6-types of digital surface points as shown in Fig. 5.8.

For M, M_i still still indicate the set of digital points with i neighbors. We have M_3, M_4, M_5, and M_6 to our image analysis. Mapping these digital surface points to the Besl-Jain classification, we would be able to see that M_3 corresponds to cases (1) and (3), $M_4(a)$ to case (2), $M_4(b)$ to cases (5) and (7), M_5 to cases (6) and (8), and M_6 to case (4).

Digital configurations contain two types of none-trivial minimal surface points in M_6. The correspondence between the shapes in the Besl-Jain classification is very interesting. Also see [10, 13, 28, 37] for more related discussion using curvatures.

14.7.2   Digital Curvatures and 3D Image Analysis

Using the digital Gaussian curvature and the digital mean curvature in the classification of 3D objects in image analysis will bring us new information. For instance in a closed digital surface, M_4 is independent to M_3; M_4 contains parabolic points (the bend ones). M_3 contains the elliptical points since the surface point is locally convex. M_5 and M_6 contain hyperbolic points, where the Gaussian curvature is negative and the surface point is locally saddle shaped.

Let S be a subset of the 2D manifold M. Now, S might not be a closed surface.

We define a feature vector called the digital curvature vector $f_S = (m_3, m_4, m_5, m_6)$, where $m_i = |M_i|$ with respect to S. We can also split the vector into four component vectors to include two cases for M_4 and M_6, respectively. Basically, the vector now contains four components: $f_S = (|M_3|, |M_4|, |M_5|, |M_6|)$.

The motivation of defining such a vector for a local region is based on the corresponding Gaussian-Bonnet theorem with the boundary curves [23]. Suppose C is a boundary curve of S that is simply connected and k_g is the geodesic curvature of C, then

$$\int_C k_g dt + \int\int_S K dA = 2\pi \tag{14.7}$$

If C is a n-polygon and α_i is the interior angle, then

$$(n-2) \cdot \pi + \int_C k_g dt + \int\int_S K dA = \Sigma_{i=1}^n \alpha_i \tag{14.8}$$

Therefore,

$$g_S = g(f_S) = |M_3| \cdot K_3 + |M_5| \cdot K_5 + |M_6| \cdot K_6) = \int\int_S K dA \tag{14.9}$$

can also be used to represent the total geodesic curvature of C:

$$\int_C k_g dt = 2\pi - g_S \tag{14.10}$$

In image processing, we can define S (or C) as a rectangular region. If S only contains one surface point, then g_S is just K. In practice, the vector f_S can be used

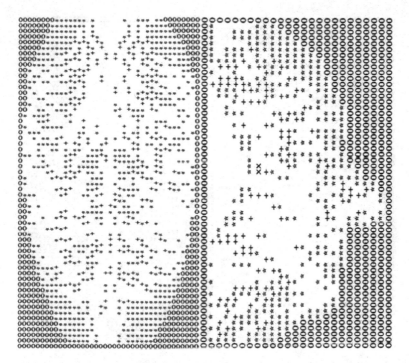

Fig. 14.10 Digital curvature on face one's surface by projection: $'*','+'$, and $'X'$ indicate M_3, M_5, M_6-points, respectively

for a closed surface or a small region centered at one point. Between two regions, there can be intersections or no intersection. The region can be 2D or 3D, depending on whether the problem can be projected into 2D easily. The most popular shapes of the domain regions are circlesspheres and rectanglescubes.

We have used this technique to analyze human facial data. If the size of the region is reduced $1, 2, 2^2, \ldots, 2^k$ times, we will get a sequence of f_S, or simply get g_S. Then, we can see the change in the curvatures. Such a method is usually called a multi-scaling method.

Let us look at the following example. The two original images are shown in Fig. 14.9 [7].

In Fig. 14.10 and 14.11, we show the initial digital curvature calculation–projections from $x-$ and $y-$ directions. In these two figures, There are three symbols, $'*','+'$, and $'X'$. $'*'$ indicates M_3-points on the face that are the positive curvature point $(\pi/2)$. $'+'$ indicates M_5-points $(-\pi/2)$, and $'X'$ indicates M_6-points $(-\pi)$; they are negative curvature points. The calculation is for the whole 3D image. The display only shows two observation angles for each image.

In Fig. 14.12, we show the projected Gaussian curvature data for each scale for face one from $64 \times 64 \times 64$ to $8 \times 8 \times 8$. Figure 14.13 is for face two. In order

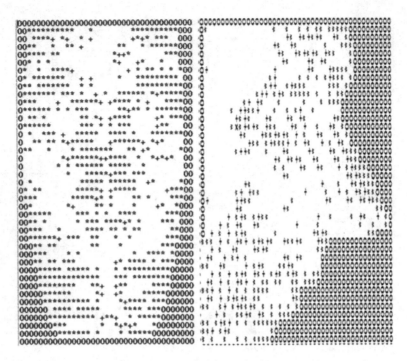

Fig. 14.11 Digital curvature on face two's surface by projection

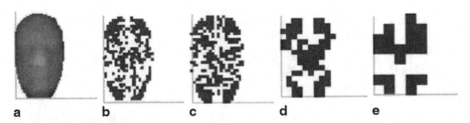

a   b   c   d   e

Fig. 14.12 Digital curvature scaling for face one

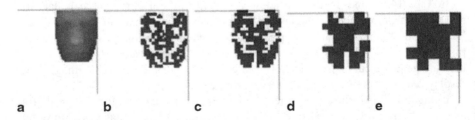

a   b   c   d   e

Fig. 14.13 Digital curvature scaling for face two

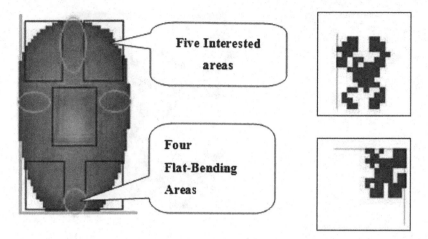

Fig. 14.14 Five areas of interests and four flat-or-bending areas are the similar characteristics of the human face

to be able to observe more easily, we display the same sized image when the scale changes.

Analyzing Figs. 14.12 and 14.13, we see that the multi-scaling method can identify some of the interesting areas of the face. For example, we can identify five areas of interest for face one and face two. They both contain two areas at the top and bottom and one area in the middle.

These five areas indicate two regions on the sides of the head, two cheeks, and the nose. The nonzero total of the Gaussian curvature for a region that remains means that the total change has not been canceled in this region.

The result shows that the first person has a flatter face than the second person since the method reaches the five interesting areas earlier. We can also identify the flat-or-bending regions in the human face as shown in Fig. 14.14. A total of four such areas can be easily found.

The advantage of Gaussian curvature-based calculation is that it is not a simple processing of pixel averages in the region. The curvature-based calculation is based on geometric and topological properties of the 3D object. For instance, we know the total Gaussian curvature will be a constant as the selected region becomes the whole 3D data array. In summary, the method described in this section can identify the five regions of interest and four flat-or-bending regions as the common similar characteristics of a human face. In addition, the calculation is much easier than that of triangulation based images. The following lemma provides evidence for this.

Lemma 14.8 *The algorithm designed to obtain scaled local Gaussian curvatures for finding five areas of interest on the image of a human face is of complexity $O(n\log n)$. The space needed is $O(\log n)$.*

Proof We first compute the curvature for higher resolution images then we reduce the resolution by half, since the $2^k - 1$-scaled Gaussian curvature data can be used to calculate the 2^k-scaled Gaussian curvature data. We can design a fast linear algorithm

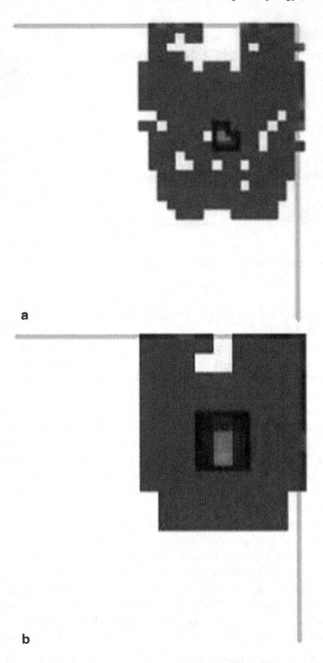

a

b

Fig. 14.15 Use digital mean curvature to find a peak point that indicates the nose on a human face: **a** Image is made by a 2×2 summation of the absolute values of digital mean curvatures, and **b** Image is made by a 4×4 summation of the absolute values of digital mean curvatures

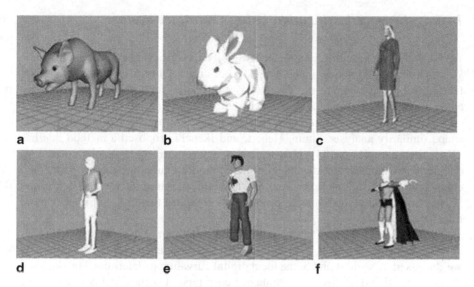

Fig. 14.16 Six 3D objects from Princeton's 3D database

due to the reduction of the size of the array by half. In such a case, the space needed would also be $O(n)$ if we just use the input array and not the intermediate data. By using the output of the result for each resolution or scale, the time complexity would be $O(n \log n)$ and the space complexity would be $O(\log n)$ (considering only the space needed to run the algorithm).

14.7.3   Investigation Based on Digital Mean Curvatures

For the mean digital curvature application, we have investigated the calculation based on the absolute average for small regions. The geometric meaning of such a treatment of digital mean curvatures is the zigzagged points on the surfaces. Mean curvature zero points indicate the critical points that change from inward to outward if it is not a flat point. The results of an example is shown in Fig. 14.15.

To summarize the findings, Gaussian curvatures describe the global features and the average characteristics such as the five regions of a human face, but mean curvatures find local features and extreme points such as the nose. In the future, we will use this method for analyzing more human facial data. We can normalize the image size and calculate the distance for a pair of faces in corresponding or nearby scales. The following section presents a method for 3D data "rough" classification using digital curvature vectors.

14.7.4   Object Classification Using Curvature Vectors

We know that we have different types of digital curvature points in a digital surface. We can list these different points as a digital curvature vector for an image. As a global feature, we may be able to use this global information for for 3D shape similarity analysis and classification of 3D objects.

There have been many research investigations that use local curvatures in 3D shape similarity analysis. Shum, Hebert, and Ikeuchi proposed a method that has $O(n^2)$ time complexity [34]. However, it is not very practical for the purpose of data retrieval. Another disadvantage of the continuous curvature method is that the calculation of curvatures return real numbers and that introduces precision errors [27]. A multi-scaling method was proposed to complete this task, but it takes more time.

The technique presented in this section is not an attempt to replace the methods developed before. We try to explore more applications using digital curvatures. As we discussed in Sects. 2 and 3, the local digital curvature is determined by the local shape of the digital surface. The number of each type of surface-point may indicate the features of a 3D object.

14.7.5   Feature Vectors of 3D Objects Based on Digital Curvatures

A feature vector only contains a number of digital surface points in each of the categories M_3, M_4, M_5, and M_6. In general, a 3D digital object may not necessarily be a 3D digital manifold. This is because we have strict definitions for 3D manifolds where each local neighborhood must be similar to a 3D Euclidean space. However, this does not affect the use of curvatures in 3D objects. For a safe claim, we can assume that the 3D object is simply connected. This method is used for calculating the boundary surface of a 3D object in this section. The feature vector is

$$fv = \frac{1}{T}(|M_3|, |M_4|, |M_5|, |M_6|) = (r_3, r_4, r_5, r_6) \tag{14.11}$$

where T is the total number of surface points. For example, an object has a total of 1679 digital points on boundary surfaces $|M_3| = 469, |M_4| = 995, |M_5| = 183, and |M_6| = 32$. The non-manifold points (2D) are the points where the neighborhood of the point is not a 2D-configuration shown in Fig. 5.8. Because of these non-manifold points, we corrected the data for $|M|$'s: 484, 995, 180, 28, and 240. It also includes a total of 8 non-manifold points. The method to delete these points is presented in [6]. Therefore, we get the ratios $r_3 = 0.288267$, $r_4 = 0.592615$, $r_5 = 0.107207$, and $r_6 = 0.016677$. The Euclidean distance of the two feature vectors is $d = \sqrt{\Sigma_{i=3}^{6}((x_i - y_i)^2)}$.

We could also use the scaling method from above section (Sect. 14.7.2) to get more features. Here, we use this method to do a rough classification for 3D shapes. The computing examples are taken from the Princeton Benchmark Website [14].

14.7.6 Similarity and Distance Analysis Based on the Feature Vectors

For n samples of solid objects, we can calculate the feature ratio vectors for each of the samples. Therefore, we will have n feature vectors $e_1, \cdots, e_n$. A simple calculation allows us to get the Euclidean distances for every pair of points. For the example in Fig. 14.16, the feature ratio vectors $e_x = (r_3, r_4, r_5, r_6)$ where $r_k = |M_k|/T$ are listed below:

$e_1 = (0.288267, 0.592615, 0.107207, 0.016677)$,
$e_2 = (0.262424, 0.508752, 0.193369, 0.044133)$,
$e_3 = (0.168149, 0.680220, 0.144854, 0.008895)$,
$e_4 = (0.152833, 0.711492, 0.122506, 0.013966)$,
$e_5 = (0.148500, 0.710425, 0.135432, 0.007128)$,
$e_6 = (0.162700, 0.688310, 0.140705, 0.010093)$.

The Distance matrix of two-vector pairs is

$$
\begin{pmatrix}
0 & & & & & \\
0.01587 & 0 & & & & \\
0.02358 & 0.04188 & 0 & & & \\
0.03271 & 0.05904 & 0.00173 & 0 & & \\
0.03430 & 0.05837 & 0.00139 & 0.00023 & 0 & \\
0.02609 & 0.0461 & 0.00011 & 0.00098 & 0.00072 & 0
\end{pmatrix}
\tag{14.12}
$$

In the matrix, a_{ij} is the distance between objects i and j. The matrix is symmetric so we only present half of it. We can easily see that objects 1 and 2 are closely related. Objects 4, 5, and 6 are similar. From the data pictures displayed in Fig. 14.16, we can see that this is correct. An interesting observation is that Object 3 does not go with three of the other objects in the second category.

In this case study, we used digital Gaussian curvatures and digital mean curvatures to analyze 3D shapes. We have found that digital curvatures may have some power in identifying the significant features of 3D objects. For instance, we could identify five regions in some facial images. We also presented a method for similarity analysis using digital curvatures in Sect. 14.7.5 and 14.7.6.

Acknowledgement We would also like to thank the NIST Face Recognition Grand Challenge (FRGC) and Princeton University for the 3D benchmark data sets.

References

1. Besl, P.J., Jain, R.C.: Three-dimensional object recognition. Comput. Surv. 17(1), 75–145 (1985)
2. Besl, P.J., Jain, R.C.: Invariant surface characteristics for 3D object recognition in range images. Computer Vision, Graphics, and Image Processing 33(1), 33–80, (1986)
3. D. Boltcheva, D. Canino, S. Merino Aceituno, J.-C. Leon, L. De Floriani, and F. Hetroy, "An iterative algorithm for homology computation on simplicial shapes," Comput.-Aided Des., vol. 43, no. 11, pp. 1457–1467, nov 2011, special Issue: Solid and Physical Modeling 2011.
4. G. Carlsson, Persistent homology and the analysis of high dimensional data, Symposium on the Geometry of Very Large Data Sets, Fields Institute for Research in Mathematical Sciences, February 24, 2005.
5. L. Chen, *Discrete Surfaces and Manifolds*, Sp Computing, Rockville, 2004.
6. Chen, L.: Genus computing for 3D digital objects: Algorithm and implementation. In: Kropatsch, W., Abril, H. M., Ion, A.(Eds.) Proceedings of the Workshop on Computational Topology in Image Context (2009)
7. L. Chen and S. Biswas, Digital Curvatures Applied to 3D Object Analysis and Recognition: A Case Study. Barneva et al, Combinatorial Image Analysis, Lectures Notes 1n Computer Science Vol 7655, Springer, pp 45–58, 2012.
8. Chen, L., Rong, Y.: Digital topological method for computing genus and the Betti numbers. Topology and its Applications 157(12) 1931–1936 (2010)
9. L. Chen, D. Cooley and J. Zhang, Equivalence between two definitions of digital surfaces, Information Sciences, Vol 115, 201–220, 1999.
10. Colombo, A., Cusano, C., Schettini, R.: 3D face detection using curvature analysis. Pattern Recognition 39(3) 444–455 (2006)
11. Damiand G., Peltier S., Fuchs L., Computing Homology Generators For Volumes Using Minimal Generalized Maps Proceedings Of 12th International Workshop On Combinatorial Image Analysis, LNCS Vol 4958, pp 63–74, 2008
12. C. J. A. Delfinado, H. Edelsbrunner, An Incremental Algorithm For Betti Numbers Of Simplicial Complexes On The 3-Sphere, Computer Aided Geometric Design 12 (1995), 771–784.
13. Dudek, G., Tsotsos, J.K.: Shape representation and recognition from multiscale curvature. Computer Vision and Image Understanding 68, 170–189 (1997)
14. Funkhouser, T., Kazhdan, M., Min, P., Shilane, P.: Shape-based retrieval and analysis of 3D models. Communications of the ACM 48(6), 58–64, (June 2005)
15. Goodman-Strauss, C., Sullivan, J. M.: Cubic polyhedra. In: Discrete Geometry: In Honor of W. Kuperberg's 60th Birthday, Monographs and Textbooks in Pure and Applied Mathematics, 253, 305–330 (2003)
16. A. Hatcher, *Algebraic Topology*, Cambridge University Press, 2002.
17. T. Kaczynski, K. Mischaikow And M. Mrozek, Computing Homology, Homology, Homotopy And Applications, Vol.5(2), 2003, Pp.233–256
18. T. Kaczynski, K. Mischaikow, M. Mrozek, *Computational Homology* Springer Series: Applied Mathematical Sciences, Vol. 157, 2004,
19. W.D. Kalies, K. Mischaikow And G. Watson, Cubical Approximation And Computation Of Homology, Banach Center Publ. 47, 115–131 (1999).
20. R. Klette, and A. Rosenfeld *Digital Geometry: Geometric Methods for Digital Image Analysis*. Morgan Kaufmann, 2004.
21. T.Y. Kong, and A. Rosenfeld (editors). *Topological Algorithms for Digital Image Processing*. Elsevier. 2006.
22. P. Kot, Homology Calculation Of Cubical Complexes In R^n, Computational Methods In Science And Technology 12(2), 115–121 (2006)
23. Kreyszig, E.: Differential Geometry. University of Toronto Press, 1959
24. L. J. Latecki, 3D Well-Composed Pictures, GRAPHICAL MODELS AND IMAGE PROCESSING Vol. 59, No. 3, pp. 164–172, 1997.

25. W. E. Lorensen and H. E. Cline, Marching Cubes: A high resolution 3D surface construction algorithm. In: Computer Graphics, Vol. 21, No. 4, July 1987
26. Meyer, M., Desbrun, M., Schroder, P., Barr, A.H.: Discrete differential-geometry operators for triangulated 2-manifolds. In: Visualization and Mathematics III, Springer, 35–58 (2003)
27. Mokhtarian, F., Bober, M.: Curvature Scale Space Representation: Theory, Applications, and MPEG-7 Standardization. Kluwer, 2003
28. Mokhtarian, F., Khalili, N., Yuen, P.: Estimation of error in curvature computation on multi-Scale free-form surfaces. International Journal of Computer Vision 48(2) 131–149 (July 2002)
29. P. S. Novikov, On the algorithmic unsolvability of the word problem in group theory, Trudy Mat. Inst. Steklov., 44 (1955); English transl. Amer. Math. Soc. Transl., 9:2 (1958), 1–122.
30. T. Pavlidis, Theodosios Algorithms for graphics and image processing, Computer Science Press, 1982.
31. S. Peltier, A. Ion, W. G. Kropatsch, G. Damiand, Y. Haxhimusa, Directly computing the generators of image homology using graph pyramids Image and Vision Computing, Vol 27, No 7, 2009, pp 846–853
32. K. Polthier. Polyhedral surfaces of constant mean curvature. Habilitationsschrift, Technische University Berlin, 2002.
33. J. H. Rubinstein. An algorithm to recognize the 3-sphere, In Proceedings of the International Congress of Mathematicians, Vol. 1, 2, pp 601–611, 1994.
34. Shum, H.Y., Hebert, M., Ikeuchi, K.: On 3D shape similarity. Proceedings of the 1996 Conference on Computer Vision and Pattern Recognition (CVPR 96), 526–531 (1996)
35. M. Siqueira, L. J. Latecki, J. Gallier, Making 3D Binary Digital Images Well-Composed, Vision Geometry XIII, Proc. SPIE 5675 pp 150–163, 2005
36. M. Siqueira, L. J. Latecki, N. Tustison, J. Gallier, J. Gee Topological Repairing of 3D Digital Images, Journal of Mathematical Imaging and Vision Volume 30, Number 3 March, 2008 pp 249–274.
37. Tanaka, H.T., Ikeda, M., Chiaki, H.: Curvature-based face surface recognition using spherical correlation. Proceedings of Third IEEE International Conference on Principal directions for curved object recognition Automatic Face and Gesture Recognition, 372–377 (1998)
38. Worring, M., Smeulders, A.W.M.: Digital curvature estimation. CVGIP Image Understanding 58(3) 366–382 (1993)
39. A. Zomorodian and G. Carlsson, Computing persistent homology, Discrete and Computational Geometry, 33 (2), pp. 247–274.

Chapter 15
Select Topics and Future Challenges in Discrete Geometry

Abstract Digital geometry is a relatively new research area. It is difficult to show the characteristics of digital geometry as a well-developed theory. On the other hand, discrete geometry used to focus on combinatorial methods such as simplicial decomposition, counting, and tillings. However, it is now also much interested in differential geometry methods. Many new problems related to digital and discrete geometry are have been discovered and have raised interests from various different research disciplinary areas. In order to synthesize some features, this chapter mainly deals with methodology issues of digital and discrete geometry in terms of future studies. We begin with detailed proofs of two basic theorems in digital and discrete geometry. In these proofs, we show the power of the digital and discrete methods in geometry. Then, we focus on future problems in BigData and the data sciences, including what digital methods can do in random algorithms, manifold learning, and advanced geometric measurements. We also present some questions for graduate students and other researchers to think about.

Keywords Discrete and digital methodology · Equivalence theorem · Jordan curve theorem · Random algorithm · BigData · Data science · Advanced measurement · Future research problems

15.1 Characteristics of Digital Geometric Methods

Digital geometric methods differ from traditional discrete geometry, performance being a key difference. Unlike triangulations and meshes, the digital method seeks to find fast algorithms beyond its special mathematical properties. Meshes in computer graphics use simplicial complexes in a space expensive storage manner. We not only need to store points, but also edges, faces, and solids. In computer graphics, the beauty of the generated picture is one of the primary concerns. However, in computer vision, speed might be the most important, such as automated driving.

This is why the digital geometric method must aim to find more efficient algorithms, even if these are not attainable immediately. The algorithmic advantages are guaranteed by the equivalence theorem of the two definitions of the digital surface.

Digital method and its extended partial graph method, have showed the advantages in fast algorithm design and even new theorem findings. In Sect. 15.2, we will prove

© Springer International Publishing Switzerland 2014
L. M. Chen, *Digital and Discrete Geometry*, DOI 10.1007/978-3-319-12099-7_15

the Equivalence theorem between Two Definitions of Digital Surfaces: Morgenthaler-Rosenfeld's definition and Chen-Zhang's parallel move based definition [2, 4, 9, 10].

Digital manifold defined in the parallel move is to has considerable meaning for such an advantage. Instead of saving all 0-cell, ... , k-cells for a digital manifold, we only need to save 0-cells. Then we generate all other cells while the program is running, dynamically.

The digitization is also related to the performance how do we digitizing an object. A good method was suggested by M. Rabin for random algorithms.

In Chap. 7, we extend this method in the fashion of discrete method in topology which is different from simplicial or cell complex in algebraic topology presented in Chaps. 9 and 13.

We use a partial graph[1], the solid object, instead of soft, the multiple interpretations in our method. This is because there are exponential cases for the interpretations.

Digital geometry and modern discrete geometry show much interests in fast constructive methods not like the classical mathematics where the existence of findings are essential in many cases.

Even though, a single simplicial complex has unique meaning, but it cannot directly used as a space that holds numbers of manifolds that has a unique meaning. This traditional method in mathematics, they some how much care about the existence.

However, in digital method, we need to find an object out. Since we only deal with the finite points object, the existence is clear. we always can find an algorithm to get it or reject it. The focus in this case is turned to be find the quick way to identify an object.

Chapter 6 is the "implementation" of such idea. One of the important theoretical result is finding the classification of the six types. We will prove the related theorem in the next section to show the equivalence of the parallel move method comparing to set theoretical method. It is obvious that our method is suit for the meaning of the simplicial complex. It is more specialized in computing the geometric object.

This development is shown in Chap 14 the simplest form of Gauss-Bonnet theorem. The following theorem was kind of lengthy in proof. The detailed technology can be seen the specialty of the data.

The cell complex does not contain the case of digital space in general (indirect) adjacency. Even though cell complexes are most general in topology. but it still not be able to cover a general digital space since 8-adjacency in 2D digital space is not a cell complex. There are two cells where their intersection is not a 0-cell, 1-cell or, 2-cell when embedding to Euclidean 2D.

A simple closed path in cell-complex or discrete manifold, might not have Jordan curve property. It could be the boundary of a smallest 2-cell. There is no internal vertex inside of a 2-cell.

In order to maintain the Jordan curve theorem for discrete manifold, a new definition of curves must be defined. The curve under the new definition must be able to

---

[1] We switched the meanings of the partial graph with the subgraph from the author's previous publication. See definition in Chap. 2.

be embedded to 2D Euclidean space for the standard meaning in general topology and combinatorial topology. In this chapter, we will give a complete proof of this theorem in discrete manifolds in Sect. 15.3.

For other related future topics, we select random algorithm issues for finding the digitization in Sect. 15.4, integral points issue and Riemann manifold learning that is related to manifold learning discussed in Sect. 12.5.

15.2   The Equivalence Between Two Definitions of Digital Surfaces

This section presents a proof of the equivalence of two digital surface definitions. It states that a simple surface point of a set $S \subset \Sigma_3$ is a regular inner surface point, and vice versa. We have used this theorem, Theorem 5.1, in Chap. 5.

The definition of the simple surface points was given by Morgenthaler and Rosenfeld (Sect. 5.1), and the definition of inner surface points is based on the parallel-move concept proposed by Chen and Zhang in direct adjacency (Sect. 5.2).

This theorem is the basis for the classification theorem (Theorem 5.2) of digital surface points, and the classification theorem is fundamental for proving the digital form of Gauss-Bonnet theorem (Theorem 14.1). That is one of most stunting results in this book.

15.2.1   The Theorem of Equivalence

In this subsection, we provide a complete proof of Theorem 5.1. This theorem is important to the classification theorem of digital surface points, Theorem 5.2. The proof is lengthy but shows some detailed technique to digital geometry [10, 5]. The second purpose to involve this proof in this book is to demonstrate that the digital geometry is a branch of mathematics, not a branch of empirical sciences.

We know most of surface points are regular. We show two examples for non-regular surface points in Fig. 15.1. So we want to exclude these cases. In other words, a simple surface point needs to be a regular point. This fact will be shown in the following proof.

The following theorem is another form of Theorem 5.1.

Theorem 15.1 *A simple surface point, under the Morgenthaler-Rosenfeld definition (Definition 5.1), of a set $S \subset \Sigma_3$ is a regular inner surface point, and vice versa.*

Proof This proof contains two parts. The first part proves that if p is a simple surface point, then p is also a regular inner surface point. The second part proves that if p is a regular inner surface point, then p is a simple surface point.

We still use N_p as the neighborhood of point p. N_p that contains 26 points.

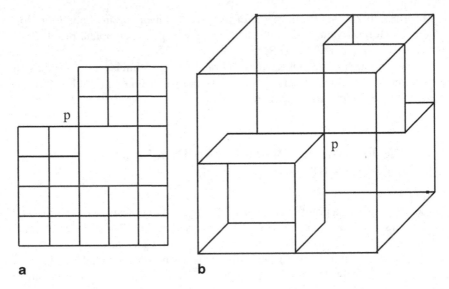

Fig. 15.1 Examples for non-regular points: **a** Non-regular point p; **b** Inner but non-regular surface point p

Part 1 Suppose that p is a simple surface point as defined in Definition 5.1 at Sect. 5. 2.1. We want to show the following:

1) $S(p) = S \cap (N_p \cup \{p\})$ does not have any 3D-cell,
2) Each line-cell containing p in S has exactly two parallel moves in $S(p)$, i.e., p is inner, and
3) Any two surface-cells containing p in S are line-connected in $S(p)$.

First, suppose p is a simple surface point in S. Obviously, p cannot be a corner point of any 3D-cell in S, hence we establish statement 1).

For 2), to begin with, p has three or more (directly) adjacent points in $S \cap N_p$. Otherwise, $\bar{S} \cap N_p$ is connected, so p is not a simple surface point in accordance with condition (2) of its definition in Sect. 5.2.1.

We define here a grid plane is a set of all points with a fixed z $P_z = \{(x,y,z)|(x,y,z) \in \Sigma_3\}$, all points with a fixed y $P_y = \{(x,y,z)|(x,y,z) \in \Sigma_3\}$, or all points with a fixed x $P_x = \{(x,y,z)|(x,y,z) \in \Sigma_3\}$.

Next, let $p' \in S$ be an arbitrary adjacent point of p. therefore, there is a grid plane (such as *plane*1, *plane*2 or *plane*3 in Fig. 15.2 which contains p in N_p and does not contain p'. Thus, all directly and indirectly adjacent points of p' are in one side of the grid plane (including the plane). Dependent upon the third condition of the definition of simple surface points, each $c_1(p)$ and $c_2(p)$, which are defined in Sect. 5.2.1, have one point in the side of the plane because they must be indirectly adjacent to p'. Also, all of the parallel-moves of line-cell $\{p, p'\}$ are in the side of the plane.

Fig. 15.2 S in N_p

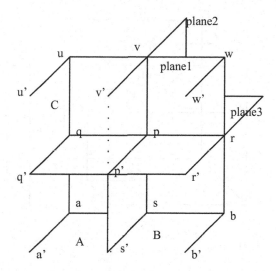

If $\{p, p'\}$ has no two parallel-moves in S, then all of the points which are in $\bar{S} \cap N_p$ and are in the side of the plane are connected; thus, $c_1(p)$ and $c_2(p)$ are connected. So, $\{p, p'\}$ has two or more parallel-moves.

Now, we prove line-cell $\{p, p'\}$ has no more than two parallel-moves. In contrast, suppose $\{p, p'\}$ has three parallel-moves; without loss of generality, we let the three parallel-moves $\{q, q'\}$, $\{r, r'\}$, and $\{s, s'\}$ of $\{p, p'\}$ be described in Fig. 15.2:

Suppose we are sure that $\{p, p', q, q', r, r', s, s'\}$ are in S. We let A be the set whose elements are in $\{a, a'\}$ but not in S, and B be the set whose elements are in $\{b, b'\}$ but not in S. Because $S \cap N_p$ does not have any 3D-cells, A and B are not empty.

Since p is a simple surface point; A and B, A and C, or B and C must be indirectly connected according to the second condition of the definition of simple surface points in Sect. 5.2.1. Using the same reasoning for A and C, and B and C, we only need to prove: p is no longer a simple surface point when A and B, or A and C are indirectly connected.

(i) If A and B are indirectly connected, then $a \in A$ and $b \in B$; otherwise, A does not connect with B in $N_p \cup p$. We know that a and b are indirectly connected. Meanwhile, every s's indirectly adjacent point in $N_p \cup p$ is below plane3, and two of them are contained in $c_1(p)$ and $c_2(p)$ respectively. Thus, all of s's indirectly adjacent points must be indirectly adjacent to a or b. Then $c_1(p)$ and $c_2(p)$ defined in Sect. 5.2.1 are indirectly connected, so p is not a simple surface point.

(ii) If A and C are indirectly connected, a must belong to A. We may suppose that $A \cup C \subset c_1(p)$ and $B \subset c_2(p)$. We now must discuss the following two cases.

(ii.a) If $b \in S$, then $c_2(p) = b'$. Thus, q cannot be indirectly adjacent to b, i.e., $c_2(p)$. According to the third condition of the simple surface points' definition, p is not a simple surface point.

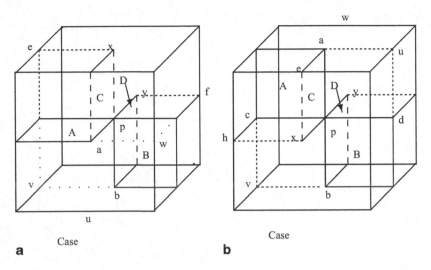

Fig. 15.3 Two cases in which surface-cells are not line-adjacent

(ii.b) If b is not in S, i.e., $b \in c_2(p)$; then we can see if q connects to b and r connects to a, they must pass the plane 2. On the other hand, if $\{u, v, w\} \subset S$, then there is a point of $c_1(p)$ in $\{u', v', w'\}$ based on no 3D-cell in S. However, the point and A cannot be indirectly connected in N_p, that is, A and C are not indirectly connected. Thus, there must be a point δ in u, v, and w such that it is in $c_1(p) \cup c_2(p)$. If δ is in $c_1(p)$ then each point in plane 2 indirectly connects with a or δ. q cannot indirectly connect with the point b. If δ is in $c_2(p)$ then r cannot indirectly connects with the point a. According to the third condition of the simple surface point definition, p is not a simple surface point.

Therefore, we have proven statement 2. We now prove statement 3 to complete part one of the proof. Statement 3 says that any two surface-cells of $S(p)$ are line-connected if p is a simple surface point.

Actually, there are only two cases for two surface-cells A and B including p in $N_p \cup p$ which are not line-adjacent (See Fig. 15.3). In the following we show that these two surface-cells A and B are line-connected in $N_p \cup p$ when p is a simple surface point and both A and B are in S.

For case (a), if one of u, v, or w is in S, then surface-cells A and B are line-connected. Otherwise, because each (p, a) and (p, b) have two parallel-moves, C and D must be in S. By the same reasoning, points e and f are in S, so we can see that the p is not a simple surface point because $\bar{S} \cap N_p$ has three indirectly connected parts.

For case (b), if u and v are in S, then A and B are line-connected. Otherwise, C and D,or $\{a, e, x, p\}$ and $\{a, p, y, w\}$ must be in S; so we have a case like case (a). Thus, p is not a simple surface point if A and B are not line-connected. We have thus proven statement 3.

Fig. 15.4 The case that is not
a regular surface point

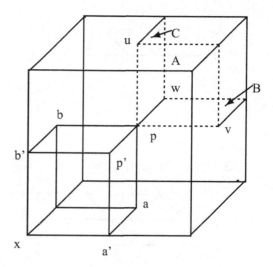

To summarize, p is a regular inner surface point if p is a simple surface point. We have proved the first part.

Part 2 Suppose p is a regular inner surface point. We want to show that p is a simple surface point, i.e., the following three statements are true:

1) $S \cap N_p$ has exactly one component adjacent to p, denote this component A_p.
2) $\bar{S} \cap N_p$ has exactly two 26-connected components, c_1 and c_2, 26-adjacent to p.
3) If $q \in S$ and q is adjacent to p, then q is 26-adjacent to both c_1 and c_2.

Our strategy for proving part two is different from part one's proof. We just enumerate all possibilities in which a regular inner surface point p can appear, and we then prove that all possible regular inner surface points are simple surface points.

We know if p is a regular surface point of S, then p is not a corner point of any 3D-cell which is in S. Because of the (point-) connectivity of S, p has an adjacent point denoted by p'. Also, $\{p, p'\}$ has two parallel-moves, denoted by $\{a, a'\}$ and $\{b, b'\}$; therefore, both a and b are also adjacent to p.

Suppose the surface-cell that is formed by $\{p, a\}$ and $\{p, b\}$ is in S, then there are three surface-cells that are line-connected to each other in $S(p)$. Therefore, they can be illustrated as shown in Fig. 5.10.

Here x must not be in S as p is not a corner of any 3D-cell. If there is a point u, v, or w that is adjacent to p; then A, B or C must be in S because any line-cell has exactly two parallel-moves. In this case, there exist two surface-cells which are not connected in $N_p \cup \{p\}$, so p is not a regular surface point. Thus, there is no such point u, v, or w which is adjacent to p, so we can see p satisfies the definition of a simple surface point (Fig. 15.4).

On the other hand, if the surface-cell formed by $\{p, a\}$ and $\{p, b\}$ is not in S. We can derive only two different cases as shown in Fig. 15.5.

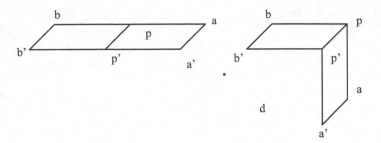

Fig. 15.5 Another case that is not a regular surface point

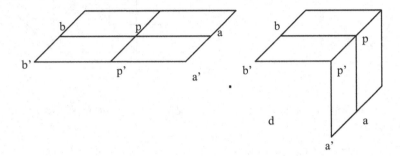

Fig. 15.6 $S(p)$ has exactly four 2-cells (surface-cells)

We now illustrate all possible developments based on the above cases, meaning that p is kept as a regular inner surface point. We only consider the points that are adjacent to p and the surface-cells that contain p. It is straightforward to see that there is only one possible case for a regular inner surface point with exactly three surface-cells.

(i) If there are exactly four surface-cells in $S(p)$, we have only the following two possible cases keeping p as a regular inner surface point (Fig. 15.6) We can see that p in either (a) or (b) is a simple surface point.

(ii) If there are five or more surface-cells in $S(p) = S \cap (N_p \cup p)$, then only two cases which contain three surface-cells can be developed to generate different results. The two cases are given in Fig. 15.7.

Each (a) and (b) in Fig. 15.8 has three possible ways for adding one more surface-cell. Some of which are overlap. We can reduce such cases to 4 distinct cases as shown in Fig. 15.14.

Next, we continuously add new surface-cells to (a), (b), (c), and (d) of Fig. 15.8. and maintain p as a regular inner surface point. We arrive at the following seven cases shown in Fig. 15.9. We can see that (e) of Fig. 15.9. already arrives at the final state, where it has only five surface-cells in $S(p)$. On the other hand, (c) also arrives at the final state because it cannot be a regular inner surface point.

We also can see that there is only one possible choice to add a surface-cell onto (a), (b), (d), (f), and (g) in Fig. 15.9. After adding a surface-cell, (a), (b), (d), (f), and (g)

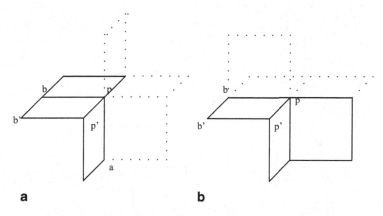

Fig. 15.7 Two cases can be developed

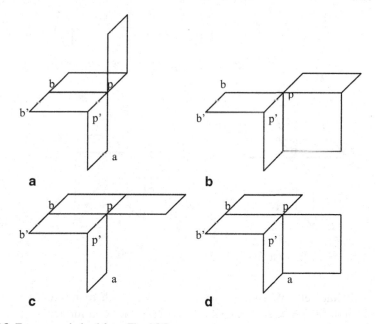

Fig. 15.8 Four cases derived from Fig. 15.7

are deduced into 2 cases shown in Fig. 15.10. Each of which have six surface-cells including p, where p is a regular inner surface point.

Finally, we shall explain, when p is a regular surface point in S, why there are no seven or more surface-cells including p in $S \cap (N_p \cup \{p\})$ that can make a simple surface point. We know p has six adjacent points in $N_p \cup \{p\}$; in other words, there are six line-cells including p. If a surface-cell A including p is in S, then A contains

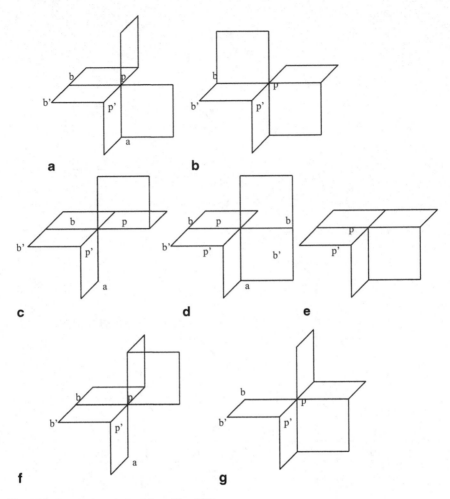

Fig. 15.9 Seven cases derived from Fig. 15.8

two of the six line-cells. When S has seven surface-cells, there must exist a line-cell which is included by three surface-cells. Therefore, S is not a surface.

From the preceding 5-step process, we obtain three types of regular inner surface points contained by five or six surface-cells. Considering Figs. 15.5 and 15.6, we have one regular inner surface point with three surface-cells and two regular inner surface points with exactly four surface-cells. There are only six possibilities for p to be a regular inner surface point. We can see that all of the three kinds of the regular inner surface points satisfy the definition of the simple surface points. We now have completed the proof of part 2.

Therefore, every regular inner surface point is a simple surface point.   □

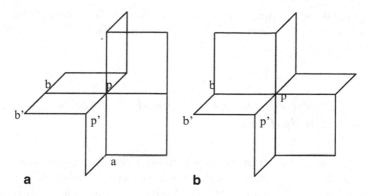

Fig. 15.10 Two cases deduced from Fig. 15.9

15.3   The Jordan Curve Theorem on Discrete Surfaces

This section focuses on the Jordan curve theorem in 2D discrete spaces, with respect to the general definition of discrete curves, surfaces, and manifolds discussed in Chap. 7 [3]. The Jordan curve theorem states that a (simply) closed curve separates a simply connected surface into two components. Based on the definition of discrete surfaces, we give three reasonable definitions of simply connected spaces in discrete spaces. Theoretically, these three definitions are equivalent.

For the Jordan curve theorem, O. Veblen in 1905 wrote a paper [35] that was regraded as the first correct proof of this fundamental theorem. The first discrete proof was given by W.T. Tutte on planar graphs in 1979 [34]. Recently, researchers still show considerable interests in the Jordan curve theorem using formalized proofs in computers [13]. In 1999, L. Chen attempted to prove the discrete Jordan curve theorem for 2D discrete manifolds without using 2D Euclidean space [7]. In this section, we will adopt some original ideas in Veblen's paper and give a proof of this theorem in discrete space.

In Chap. 7, we have defined a discrete surface. More importantly, this discrete surface can be naturally embedded to Euclidean plane. Or a closed discrete surface, can be easily embedded to a 3D or higher dimensional Euclidean spaces.

Let us first review some concepts for curves in Sect. 7.2: (1) A simple path is called a (discrete) pseudo-curve, (2) a simple semi-curve can be a curve or a surface-cell (2-cell), (3) A simple curve must not contain any proper subset that is a 2-cell.

It is obvious, if we define a path (a pseudo-curve) is a discrete curve, there is no Jordan curve theorem in discrete space. This is because that the inner part of a 2-cell is empty in graphs structures.

In Sect. 7.5, we defined regular points or ordinary points in a discrete manifolds. We proved that for a discrete surface S (Lemma 7.6): If p is a inner and regular point of S, then there exists a simple cycle containing all points in $S(p) - \{p\}$ in S where $S(p)$ is the neighborhood of p in graph G. This lemma is particularly important in our proof. In addition, we also have (Lemma 7.5): For a discrete surface S, let a point

$p \in S$, if p has only two adjacent points p', p'' in S, then there are two surface-cells A, B such that $A \cap B$ contains p', p, p''. If p', p'' are adjacent in S, then p', p, p'' form a surface-cell.

15.3.1 Discrete Deformation and Simply Connected Discrete Surfaces

In topology, the formal description of the Jordan curve theorem is: A simply closed curve J in a plane Π decomposes $\Pi - J$ into two components [24, 28]. In fact, this theorem holds for any simply connected surface. A plane is a simply connected surface in Euclidean space, but this theorem is not true for a general continuous surface. For example, the boundary of a donut.

What is a simply connected continuous surface? A connected topological space T is simply connected if for any point p in T, any simply closed curve containing p can be contracted to p. The contraction is a continuous mapping among a series of closed continuous curves [28]. So, we first need the concept of "discrete contraction."

In order to keep the concepts simple to understand, we first define the gradual variation between graphs. Then we define discrete deformation among discrete pseudo curves. And finally, we define the contraction of curves is a type of discrete deformation. See [5, 6] for more details of the definitions.

In this section, we assume the discrete surface is both regular and orientable (Chap. 7).

Definition 15.1 Let G and G' be two connected graphs. A mapping $f : G \to G'$ is gradually varied if for two vertices $a, b \in G$ that are adjacent in G, then $f(a)$ and $f(b)$ are adjacent in G' or $f(a) = f(a')$.

So gradual variation defined in Chap. 11 is the counterpart of continuation of Euclidean space. The special case of this definition is that two paths are subgraphs of a graph G. We want to change a path to another "continuously" is the same as to change one "gradually" to another one. Herman defined "elementarily N-equivalent" for defining simply connected space[19].

Intuitively, "continuous" change from a simple path C to another C' is that there is no "jump" between these two paths. If $x, y \in S$, $d(x, y)$ denotes the distance between x and y. $d(x, y) = 1$ means that x and y are adjacent in S. It is important to point out that in a 2-cell (or any other k-cell), from a point p to another point q in the cell, $p \neq 1$, the distance $d(p, q)$ can be viewed as 1. In other words, a cell can be viewed as a complete subgraph on its vertices.

Definition 15.2 Two simple paths $C = p_0, ..., p_n$ and $C' = q_0, ..., q_m$ are gradually varied in S if $d(p_0, q_0) \leq 1$ and $d(p_n, q_m) \leq 1$ and for any non-end point p in C, then

(1) p is in C', or p is contained by a 2-cell A (in $G(C \cup C')$) such that A has a point in C'.

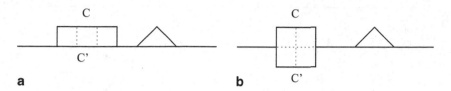

Fig. 15.11 Gradually varied curves: **a** C and C' are gradually varied; **b** C and C' are not gradually varied

(2) Each non-end-edge in C is contained by a 2-cell A (in $G(C \cup C')$) which has an edge contained by C' but not C if C' is not a single point.
and vise versa for C'.

For example, C and C' in Fig. 15.11a are gradually varied, but C and C' in Fig. 15.11b are not gradually varied. We can see that a 2-cell, which is a simple path, and any two connected parts in the 2-cell are gradually varied. So we can say that a 2-cell can be contracted to a point gradually.

Assume $E(C)$ denotes all edges in path C. Let $XorSum(C, C') = (E(C) - E(C')) \cup (E(C') - E(C))$. $XorSum$ is called $sum(modulo2)$ in Newman's book [28].

Attach a 2-cell to a simple path C, if the intersection is an arc (connected path) not a vertex, we can see cut the intersection (keep the first and last vertices of the intersection which is an arc), the simple path will go another half of the arc of the cell. The new path is also a simple path, and it is gradually varied to C. Therefore,

Lemma 15.1 *Let C be a pseudo-curve and A be a 2-cell. If $A \cap C$ is an arc containing at least an edge, then $XorSum(C, A)$ is a gradual variation of C.*

It is not difficult to see that $XorSum(XorSum(C, A), A) = C$ and $XorSum(XorSum(C, A), C) = A$ under the condition of the above lemma.

Definition 15.3 Two simple paths (or pseudo-curves) C and C' are said to be homotopic if there is a series of simple paths $C_0, ..., C_n$ such that $C = C_0$, $C' = C_n$, and C_i, C_{i+1} are gradually varied.

We say that C can be discretely deformed to C' if C and C' are homotopic. The following lemma is guarantee that we deform a curve by just making changes a cell a time.

Lemma 15.2 *If two (open, not closed) simple paths C and C' are homotopic then there is a series of simple paths $C_0, ..., C_m$ such that $C = C_0$, $C' = C_n$, and $XorSum(C_i, C_{i+1})$ is a 2-cell excepting end-edges of C and C'.*

15.3.2   Cross-Over of Simple Paths

To prove the Jordan curve theorem, we need to describe what the disconnected components are by means of separated from a simple curve C? It means that any path from a component to another must include at least a point in C. It also means

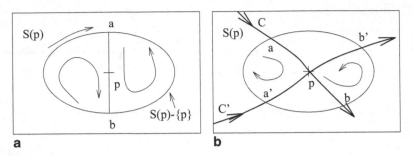

Fig. 15.12 $S(p)$ and Cross-over at p: **a** Two adjacent points a and b of p in $S(p)$, and **b** an example for two cross-over paths

that this linking path must cross-over the curve C. In this subsection, we want to define it.

Because a surface-cell A is a closed path, we can define two orientations (normals) to A: clockwise and counter-clockwise. Usually, the orientation of a 2-cell is not a critical issue. However, for the proof of the Jordan curve theorem it is necessary.

In other words, a pseudo-curve which is a set of points has no "direction," but aa path, has its own "travel direction" from p_0 to p_n. For two paths C and C', which are gradually varied, if a 2-cell A is in $G(C \cup C')$, the orientation of A with respect to C is determined by the first pair of points $(p, q) \in C \cap A$ and $C = ...pq...$. Moreover, if a 1-cell of A is in C, then the orientation of A is fixed with respect to C.

According to Lemma 7.6, $S(p)$ contains all adjacent points of p and $S(p) - \{p\}$ is a simple cycle—there is a cycle containing all points in $S(p) - \{p\}$.

We assume that cycle $S(p) - \{p\}$ is always oriented clockwise. For two points $a, b \in S(p) - \{p\}$, there are two simple cycles containing the path $a \to p \to b$: (1) a cycle from a to p to b then moving clockwise to a, and (2) a cycle from a to p to b then moving counter-clockwise to a. See Fig. 15.12a.

It is easy to see that the simple cycle $S(p) - \{p\}$ separates $S - \{S(p) - \{p\}\}$ into at least two connected components because from p to any other points in S the path must contain a point in $S(p) - \{p\}$. $S(p) - \{p\}$ is an example the Jordan curve.

Definition 15.4 Two simple paths C and C' are said to be "cross-over" each other if there are points p and q (p may be the same as q) such that $C = ...apb...sqt...$ and $C' = ...a'pb...sqt'...$ where $a \neq a'$ and $t \neq t'$. The cycle $apa'...a$ without b in $S(p)$ and the cycle $qt...t'q$ without s in $S(q)$ have different orientations with respect to C.

For example, in Fig. 15.12b, C and C' are "cross-over" each other. When C and C' are not "cross-over" each other, we will say that C is at a side of C'.

Lemma 15.3 *If two simple paths C and C' are not cross-over each other, and they are gradually varied, then every surface-cell in $G(C \cup C')$ has the same orientation with respect to the "travel direction" of C and opposite to the "travel direction" of C'.*

We also say that C and C' in the above Lemma are side-gradually varied.

Intuitively, a simply connected set is such a set so that for any point, every simple cycle containing this point can contract to the point. According to the nature of the word "contraction," we can give the mathematical definition of "contraction" for discrete spaces. In fact, the contraction procedure relates to some substance. This substance gradually loses the size (which can be space occupied), and when a part was lost, it will never come back again in this contraction.

Definition 15.5   A simple cycle C can contract to a point $p \in C$ if there exist a series of simple cycle, $C = C_0, ..., p = C_n$: (1) C_i contains p for all i; (2) If q is not in C_i then q is not in all C_j, $j > i$; (3) C_i and C_{i+1} are side-gradually varied.

We now show three reasonable definitions of simply connected spaces below. We will provide a proof for the Jordan curve theorem under the third definition of simply connected spaces. The Jordan theorem shows the relationship among an object, its boundary, and its outside area.

A general definition of a simply connected space should be:

Definition 15.6   Simply Connected Surface Definition (a) $< G, U_2 >$ is simply connected if any two closed simple paths are homotopic.

If we use this definition, then we may need an extremely long proof for the Jordan curve theorem. The next one is the stardard definition which is the special case of the Definition 15.6.

Definition 15.7   Simply Connected Surface Definition (b) A connected discrete space $< G, U_2 >$ is simply connected if for any point $p \in S$, every simple cycle containing p can contract to p.

This definition of the simply connected set is based on the original meaning of simple contraction. In order to make the task of proving the Jordan theorem simpler, we give the third strict definition of simply connected surfaces as follows.

We know that a simple closed path (simple cycle) has at least three vertices in a simple graph. This is true for a discrete curve in a simply connected surface S. For simplicity, we call an unclosed path an arc. Assume C is a simple cycle with clockwise orientation. Let two distinct points $p, q \in C$. Let $C(p, q)$ be an arc of C from p to q in a clockwise direction, and $C(q, p)$ be the arc from q to p also in a clockwise direction, then we know $C = C(p, q) \cup C(q, p)$. We use $C^a(p, q)$ to represent the counter-clockwise arc from p to q. Indeed, $C(p, q) = C^a(q, p)$. We always assume that C is in clockwise orientation.

Definition 15.8   Simply Connected Surface Definition (c) A connected discrete space $< G, U_2 >$ is simply connected if for any simple cycle C and two points $p, q \in C$, there exists a sequence of simple cycle paths $Q_0, ..., Q_n$ where $C(p, q) = Q_0$ and $C^a(p, q) = Q_n$ such that Q_i and Q_{i+1} are side-gradually varied for all $i = 0, \cdots, n - 1..$

In fact, it is easy to see that Simply Connected Surface Definition (b) and (c) are special cases of Definition (a). $C(p, q) = Q_0$ and $C^a(p, q) = Q_n$ are two arcs of C.

Now we want to prove that the simply connected surface definition (b) and the definition (c) are equivalent (In [5], we thought that such a proof would be hard).

Proposition 15.1 *Definition (b) and Definition (c) are equivalent.*

Proof It is easy to see that Definition (b) is a special case of Definition (c). This is because that we can define an edge e in C as the one arc and the rest of C is another arc.

Now we prove that Definition (c) can be induced from Definition (b). We first select the contracting point x to be p. Let the contracting sequence $C_0, C_1, \ldots , C_i$ contain another point q if C is not a 2-cell, but $C_{i+1}, \ldots , C_n = p$ do not contain q (C_{i+1} does not contain q is the key. If q is not in the same 2-cell of p, we can always find such a q).

So the paths from p to q, $C_k(p, q)$, and q to p, $C_k(q, p)$ (both arcs are in the clockwise direction) have their own corresponding gradually varied paths, respectively in $C_k, k = 0, \ldots , i$).

Note that, all $C_t, t = i + 1, \ldots , n - 1$ are closed path. We know q is not in C_{i+1}, but q has a corresponding point in C_{i+1}, denoted as q^{i+1}. In other words, q changed to q^{i+1}, in the contraction process.

Therefore, from p to q^{i+1}, there are two paths from p to q^{i+1}, $C_{i+1}(p, q^{i+1})$ and from q^{i+1} to p, $C_{i+1}(q^{i+1}, p)$. Thus, $C_i(p, q)$ and $C_{i+1}(p, q^{i+1})$ are gradually varied, so are $C_i(q, p)$ and $C_{i+1}(q^{i+1}, p)$. In the same way, we can find $q^{i+2}, \ldots$ $, q^{n-1}$. Finally, $C_0(p, q), \ldots , C_{i+1}(p, q^{i+1}), \ldots , C_{n-1}(p, q^{n-1}), C_{n-1}(q^{n-1}, p), \ldots$ $C_{i+1}(q^{i+1}, p), C_0(q, p)$ are such a gradually varied sequence. $\square$

Basically, the deformation procedure does not really care about cross-over points. We does not allow cross-over points in the definition is to make the proof easier.

15.3.3 The Jordan Curve Theorem

Since a simple cycle could be a surface-cell, it can not separate S into two disconnected components. So for the strict case of Jordan curve theorem, we must use a discrete curve not a simple cycle.

In the case of allowing the central pseudo points, (it is true for embedding a surface into a Euclidean space.) we will have the general Jordan Curve Theorem. We will prove this case at the last of this section.

However, for a closed discrete curve, we have

Theorem 15.2 (The Jordan Curve Theorem in Discrete Space) *A discrete simply connected surfaces S defined by Definition 15.8 (Definition (c)), has the Jordan property: For a closed discrete curve C on S, if C does not contain any point of ∂S, C divides S into at least two disconnected components. In other words, $S - C$ consists of at least two disconnected components.*

Proof Suppose that C is a closed curve in a simply connected surface S. C does not reach the border of S, i.e. $C \cap \partial S = \emptyset$.

Assume point $p \in C$, then suppose that q and r are two adjacent points of p in C with form of ...qpr, ..., where the direction of ... q to p to r ... to p is

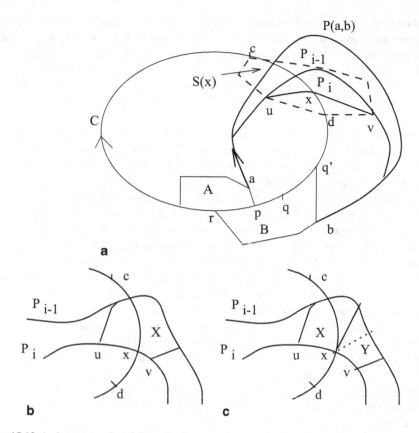

Fig. 15.13 A close curve C and the paths from a to b

clockwise. See Fig. 15.13 $\{p,r\}$ is a line-cell, then there are two 2-cells containing $\{p,r\}$. Denote these 2-cells by A and B with clockwise orientation.

Our strategy is to prove that if there is a point a in A which is not in C, and a point $b \in B$ and $b \notin C$, then any path from a to b must contain a point in C. Then we can see that $S - C$ are not (point-) connected and we have the Jordan curve theorem.

First, we want to prove that there must exist a point in $A - C$. If each point in A is in C, since A is a simple cycle, then $C = A$. However, C is not a surface-cell, so the statement can not be true. Thus, there is a point $a \in A - C$. For the same reason there is a point $b \in B - C$. We assume that a is the last such point in A starting with p, and b is the first such point in B starting with p (see Fig. 15.13a).

We always assume clockwise direction here for cell A and B unless we indicate otherwise.

Let us make a summery of above idea: Suppose that C is a closed curve. (If C is a closed simple path, we allow C is a 2-cell and we allow to assign a pseudo-point in the center of the 2-cell, we still can prove this theorem.) The idea of the proof is to find two points in each sides of the curve C. This is because that for any 1-cell (r, p) in C, there are two 2-cells A, B sharing (r, p) by the discrete surface definition. A must contain a vertex a and B must contain b, and they are not in C. a, b are adjacent to some points in C, respectively. We are going to prove that from a to b, any path must cross-over C. That is the most important part of the Jordan curve theorem.

We assume, on the contrary, there is a simple path from a to b does not cross-over C, called $P_{(}a, b)$ in Fig. 15.13 a. But we know there is $P(b, a)$ in $A \cup B$ (i.e. $P(b, a) = b \cdots rp \cdots a$) does cross-over C. $P_{(}a, b) \cup P_{(}a, b)$ is a cycle in clockwise (Fig. 15.13a).

We know $S(r)$ is the neighborhood of r in S. So $S(r)$ contains all 2-cells containing r. The boundary of $S(r)$ is a simple closed curve. (This is because we always assume that r is a regular point). a is on the boundary of $S(r)$. (The boundary of $S(r)$ is $S(r) - \{r\}$). $A \cup B$ is a subset of $S(r)$.

We now prove that $P_{(}a, b)$ is not a subset of $S(r)$; otherwise, it must cross-over C. (a 2-cell containing r must have an edge on C, or all points of the 2-cell are on the boundary of $S(r)$ except r). If $P(a, b)$ does not contain r, must be a part of boundary of $S(r)$ which is a cycle. r has two adjacent points on C, (If they are not pseudo points, meaning here it can be eliminated or added on an edge that does not affect to the 2-cell) so these two points are also in the boundary of $S(r)$. So there are only two paths from a to b on the boundary of $S(r)$. These two points are not on the same side of the cross-over path containing r. (The boundary of $S(r)$ was separated by the cross-over path containing r.) $P_{(}a, b)$ must contain such a point that is on C.

Therefore we proved $P_{(}a, b)$ is not a subset of $S(r)$. Then $P_{(}a, b) \cup P(b, a)$ is a simple closed curve. ($P(b, a)$ passes r). By the definition of the simply-connected surface, there are finite numbers of paths $P(a, b) = P_0(a, b), \ldots, P_{n-1}(a, b)$, such that so that $P_i(a, b)$ and $P_{i+1}(a, b)$ are (side-)gradually varied.

In addition, $P_{n-1}(a, b)$ is gradually varied to $P_n(a, b) = P^a(b, a)$ (a reversed $P(b, a)$ that passes r).

We now can assume that there is a smallest i such that $P_i(a, b)$ cross over C, but $P_{i-1}(a, b)$ does not (Fig. 15.13 a). We will prove that is impossible if $P_{i-1}(a, b)$ does not cross over C.

Let point x in $P_i(a, b) \cap C$ and $x \notin P_{i-1}(a, b)$. There are two cases: (1) cross over at a single point x on C, or (2) cross over at a sequence of points on C. We will prove these two cases, respectively.

Case 1 Suppose that $x =$ "p''" $=$ "q''" in Definition 15.4 (See Fig. 15.12 a, b)). It means two curves $P_i(a, b)$ and C share just one point x. and assume $P_i(a, b) =$ $...uxv...$ and $C = ...cxd...$, where $v \neq d$.

We know that u, v, c, d are in the boundary of $S(x)$, a simple cycle $S(x) - \{x\}$ (See Lemma 7.6). There is a 2-cell X (in between P_{i-1} and P_i) contains (u, x). See Fig. 15.13b) X has a sequence of points $S1$ in P_{i-1} and a sequence of points $S2$ in P_i. X has at most two edges $e1, e2$ not in $P_{i-1} \cup P_i$; $S1, e1, S2, e2$, are the boundary of X. $e1$ is the edge linking $S1$ to $S2$, and $e2$ is the edge linking $S2$ to $S1$ counterclockwise.(Again, $e1$ may or may not be directly incident to u, and $e1$ may be an empty edge if P_{i-1} intersects P_i at point u. $e2$ may also in the same situation.) We might as well assume that x is the first point on $P_i(a, b)$ (from a to b in path

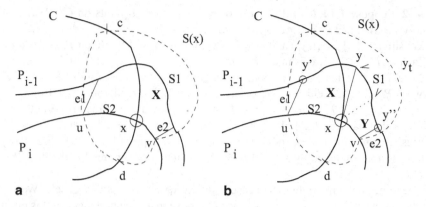

Fig. 15.14 Extended figures of Fig. 15.13 **b** and **c**, respectively

P_i)that is in C. Thus, $c, d \notin X$ (If c is in X c must be in P_{i-1}. if d is in X, x is not only the cross over point).

If X contains v, we will have a cycle $u \cdot d \cdot v(e2)(S1)(e1)$ in the boundary of $S(x)$ $(e2)(S1)(e1)$ contains only points in P_{i-1} and u,v (that are possible end points of $e1$, $e2$). c is on the boundary of $S(x)$ too. Where is c? It must be in the boundary curves (of $S(x)$) from u to d or the curve from d to v. Then c, d in $S(x)$ must in the same side of uxv which is part of P_i. Therefore, C and $P_i(a, b)$ do not cross-over each other at x. See Fig. 15.13b and the following extended figure.

If X does not contain v, then there must be a 2-cell Y (in between P_{i-1} and P_i) containing (x, v). We can see that X and Y are line-connected in $S(x)$ (See Fig. 15.13c and above extended figure). This is due to the definition of regular point of x, all surface-cells containing x are line-connected. Meaning there is a 2-cell paths they share a 1-cell in adjacent pairs.

Since X and Y are line-connected, we can assume:

a) X and Y share a 1-cell, i.e. $X \cap Y = (x, y)$. Then y is on $P_{i-1}(a, b)$. Let $e2$ be the possible edge from v to $P_{i-1}(a, b)$. ($e2$ could be empty as $e1$) and $u(e1)..y...(e2)v$ is on the boundary cycle of $S(x)$. Except u and v, $u(e1)..y...(e2)v$ is on $P_{i-1}(a, b)$. $u...d...v$ is part of the boundary cycle of $S(x)$. In addition, c (that is not in $P_{i-1}(a, b)$) must be in the boundary curves (of $S(x)$) from u to d or the curve from d to v. Again, c, d in $S(x)$ must in the same side of uxv which is part of P_i. Therefore, C and $P_i(a, b)$ do not cross-over each other at x (See Fig. 15.13c also see the above extended c).

b) X and Y share the point x, and there are line-connected 2-cells as a path in between X and Y. $X \cap Y = x$. Let us assume that $e1$ incident to $P_{i-1}(a, b)$ at y' (y' is u if $e1$ is empty) and $e3$ incident to $P_{i-1}(a, b)$ at y''. We will have a set of points $y' = y_0, y_1, ..., y_k = y''$ in $P_{i-1}(a, b)$. Each y_t is contained in a 2-cell containing x. All $y_0, y_1, ..., y_k$ are in the boundary cycle of $S(x)$. c that is not in $P_{i-1}(a, b)$. c must be in the boundary curves (of $S(x)$) from u to d or the curve from d to v. Thus, c, d in $S(x)$ must in the same side of uxv which is part of P_i. C and $P_i(a, b)$ do not cross-over each other at x (See Fig. 15.13 c and 15.14c).

Case 2 Suppose $P_i(a,b)$ and C cross over a sequence of points on C: $P_i(a,b) = ...ux_0x_1...x_mv...$ and $C = ...cx_0x_1...x_md...$, where $v \neq d$.

We still have $e1 = (u, y_0)$ and $e2 = (v, y_k)$ where y_0 and y_k are on P_{n-1} for some k. Each y_t, $t = 0, 1, ..., k$, is in a 2-cell that containing some x_j, $j = 0, 1, ..., m$.

Note that: If u does not have a direct edge linking to P_{n-1}, u will be in a 2-cell between P_n and P_{n-1}, either u is a pseudo point on P_n for the deformation from P_{n-1} to P_n, or P_{n-1} and P_n intersects at u. That u is a pseudo point means here it has a neighbor that has an edge link to P_{n-1}, or the neighbor's neighbor, and so on. We can just assume here u is the point that is adjacent to a point in P_{n-1}. In the theory, as long as u is contained by a 2-cell such that all the points in the 2-cell are in P_{n-1} or P_n.

The same way will apply to this case just treat $x_0, \ldots ,x_m$ to x in Case 1. We first get the union of $S(x_0), \ldots ,S(x_m)$. We want to prove that: The boundary of this union will be simple cycle too.

Using mathematical induction we can prove it. After that, we can prove the rest of theorem using the same method presented in Case 1. See Fig. 15.15.

The following is the detailed proof: Let $S(x_0, ..., x_k) = S(x_0) \cup ... \cup S(x_k)$.

First, we will prove that the boundary of $S(x_0) \cup S(x_1)$ is a simple cycle (it is a simple closed curve too). We know that (x_0, x_1) is an edge in $C \cap P_i(a,b)$. Also, there are two 2-cells A, B in $S(x_0)$ containing (x_0, x_1).

x_1 is a boundary point in $S(x_0)$, so no other 2-cell will contain x_1. In the same way, $S(x_1)$ also contains A, B, and x_1 is only contained in two 2-cells in $S(x_1)$. Therefore, $S(x_0) \cap S(x_1) = A \cup B$ and $A \cap B = (x_0, x_1)$.

Note that A and B are adjacent 2-cells. On the other hand, x_1 is on the boundary curve (that is closed) of $S(x_0)$. So x_1 has two adjacent points on this cycle, y_1 and y_2. (We assume that y_1 and y_2 are not pseudo points, so) y_1 and y_2 are both on the boundary of $S(x_0) \cup S(x_1)$. (If y_1 or y_1 is pseudo points, we can ignore y_1 or y_2 to find the a actual point that adjacent to x_1.) (x_1, y_1) has two 2-cells containing (x_1, y_1) in $S(x_0) \cup S(x_1)$. For instance, in Fig. 15.15a, A and A_1 contain (x_1, y_1) and B and B_1 contain (x_1, y_2). Thus, the boundary of $S(x_0) \cup S(x_1)$ is a closed curve that is formed by the arc from y_1 to y_2 in the boundary of $S(x_0)$, plus the arc from y_2 to y_1 in the boundary of $S(x_1)$.

Second, we assume the boundary of $S(x_0, ..., x_{k-1})$ is a closed curve, when we consider the arc $x_0, ..., x_{k-1}, x_k$ in C, we can prove the boundary of $S(x_0, ..., x_k)$ is also a closed curve. Please note that this property may contain a degenerate or pathological case. We can add some 2-cells in S, but not on C. So that if x_i, x_j in C is not adjacent, then any path linking x_i and x_j not intersecting C, contains at least two other points. In other words, from x_i to x_j, a path must pass three 2-cells (including the one that contains x_i or x_j). This is also to say that x_i and x_j are separated by a 2-cell that does not contain x_i, x_j. This is very easy to make since we can always refine a 2-cells into several two cells using Veblen points.

We know that we have two closed curves: Suppose that Q is the boundary of $S(x_0, ..., x_{k-1})$, and R is the boundary of $S(x_k)$. (x_{k-1}, x_k) is in $S(x_k)$, and (x_{k-1}, x_k) is in $S(x_0, ..., x_{k-1})$. There are two 2-cells A, B containing $(x_k - 1, x_k)$ in $S(x_k) \cap S(x_0, ..., x_{k-1})$.

Fig. 15.15 The union of neighborhoods of a sequence of adjacent points $S(x_0, ..., x_k)$ and its boundary

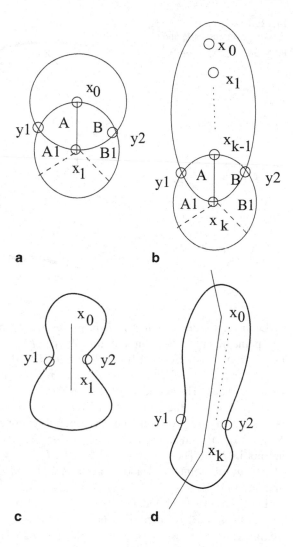

x_{k-1} is on the boundary cycle of $S(x_k)$, then x_{k-1} must have two adjacent points in R, y_1, and y_2. (x_{k-1}, y_1) and (x_{k-1}, y_2) are two edges in $S(x_k) \cap S(x_0, ..., x_{k-1})$. In the same way above, we will have the cycle passing y_1 and y_2 that is the boundary curve of $S(x_0, ..., x_k)$.

Thus, we have proved that the boundary curve of $S(x_0, ..., x_k)$ is a simple closed curve. In the rest of the proof, we will treat $S(x_0, ..., x_m)$ to be $S(x)$ in Case 1. See Fig. 15.15.

We now use denote $X = \{x_0, ..., x_m\}$ and X is an arc in C (Please note that in Case 1, X was used as a 2-cell. Now X is an arc in C).

In the rest of the proof, we will prove: if $P_{i-1}(a, b)$ and C are not cross over each other, then, $P_i(a, b)$ and C will not be cross over each other. Therefore, any $P(a, b)$ must cross over C. This completes the proof of the discrete Jordan curve theorem.

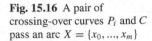
Fig. 15.16 A pair of crossing-over curves P_i and C pass an arc $X = \{x_0, ..., x_m\}$

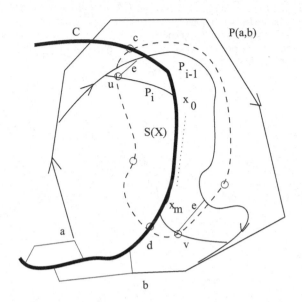

Let us first state again that $P_i(a, b)$ passes $x_0...x_m$ but $P_{i-1}(a, b)$ does not contain any point of $\{x_0, ..., x_m\}$. In addition, $P_{i-1}(a, b)$ and $P_i(a, b)$ is gradually varied, i.e. $P_i(a, b)$ was deformed from $P_{i-1}(a, b)$ directly. We also know that $S(X) = S(x_0, ..., x_m)$ is the neighborhood of the arc in C, i.e. the arc $x_0, ..., x_m$ is a part of the closed curve C. The boundary of $S(X) = S(x_0, ..., x_m)$ is a closed curve too.

u, v, c, d are on the boundary of $S(x_0, ..., x_m)$ (Assume u, v, c, d are not pseudo points, otherwise, we can find corresponding none-pseudo on the boundary of $S(x_0, ..., x_m)$.) $u, (x_0, ..., x_m), v$ is a part of P_i We also know that c and $(x_0, ..., x_m)$ are not in P_{i-1}. There will be two 2-cells, U and V, are in between $P_i(a, b)$ and $P_{i-1}(a, b)$ (all points of U and V are in $P_i(a, b) \cup P_{i-1}(a, b)$) such that $(u, x_0) \in U$ and $(x_m, v) \in V$.

Let $P_{i-1} \cap U = S1$ and $P_i \cap U = S2$. Let $e1$ be the edge in U linking $S1$ to $S2$ (in most cases, $e1$ incident to u, but not necessarily), and let $e2$ be the edge in U linking $S2$ to $S1$ (possibly starting at x_0). So, $(e2)(S1)(e1)(S2)$ are the boundary of U, counterclockwise.

Subcase (i) If U contains v ($U = V$), all points in U's boundary are contained in $S(\{x_0, ..., x_m\})$ by the definition of $S(x_0)$. we will have a cycle $u \cdot d \cdot v(e2)(S1)(e1)$ in the boundary of $S(X = \{x_0, ..., x_m\})$. c is on the boundary of $S(X)$ too. But $c \notin P_{i-1}$. It must be in the boundary curves (of $S(X)$) from u to d or the curve from d to v. Then c, d in $S(X)$ must in the same side of uXv which is part of P_i. Therefore, C and $P_i(a, b)$ do not cross-over each other at X (See Fig. 15.16).

Subcase (ii) If U does not contain v, then there must be a 2-cell V (in between P_{i-1} and P_i) containing (x_m, v).

Let $e1 = (p1, p2)$ be the edge in U incident to a point in P_{i-1} and a point in P_i, respectively. (In most cases, $e1$ incident to u, i.e. $u = p2$, but not necessarily).

Fig. 15.17 An edge only
starts at x_k linking to P_{i-1};
$x_0, ..., x_{k-1}$ do not have any
edge to P_{i-1}

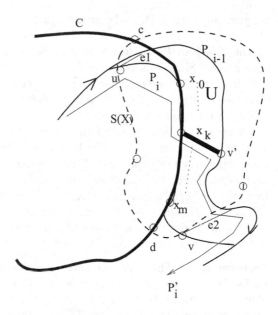

And let $e2 = (r2, r1)$ be the edge in V incident to a point in P_i and a point in P_{i-1}, respectively. $r2$ is usually v.

Note that: c must not be in U. Gradual variation (direct deformation) means that each point in each 2-cell of U and V in between P_i and P_{i-1} must be in $P_i \cup P_{i-1}$. Formally, $P_i \, XoRSum \, P_{i-1}$ is a set of 2-cells; every point in these 2-cells is in $P_i \cup P_{i-1}$.

We can see that U and V are line-connected in $S(X)$ by the definition of line-connected paths, meaning there is a path of 2-cells where each adjacent pair shares a 1-cell (See Figs. 15.15d and 15.16).

From $r1$ to $p1$, there is an arc in P_{i-1}. To prove that all points in this arc are in the boundary of $S(X)$ we need to prove each point on the arc must be in a 2-cell that contains a point in $\{x_0, ..., x_m\}$, and this 2-cell is other than (except this 2-cell is) U or V. It gives us some difficult to prove it. The above discussion seems not very productive.

We found a more elegant way to prove this case by finding another simple path (or pseudo curve) that cross-over C. The method is the following: If $U \neq V$, there must be a x_k in $\{x_0, ..., x_m\}$, x_k has an edge linking to P_{i-1}. (Otherwise, $u, x_0, ..., x_m, v$ are in a 2-cell that contains some points in P_{i-1}. Therefore, $U = V$.) We can also assume that k is not m, otherwise, v is in P_{i-1}, so $U = V$. See Fig. 15.17.

We select the smallest k having an edge linking x_k to P_{i-1}, $0 \leq k \leq m - 1$. x_k is in both P_{i-1} and C. We might as well let (x_k, v') is such an edge, and v' is a point in P_{i-1}. Therefore, we will have the new path (simple path), P_i'. This new path has two parts: The first part is the same as P_i before and including the point x_k, and the second part is the partial path (curve) of P_{i-1} after point v'. This path $P_i' = ..., u, x_0, ..., x_k, v', ...$ does cross-over $C = ..., c, x_0, ..., x_k, x_{k+1}, ..., x_m, d, ...$ It is obvious that P_{i-1}, P_i', P_i

are gradually varied. This is because we just inserted a path in between of P_{i-1} and P_i.

This new path P_i' has such a good property that is $\{x_0, ..., x_{k-1}\}$ do not have an edge in P_{i-1}. Since v' is in P_i', the 2-cell V' (in P_{i-1} $XoRSum$ P_i') contains v' also contains (x_{k-1}, x_k) and (x_k, v') (in $S(x_k)$). Since no edge from $x_0, ..., x_{k-1}$ to P_{i-1}, U containing u is just V'. We will have just **Subcase (i)** using P_i' to replace P_i.

The entire theorem is proven.   □

Theorem 15.2, the discrete Jordan curve theorem, has a little difference from the classical description of The Jordan curve Theorem. This is because that discrete curve has its own strict property: C does not contain any 2-cell. In order to satisfy the classical form. We need to use central pseudo points, we call it the Veblen point, for each type of cells, especially 1-cells (line-cells) and 2-cells (surface-cells) So we will allow the simple path (semi-curve) in the proof of the Jordan curve theorem. In fact, a little modification will assist the proving of the theorem. The rest of work is just to prove that there are only two (connected) components in $S - C$.

Theorem 15.3 *(The Jordan Curve Theorem for Generalized Simple Closed Paths) Let S be a discrete simply connected surfaces, (S can be closed or a discrete plane embedded in 2D Euclidean Space). A closed simple path (0-cell connected semi-curve) C which does not contain any point of ∂S divides S into two components (in terms of allowing central pseudo points for each cell). In other words, $S - C$ consists of two components. These two components are disconnected.*

Proof In this proof, we can put the central pseudo points for each 1-cells and 2-cells to assist our proof. The reason is that if we embed 1-cells and 2-cells into Euclidean plane or higher dimensional space. We can always find the central points for each cell. The idea of the central pseudo points is at least valid in Euclidean space. In fact, the central pseudo points also have two normal directions for a 2-cell. It also has two directions for a 1-cell. For instance, $\{a, b\}$ is an edge, $a \rightarrow b$ and $b \rightarrow a$ are two directions.

In the proof of Theorem 15.2, we know that we have two 2-cells A and B at the different side of the cycle C. We proved that point $a \in A$ and $b \in B$ are not connected in $S - C$.

C has the orientation of clockwise (or counterclockwise as we first made). (p, r) is clockwise in A, but (p, r) is counterclockwise in B. So we call A is clockwise, and B is counterclockwise. For each edge e_i (e.g. (p, r)) in C, we will have two 2-cells containing e_i, denoted by A_i and B_i. There must be one in clockwise and another is in counterclockwise.

We always assume that A_i is clockwise and B_i is counterclockwise. We now add all the central pseudo points to all 1-cells and 2-cells in S. And immediately remove all central pseudo points from 1-cells in C. (This operation is to stop a path will go through the central pseudo points on C.)

We also know in our assumption: each 2-cell must have at least three edges (1-cells) in its boundary. This is because S is a simple graph. (We can always add a point to make it in Euclidean space.) We also assume that C does not reach the border of S meaning that $C \cap \partial S = \emptyset$.

Case 1: A special case C is the boundary of a single 2-cell, denoted as A. A simple path could be just the boundary of a 2-cell. In this case, we have a central pseudo point in the cell A. So this theorem is virtually true if we can prove that $S - C$ is point-connected. That is to say that except the central pseudo point in A, $S - C$ is a point-connected component. In other word, $S - A$ is one point-connected component.

We can prove this is because of the following facts: For each cell B in $S - A$, if B has an edge in C. The boundary of B is a simple closed path. If the boundary of B is C, then $S = A \cup B$ since every edge has shared by two 2-cells (A and B) already. We have two components in $S - C$: the central pseudo point of A and the central pseudo point of B.

Case 2: The general case We know 2-cells that are joint with an edge in C have two types: the clockwise type, denoted as A_i, $i = 1, \cdots, m$, $m > 0$, and the counterclockwise type, denoted as B_j, $i = 1, \cdots, n, n > 0$.

We can prove that all A_i are connected without using points in C. This is because that any point p in C is contained by two 1-cells $e1$ and $e2$ in C. These two 1-cells are contained by A_i and A_j, respectively. If $A_i = A_j$, it is connected. If A If A_i and A_j share an edge, then, the central pseudo points of A_i and A_j are connected. If A_i and A_j do not share an edge, we know A_i and A_j are in $S(p)$, there must be a cycle contains some edges in A_i and some edges of A_j, and $e1 \cup e2$. So A_i and A_j are connected (meaning through their central pseudo points) do not pass $e1 \cup e2$. Therefore, all A_i's (meaning using their central pseudo points) are connected. In other words, $e1 \cup e2$ split $S(p)$ into two parts, one called $Part_A(p)$ include A_i and A_j, and another one, $Part_B(p)$ include some B's. A_i and A_j are point connected in $Part_A(p)$ without passing any point in $e1 \cup e2$. All cells that are not A_i or A_j in $Part_A(p)$ will also assign as the clockwise type, i.e. A_k for some k. So all A_i's are connected.

In the same way, we can prove that all B_i's are connected. ($C \cap \partial S = \emptyset$.)

We now prove that any point x in $S - C$, must be connected to the component containing A_i or to the component containing B_i. We know that any two points are point-connected by a path in S. Let $c \in C$, $P(x, c)$ is such a path connecting x and c. Note that every point c' in C is contained by $S(c') = Part_A(c') \cup Part_B(c')$. Since x is not in C, $P(x, c)$ (which has finite numbers of points) must contain the first point in C, we assume it is c'. In many cases, $c' = c$. Let $P(x, c) = x \cdots x'c' \cdots c$. Then x' must be not in C. Thus, x' must be in $S(c') - C$. x' must be in some A_i or B_j because $S(c') = Part_A(c') \cup Part_B(c')$.

In other words, there must be a first point in $P(x, c)$, x', that is adjacent a point $c' \in C$ (may or may not be point c). (x', c') must belong to an A_i or B_j. So if (x', c') belong to A_i, x is a point connected to the central pseudo points of A_i. We call it component $\mathcal{A}$. All points in $\mathcal{A}$ are connected since A_i are connected for all i.

If (x', c') belong to B_j, x is a point connected to the central pseudo points of B_j. We call it component $\mathcal{B}$. All points in $\mathcal{B}$ are connected since B_j are connected for all j. We also know $S - C = \mathcal{A} \cup \mathcal{B}$ since x was selected from $S - C$.

According to the proof of on Theorem 15.2, there is a in some $A_i - C$ is not connected to b in some $B_j - C$ in $S - C$. a is in $\mathcal{A}$, and b in $\mathcal{B}$. Therefore, any point in $\mathcal{A}$ is not connected to any point in $\mathcal{B}$ in in $S - C$ (Otherwise, a will be connected to that point in $\mathcal{A}$, then will be connected to any point in $\mathcal{B}$).

We now complete the proof of Theorem 15.3, the general Jordan Curve Theorem.

$\square$

Therefore, we can allow the simple path (pseudo-curves) for this theorem. This is the general case of Jordan Curve Theorem. We know that 2D Euclidean space can be partitioned into triangles and it is simply connected in this discrete space in terms of simplicial complexes. So we can prove the Jordan Curve Theorem for 2D Euclidean plane. The only problem is that we need to assume that we have a refinement process that will make the triangulation (joint with the simple curve C) infinitively approximates C. Some philosophical issue may be occur when the curve is an area filling curve. However such a question could ask for any of proofs of this theorem.

A detailed technique to make such a refinement is called the midpoint subdivision method. Please see L. Chen, A Concise Proof of Discrete Jordan Curve Theorem, *http://arxiv.org/abs/1411.4621.*

15.4   Randomized Algorithms of Closest Pair Problem in Geometry

Given a set of points in a metric space, find two points that is closest. This problem is called the closest pair problem. This problem also has the relationship with nearest neighbor search (closest point search) when a quarry point (a vector) is given, we want to get a point that is closest to the query point. This problem also relates to the famous classification method called k-nearest neighbors (k-NN). We have discussed in Chap. 9. When the data set is very large, it is important to consider the to use multiple computer to complete the search, that is cloud computing and BigData technology. Such as using the map-reduce mechanism to distribute a small set of points to each individual machine and each individual machine will return a k nearest points of each member of the small set.

In designing closet pair algorithm, Rabin discovered a fundamental technology related to geometric computing. This section will introduce his method[31].

Rabin's the closest pair problem is the first randomized algorithm. It was discovered even before randomized prime number testing. This algorithm has historical importance in computational complexity not only in computational geometry. In fact, Rabin's algorithm is an digitization algorithm. It can also be viewed as a digital geometry method.

For a set with n elements, the brute-force method meaning calculate the distances for all pairs needs $O(n^2)$ time. If we select the divide-and-conquer method it uses $O(nlogn)$ [12].

Rabin's randomization algorithm only takes $O(n)$ time. In the proof of the algorithm, the technique of building the recurrence equation for the algorithm analysis is somewhat similar to the proof of the Chan's Convex hull algorithm $O(n \log (h))$.

We call this type of digitization the Rabin's digitization. Rabin's algorithm is a digitization algorithm. We can modify his idea to get a definitive algorithm:

Algorithm 15.1 Rabin's algorithm for digitization

Step 1   Random pick N pair ($N = \sqrt{(n)}$) find the minimum distance d
Step 2   Use d to be the length of the square (or cube) for unit grids in Euclidean space. The array of the grid is $((x_{max} - x_{min})/d) \times ((y_{max} - y_{min})/d)$
Step 3   Scan through n points and get all point indexes in the grid. Mark all filled gird meaning those grid square must contain at least one point.
Step 4   Build two hash tables for points and build a graph with neighboring link

Linking all filled squares together will cost $O(n)$. To give a complete analysis of this algorithm is beyond the scope of this book. We can just assume that there are half point is just in the square containing one element. for all other squares, we select $\sqrt{(N)}$ to get smaller $d_n ew$ for digitizing the points. So the total complexity is

$$O(n) + O(n/2) + O(n/4) + ... = O(n)$$

The expect time is $O(n)$ and worst case of the algorithm is $O(nlogn)$. For each filled square, we check the distance pairs inside of the square and the distance pairs in its adjacent squares.

Some other resources for this algorithm are available in [26] and a detailed algorithm was presented in [22] S. Suri requires each square or cube contains only $O(1)$ points for this algorithm the algorithm analysis will be simpler [33]. This may be the original thought of Rabin. We can modify his idea to get a definitive algorithm:

Algorithm 15.2 Modified Rabin's algorithm for digitization

Step 1   Random pick N pair ($N = \sqrt{(n)}$) find the minimum distance d.
Step 2   Use d to be the length of the square (or cube) for unit grids in Euclidean space. The array of the grid is $((x_{max} - x_{min})/d) \times ((y_{max} - y_{min})/d)$.
Step 3   Scan through n points and get all point indexes in the grid. Mark all filled gird meaning those grid square must contain at least one point.
Step 4   Build two hash tables for points and build a graph with neighboring link.
Step 5   If a $d \times d$ contains more than $\sqrt{(n)}$ splite that in half or select t to find dist to split.
Step 6   Repeat and until all cells contain less than $\sqrt{(n)}$.

This algorithm will be $O(n(logn))$ such a digitization is better than a quadtree method. Identify all square that has at least an element, scan all points to identify the locations of the square. Mark it as filled square.

15.4.1 Relationship to Cloud Computing

For m-dimensional problem it will be the same. This is an idea of digitization. Using the property of digitization, the neighboring is the limited at least in low-dimensions.

It can apply to the problem in cloud density, counting, and splitting quad tree and octree for wireless communications such as balance the resource use for random and moving stations.

There two ways to do a square covering for quadtree type. Split a square or cube until only one point is inside of the cube. This splitting process will save most of space. Possibly the best in terms of space storage. This process will be $O(n \log n)$ time.

Use Rabin's method but for all squares that contains more than one point, we only split it use the same manner. So this method will be much faster. The average time will be $O(n)$ or at most $O(n \log \log n)$ [14].

It is also true that we can set a constant limit for the points a square can hold too. Finding a minimum numbers of squares in different sizes to cover all spatial points will be interesting too.

15.5 BigData, Manifolds, and Advanced Measurement in Geometry

BigData and Data Science contain the huge Opportunity for Scientists, Engineers and IT Business. It also provides tremendous opportunity to mathematicians and computer scientists to discover new mathematics and new algorithms. In this section, we will attempt to outline the different aspects, a mathematician or computer scientist may be interested.

15.5.1 What is the Bigdata Technology

BigData technology is about the data sets from many sources and collections such as different format of data. It also has the properties of massive storage, and it requires fast analysis through a large number of computing devices including cloud computers. It may yield revolutionary breakthroughs in science and industry. BigData is a phenomenon in the current appearance of problems regarding data sets. The characteristics of BigData are: (1) Large data volume, (2) Use of cloud computing tech, (3) High level of security, (4) Potential business values, (5) Many different data sources.

Modern Big-Data computing is also called Petabyte age: Petabyte (PB) means $1MB \times 1GB$. For instance, Google give each person 1 G of 1 billion People in the World, the data volume will be $1G \times 1G = 1000PB$.

The software tool for BigData is called Apache Hadoop, which is an open-source software framework that supports data-intensive distributed applications and it enables applications to work with thousands of computation-independent computers and petabytes of data. Hadoop was derived from Google's MapReduce and

Google File System (GFS). MapReduce is a technological framework for processing parallelize-able problems across huge datasets using a large number of computers (nodes), in the meantime, MapReduce can take advantage of locality of data, processing data near the storage in order to reduce the distance transmission costs.

MapReduce consists of two major steps: "Map" and "Reduce." They are similar to original *Fork* and *Join* operations in distributed systems, but considering very large numbers of computers that can be constructed based on Internet based cloud. In the Map-step, the master computer (a node) first divides the input into smaller sub-problems and then distributes them to worker computers (worker nodes). A worker node may also be a sub-master node to distribute the sub-problem in even more smaller problems that will form a multi-level structure of a task tree. The worker node can solve the sub-problem, and report the result back to its up level master node. In the Reduce-step, the master node will collects the results from the worker nodes, and then it combines the anwsers as an output (solution) of the original problem.

Data Science is a new terminology for BigData. How to make a Petabyte problem to be parallelize-able is the key to use Hadoop. What is Data Science? Data contains science. However, data science has a different approach than that of classical mathematics, which uses mathematical models to fit data and to extract information. Moreover, some mathematical and statistical tools are expect to find some fundamental principle behind the data.

For instance, to find rules and properties of the data set and among different data sets—the relationship of connectivity between data sets. The new research would be more likely to include partial and incomplete connectivity, which is also a hot topic in the current research of social networks. Previously developed technology such as numerical analysis, graph theory, uncertainty, and cellular automata will play some role.

However, developing new mathematics is more likely to be key for the scientists. A good example for face recognition, finding a person in a data base with 10 million pictures, the pictures are randomly taken. To find a best match of the new pictured person needs tremendous calculations.

It is related to person's orientation in each picture. Let's assume that we have 100 computers but they are not available for all time. One can build a tree structures of these 100 computers. When a computer is available, it will get task from its father node. When it is not available, it will return its job to its father node.

Not every problem with Massive data set can be easily split into sub problems, it depends on connections of its graph representation. For instance, a NP hard problem such as the traveling salesman problem, the sub-problem with less nodes does not help much for the whole problem. However, for a scoring problem, a solution of sub problem would be helpful. If a merge-sort algorithm is used, Map-step can give a sub problem to its worker nodes, "Reduce" step only takes a linear time to merge them.

15.5.2 α *Shape, Digital* α *Shape, and Homology Computing*

Finding topological structure of the spatial data, has the property of divide-and-conquer or map-reduce meaning that a problem can be split into sub problems than get the merge when the sub problem was solved individually. This is because that the structure of a topological space in the cell-complex format can be partitioned into subspace. However the merge process (reduce process) may need not only just combining. For instance it may or may not be connected. We still need to recognize the structure in the father node.

In the cloud data persistent analysis in Chap. 9, the α shape (growing the size of the ball, some author used it as dual-shape of α shape.) will get a dynamic structure of the topological homology groups.

The following strategy of the process is natural: partition the space, Euclidean E_n into m nodes, each node can be a sub-father node. Let S is a set of points, a procedure will send each point to the node that only handle the corresponding partition, a regular n-cube subspace. When all data points are distributed, each node (computer) will start its own calculation on the α shape and homology associating with the value α. In the merge (reduce) process, there are two ways: (1) Use the father node to check the boundary of partition and the manifold or complex to be merged. (2) Instead of use the father node, just use son-node to merge its neighboring partitions (nodes). And report to the father node. This merge can be done in a hierarchy manner. Linking two manifold or complex along the edges of regular n-cube subspace was not trivial but has a fast algorithm. It just the same as the quad-tree segmentation merge process. Chen designed an algorithm for this merge that was also cited in a Canon's patent application [2].

The homology group may add the fact on the boundary, they may generate more k-holes. To be evenly distributing the tasks to each node-computer, we always use the following assignment schema: if u,v are two neighboring nodes, we now just assume that $u \cap v$ is a $(n-1)$-hyperplane for now. v take care the merge if u is coded before v, meaning that the index of u given by the father node is smaller than v's. For instance, the coordinate of u is smaller than v.

The (dual) α shape uses n-balls to extend the data volume it is time expensive since the calculation of the intersection of n-balls is not simple. We can use digital n-cubes to replace the n-balls in the homology computing. A linear algorithm for digital α shapes is given in [6].

A challenge question is that how do we computationally attach a complex to a complex to get the correct homology groups.

15.5.3   *Advanced Manifold Learning*

General manifold learning and dimension reduction is still the open problem. It is highly related to the homology of data set. The existing methods, even well developed, such as isomaps and Laplacian eigenmaps methods. They are not a definitive mathematical approaches to get the solutions to most of data sets. They are still the specialized methods for some specific data sets.

The current research has started to calculate Riemann manifold in 3D and high Dimensional Euclidean space [25]. Especially to find Riemann metric and curvatures based on discrete methods. In fact, one must know the Riemann manifold in order to get the first or second form of Riemann manifolds.

Guillemard has made extensive work on the manifold learning in his PhD dissertation [16]. The α shape was combined with dimension reduction in the dissertation.

Another related problem is to find a subspace such as a plane that contains most of the points for a data set given. This could be another version of manifold learning.

This problem is called subspace recovery: Given m points in R^n. If many of them are contained in a t-dimensional subspace T can we find it using an efficient algorithm?

Most of researchers use statistical method to find the subspace [32].

This problem also has a definitive version in computing theory: Is there a polynomial algorithm that can find a subspace that contains N elements?

In [18], Hardt and Moitra proved a theorem related to find a subspace: If a set of m points in R^n has strictly more than $\frac{dm}{n}$ points in a d-dimensional subspace (with a condition), then there is a deterministic polynomial time algorithm to find a t-subspace, $t \leq d$, that has more than $\frac{dm}{n}$ points.

To understand this problem, we can assume that A plane has 100 points, we want to determine if a line that contains at least 50 points. This problem is not very easy to solve since we can make the 50 points on a line intentionally. And the other points just random arranged. Using statistics, we use the least square method to do regression. But for a determinative solution, it is not so obvious [18, 21].

15.5.4   Integral Points Counting

Counting the integral points in a polyhedra is a way to estimate the volume of the polyhedra. This is an important measure for a geometric object. As we know that there is no simple extension of Pick's theorem in 3D, the problem of counting the integral points is not easy to solve.

One can use an algorithm to count the integer points. The algorithm is simple if one can determine whether or not a point is inside of the polyhedra.

However, in this case, one may need to determine all integer neighbors of the 3D polygons in the polyhedra [23].

Another algorithm is called the odd-even test, but this algorithm needs to scan through every integer point in the space. It may be very slow if the polyhedra is relatively much smaller than the space we considered.

The idea of the odd-even test is also simple: Let's use 3D space as example, we scan an integer point p along z-axis. The scanning process starts at the smallest value point in z-axis. To check how many polygons (on the boundary of the polyhedra) have passed to the recent location at p. If the number is odd, that means p is inside of the polyhedra; otherwise, p is outside. As we said this process needs check every integer point in the space.

A more sophisticated method was given in [1]. For a m-D polytope given by its vertices or by its facets, the complexity appears to be $O(n^{O(m \log m)})$ where n is the input size.

As a discussion question, still using 3D, the following method may obtain a fast solution in digital approximation: (1) Assume we have vertices as the input. Use Bresenham algorithm for each digital lines and we can get digital polygon of a polygon. So we will get the integral boundary of the polyhedra. (2) Locate a point inside of the polyhedra. (3) Use breadth first search to get the integer points in the polyhedra.

Since only two points are need to be checked when scanning through the 3D grid array.

Another algorithm use depth-first-search (DFS) and breadth-first-search (BFS) to determine the inside integral boundary of polyhedra.

There are three digital planes are important. one is Bresenham plane the closet to the true (original) plane. One is in the left closest to the original plane (< 1) and another one is the right closet to the original plane.

Mark red to the left, and green to the Bresenham, and blue to right.

So if one red is inside of polyhedron its counterpart at same reference point will be not. Most likely, the red neighbor is in the P.

It is possible one point marked as g or R, b and g. not r and b. If g=b=r means that this point is on the exact position of digital point (define left-right plane equation >0 or <0).

Assume vertices of each polygon on the lattice points. each polygon is convex. (otherwise it is easy to split to that). Kaufman has an algorithm to determine all integer points inside of the convex polygon [23].

Link all inside points will be the answer. Use odd-even to decide if a point is in or not in the polyhedron by scanning each line.

The boundary edges of each polygon in 3D. only affect its neighbor digital points. so the red, blue, green for each polygon are limited. only need to decide among those edge surroundings. only two polygons are concerned except the points at ending points. all immediate neighbors are sealed.

Klette also realized that there are an envelop of cubics containing a 3D polygon. So we mark the all cubics, the edge of the cubic plane.

We continue to use our treatment. The neighbor points at edge will be determined with the neighboring polygon. r connected with g will change the g to r or vise versa. We will keep all g or r in one side of the orientation.

After all, we will have either r is the closed inside or g is. to determine that. we can link a r point to boundary of 3D domain edge if pass odd numbers of polygons (except the edge) is inside. if pass even numbers of polygons (g points). r is in outside.

This algorithm is linear (numbers of polygons and all vertices. or size of outer envelop with insiders and length of the array). This algorithm will be the Optimum.

Find a point is inside a polygon or outside one just need to calculate the left-right for each polygon line. get actual plane value to get r or b integral points. all other things follows.

If all those points are closed (a closed connected objects in 26-connectivity) not need not a pure manifold. Then we can use above scan algorithm to make it.

15.5.5 Advanced Measurement in Geometry

How to calculate actual length, surface area, and volume in Riemann manifolds? One way is to transform or embed the Riemann manifold into Euclidean space. The isometric embedding mean that the transformation maintains the length value no change in two spaces. J. Nash proved a theorem indicating such a smooth isometric embedding exists [17]. In particular, Nash proved: A continuously differentiable (C^1) n-Riemannian manifolds can be isometrically embedded to $2n + 1$ Euclidean space. Since a C^k manifold will have a continuous decomposition and C^1 will have C^k approximation. This theorem is enough for us for discussion.

We know there are discrete decompositions for manifolds in Euclidean space. It is also true that for any Riemannian manifolds, we will have the digital form. So we can use simplicial complexes to get the approximation and use the method of polyhedras discussed in above sections to calculate the volume. For digital k-cell decomposition of a manifold, the volume of the manifold is not near the value of total integral points inside of the manifold.

On the other hand, some estimation methods for length and areas are also useful. For instance, it is not hard to know that a circle in 2D has the shortest length if the area of a region is fixed. In other words, give a closed curve C, the area that is bounded by C will reach the maximum if C is a circle. So we have the following inequality called isoperimetric inequality [20, 29]:

$$4\pi Area_C \leq (length(C))^2$$

Another inequality is to measure the opposite direction of an inequality about the length vs area. Let's define the systole as the least length of a noncontractible loop in a space X. Loewner proved a theorem for systole inequality for every metric. It states for a torus T, (Note: A torus is also called a 2-torus; 1-torus is a circle; n-torus is $C \times C \times \cdots \times C$ where C is a circle.) we have,

$$systole^2 \leq \frac{2}{\sqrt{3}} Area_T$$

Pu proved another inequality for the real projective plane. A general result was proved by Gromov [30, 15]:

$$systole^n \leq c_n \cdot volume_M,$$

systole here is the length for closed curves of an essential n-manifold M. c_n is a constant only depending on the dimension of M.

Isometric embedding a Riemann manifold to Euclidean space, then we can get its measurement digitally or discretely. The genus calculation can also be found use the method discussed in Chap. 14. Calculation of systole in digital case or discrete case is an approximation measurement. The question for researchers in this field could be find the error estimation of such a digitization or discretization.

Unlike Riemann geometry, one might only know the relationship of some measures such as length or area in differential forms. In digital or discrete geometry, we can actually know all real values for length or area etc. Even though one does not know how to use the mathematical function to discribe a digital or discrete manifold, we know actual manifolds in space. That is kind of tread off.

Digital method dose not need to do a triangulation or a simplicial decomposition. It just need to count the integer number to get a measurement estimation.

15.5.6   Discussion of Area Measurement of Shapes in Grid Space

The area of a shape in grid space is the total points included in the shape including the edge of the shape. This intuitive definition is most used.

Why we can not just use Pick's formula as the definition of area of a geometric shape in 2D. This is because, the meaning of a grid point may not be just treated as a point in Euclidean space.

A grid point may mean many things: Could be a single point, an area near the point, or a neighborhood of the point, a open or close set.

If we use Pick's formula as the area, then Let U the the $m \times n$ region with area $(m - 1) \times (n - 1)$.

A is a connected region (simply connected) polygon. $Pick_{area}(U) \neq Pick_{area}(A) - Pick_{area}(U - A)$ since there some gap between point $Pick_{area}$ A and $Pick_{area}$ $(U - A)$.

Therefore, the most reasonable measurement of areas of grid point sets is just the count of points in the set.

Digital points are more like a physical point or quantum points. It is not just can be treated as the Euclidean point. It may have length, area, or volumes. It is depended on the need of the calculations. When calculate the volume, one may need to consider the half point location as well (just like the matching cubes).

The boundary of an area of digital object is not the boundary points, it is a continuous curve (we do not know precisely) that best fit the points. A mathematical curve observed in real world, we cannot touch, we only can feel its digital or discrete approximation.

15.6 Historical Concerns: From Geometry of Numbers to Applied Discrete Geometry

Although digital geometry, discrete geometry, geometry, and topology have their differences, they are closely related in computation. The main difference between geometry and topology is the measurements. Strictly speaking, geometry must have a metric but topology does not have to have a metric. In other words, geometry must consider the distance between two points. However, topology does not focus on this distance measure. At the same time, the common ground is that they both deal with neighborhoods, meaning that they both deal with continuity and differentiability. Discrete geometry focuses on finite points, finite line-cells, and solid-cells. It provides a way of decomposing a continuous space to finite space. Digital geometry is for computer graphics and image processing. Even though the theory of the geometry of numbers was discovered by Minkowski more than 100 years ago and it is related to digital geometry, modern image processing and massive data analysis requires efficient algorithms. These algorithms must be able to handle new geometrically related infrastructures such as cloud data computing, which uses a large number of computers within one network connection. Therefore, discrete geometry is highly applicable in current hot topics such as BigData and data sciences. On the other hand, in theory, we are still searching for concise proofs of the four-color problem and the topological proof of the Poincare conjecture. This is not to repeat research already done, but to seek to better understand mathematics in a more natural way. In addition to their contributions to modern applied sciences and classical mathematics, geometric and typological computing also play important roles in other cutting-edge fields. In quantum computing, research includes the extensive study of elliptical curves for cryptographies and topological quantum computers for non-abelian states. Discrete geometry and digital geometry will continue to be highly significant in the near future.

References

1. A. Barvinok, Integer points in polyhedra. Zurich Lectures in Advanced Mathematics. European Mathematical Society (EMS), Zrich, 2008.
2. L. Chen, The λ-connected segmentation and the optimal algorithm for split-and-merge segmentation, Chinese J. Computers, 14(2), pp. 321–331, 1991.
3. L. Chen, Generalized discrete object tracking algorithms and implementations," In Melter, Wu, and Latecki ed, Vision Geometry VI, SPIE Vol. 3168, pp. 184–195, 1997.
4. L. Chen "Point spaces and raster spaces in digital geometry and topology," in Melter, Wu, and Latecki ed, *Vision Geometry VII*, SPIE Proc. 3454, pp. 145–155, 1998.
5. L. Chen, Discrete Surfaces and Manifolds, SP Computing, Rockville, 2004.
6. L. Chen, Digital Functions and Data Reconstruction, Springer, NY, 2013.
7. L. Chen, L. Chen, Note on the discrete Jordan curve theorem, Vision Geometry VIII, Proc. SPIE Vol. 3811, 1999. pp. 82–94. Its revised version is at *http://arxiv.org/abs/1312.0316*.

8. L. Chen and J. Zhang, Digital manifolds: A Intuitive Definition and Some Properties, *The Proc. of the Second ACM/SIGGRAPH Symposium on Solid Modeling and Applications*, pp. 459–460, Montreal, 1993.

9. L. Chen and J. Zhang, "Classification of Simple digital Surface Points and A Global Theorem for Simple Closed Surfaces", in Melter and Wu ed, *Vision Geometry II*, SPIE Vol 2060, pp. 179–188, 1993.

10. L. Chen, H. Cooley and J. Zhang, The equivalence between two definitions of digital surfaces, *Information Sciences*, Vol 115, pp. 201–220, 1999.

11. L Chen, H. Zhu and W. Cui, Very Fast Region-Connected Segmentation for Spatial Data: Case Study, IEEE conference on System, Man, and Cybernetics, 2006.

12. T. H. Cormen, C.E. Leiserson, and R. L. Rivest, Introduction to Algorithms, MIT Press, 1993.

13. J.-F. Dufourd, An intuitionistic proof of a discrete form of the Jordan curve theorem formalized in Coq with combinatorial hypermaps, Journal of Automated Reasoning 43 (1) (2009) 19–51.

14. S. Fortune and J. Hopcroft, A note on Rabin's nearest-neighbor algorithm, Inform. Process. Lett. No 8, pp. 20–23, 1979

15. Gromov, M. Systoles and intersystolic inequalities. (English, French summary) Actes de la Table Ronde de Gomtrie Diffrentielle (Luminy, 1992), 291-362, Smin. Congr., 1, Soc. Math. France, Paris, 1996.

16. M. Guillemard, Some Geometrical and Topological Aspects of Dimensionality Reduction in Signal Analysis. PhD thesis, University of Hamburg, 2011, *ftp://ftp.math.tu-berlin.de/pub/numerik/guillem/prj2/mgyDiss.pdf*.

17. Q. Han, Qing, J-X Hong (2006), Isometric Embedding of Riemannian Manifolds in Euclidean Spaces, American Mathematical Society, 2006.

18. M. Hardt and A. Moitra. Algorithms and hardness for robust subspace recovery. In Proc.26 COLT, pages 354–375. JMLR, 2013.

19. G.T. Herman, *Geometry of Digital Spaces*, Birkhauser, Boston, 1998.

20. M.G. Katz, M (2007), Systolic geometry and topology, Mathematical Surveys and Monographs, 137, Providence, R.I.: American Mathematical Society, pp. 19,

21. S. Khot and D. Moshkovitz. Hardness of approximately solving linear equations over the reals. STOC, pages 413–420, 2011.

22. J. Kleinberg, E. Tardos. Algorithm Design. Addison Wesley, 2005.

23. R. Klette, and A. Rosenfeld *Digital Geometry: Geometric Methods for Digital Image Analysis*. Morgan Kaufmann, 2004.

24. S. Lefschetz "Introduction to Topology," Princeton University Press New Jersey, 1949.

25. T. Lin and H. Zha, Riemannian Manifold Learning, IEEE Trans. Pattern Analysis and Machine Intelligence, vol. 30, no. 5, pp. 796–809, May 2008.

26. R. Lipton, Rabin Flips a Coin, *http://rjlipton.wordpress.com/2009/03/01/rabin-flips-a-coin/*

27. D. G. Morgenthaler and A. Rosenfeld, "Surfaces in three-dimensional images," *Inform. and Control*, Vol 51, pp. 227–247, 1981.

28. M. Newman, *Elements of the Topology of Plane Sets of Points*, Cambridge, London, 1954.

29. R. Osserman, The isoperimetric inequality. Bull. Amer. Math. Soc. 84 (6): 1182–1238. 1978.

30. P. M., Pu, Some inequalities in certain nonorientable Riemannian manifolds. Pacific J. Math. 2 (1952), 55–71.

31. M. O. Rabin, Probabilistic algorithms, in Algorithms and Complexity: New Directions and Recent Results J. F. Traub, Ed., Academic Press, New York, 1976, pp. 21–39

32. M. Soltanolkotabi and E. Candes. A geometric analysis of subspace clustering with outliers. Ann. of Statistics. pp. 2195–2238, 2012.

33. S. Suri, The closest pair problem, University of California at Santa Barbara, *http://www.cs.ucsb.edu/ suri/cs235/ClosestPair.pdf*

34. W. T. Tutte, Graph Theory, Addison-Wesley, Reading, Mass., 1984.

35. O. Veblen, Theory on Plane Curves in Non-Metrical Analysis Situs, *Transactions of the American Mathematical Society* 6 (1): 83–98, 1905.

Glossary

Algorithm A set of steps of instructions for solving a problem.

Approximation Similar to interpolation, but approximation does not require the new function passes the sampling values at each sample points. Because of that, approximation is usually easy to get smoothness.

Boundary of domain The edge of domain.

Classification The process to categorize a set of elements to different classes.

Curve A one dimensional shape in which each point has neighborhood that is homemorphic to E_1.

Digital image A digital function from a 2D domain to integers.

Digital k-cell A k dimensional cell in digital space.

Digital k-manifold A digital object that is a collection of k-cells with some special properties.

Digital space A grid space in which every element is a integer point.

Discrete space A graph-based space with geometric metric. It contains definition of k-dimensional cells.

Domain The mathematical term for the region that holds locations. Range or co-domain is usually used as the values.

Gradually varied function Gradually varied function is a type of discrete functions where the values of each pair of neighbors are the sample or only have small change. Gradually varied functions use A1,. . .,Am to represent the value and level changes. It is essentially to translate the discrete function into digital functions in terms of algebra. This method has the limitation to deal with derivatives.

Interpolation Extend the sample data (usually a few points) into entire region or domain. It will be the function (or surface for 2D) for the region. Interpolation requires the new function passes the sampling values at each sample points.

Manifold A k dimensional shape in which each point has neighborhood that is homemorphic to E_m.

Sample Points A group of samples collected from a region that is usually a rectangle, but could be any other shapes. The sample point has two components: one is the location as (x,y) in 2D or (x,y,z) in 3D; another is the value in real numbers.

Segmentation Partition a image into different components.

Surface A two dimensional shape in which each point has neighborhood that is homemorphic to E_2.

© Springer International Publishing Switzerland 2014   317
L. M. Chen, *Digital and Discrete Geometry*, DOI 10.1007/978-3-319-12099-7

Index

© Springer International Publishing Switzerland 2014
L. M. Chen, *Digital and Discrete Geometry*, DOI 10.1007/978-3-319-12099-7

Printed in the United States
By Bookmasters